The Role of Carotenoids in Health and Disease

The Role of Carotenoids in Health and Disease

Guest Editors

Nikolay E. Polyakov
Lowell D. Kispert

MDPI

Basel • Beijing • Wuhan • Barcelona • Belgrade • Novi Sad • Cluj • Manchester

Guest Editors
Nikolay E. Polyakov
Institute of Chemical Kinetics and Combustion
Russian Academy of Sciences
Novosibirsk
Russia

Lowell D. Kispert
Department of Chemistry
University of Alabama
Tuscaloosa
USA

Editorial Office
MDPI AG
Grosspeteranlage 5
4052 Basel, Switzerland

This is a reprint of the Special Issue, published open access by the journal *International Journal of Molecular Sciences* (ISSN 1422-0067), freely accessible at: https://www.mdpi.com/journal/ijms/special_issues/YKFB4DDCZ1.

For citation purposes, cite each article independently as indicated on the article page online and as indicated below:

Lastname, A.A.; Lastname, B.B. Article Title. *Journal Name* **Year**, *Volume Number*, Page Range.

ISBN 978-3-7258-7038-7 (Hbk)
ISBN 978-3-7258-7039-4 (PDF)
https://doi.org/10.3390/books978-3-7258-7039-4

Contents

About the Editors

Nikolay Polyakov

Nikolay Polyakov is an academic researcher from the Institute of Chemical Kinetics and Combustion, Russian Academy of Sciences, Novosibirsk, Russia. Dr. Polyakov is now the head of the Laboratory of Magnetic Phenomena at the same institute. He is interested in understanding the role of free radicals in biology and medicine. He has contributed to research in topics including carotenoids; drug delivery systems; spin chemistry; free radicals; antioxidant activity; electron transfer; and membrane biophysics. He has an h-index of 33 and has co-authored 163 publications. Dr. Polyakov holds a PhD in Physics and a doctoral degree in Chemistry from the Institute of Chemical Kinetics and Combustion, Russia. After his PhD, he worked as a postdoctoral fellow at the National Industrial Research Institute of Nagoya, Japan. From 1997 to 2014, he was a visiting Professor at the University of Alabama (USA) and the University of Utah (USA). He is also an Honored Professor of Zhejiang University of Technology (China).

Lowell Kispert

Lowell Kispert is a Professor Emeritus at the Department of Chemistry and Biochemistry, University of Alabama, Tuscaloosa, USA. He is interested in understanding the energy transfer between carotenoids and chlorophyll, as well as the structure of carotenoids' excited states, radical cations, and triplet states. He holds a PhD in Chemistry from Michigan State University, USA. He has contributed to research in topics including carotenoids; electron transfer; carotenoid radicals; electron paramagnetic resonance; and DFT. He has an h-index of 43 and has co-authored 249 publications.

Preface

Carotenoids are naturally occurring pigments found in most fruits and vegetables, plants, algae, and photosynthetic bacteria, and have a range of functions in human health. They primarily exert antioxidant effects, but individual carotenoids may also act through other mechanisms. Recently, significant research has focused on the reactions between carotenoids and free radicals and the ability of carotenoids to prevent the development of diseases caused by toxic free radicals. One factor contributing to the development of various diseases, including infarction, cerebral thrombosis, and tumors, has been attributed to the action of free radicals and toxic forms of oxygen. No less significant are the membrane-stabilizing and immunostimulating functions of carotenoids, as well as their pro-vitamin A activity. At the same time, the widespread practical application of carotenoids as antioxidants or food colorants is substantially hampered by their hydrophobic properties, instability in the presence of oxygen, and high photosensitivity. Applying carotenoids in pharmaceutical contexts requires a chemical delivery system that overcomes the problems related to the parenteral administration of highly lipophilic, low-molecular-weight compounds.

This Reprint—entitled "The Role of Carotenoids in Health and Disease" and published by the *International Journal of Molecular Sciences*—includes contributions that assess state-of-the-art research as well as future developments in the field of carotenoid studies. Topics include, but are not limited to, the health benefits of carotenoids, the antioxidant activity of carotenoids, carotenoids in eye diseases, supramolecular complexes of carotenoids, carotenoid–membrane interactions, etc. We believe that this Reprint will be interesting to specialists in life sciences, including supramolecular and physical chemistry, biophysics, medicine, nutrition, and pharmacology.

Nikolay E. Polyakov and Lowell D. Kispert
Guest Editors

International Journal of
Molecular Sciences

MDPI

Editorial

Special Issue "The Role of Carotenoids in Health and Disease"

Nikolay E. Polyakov [1,*] and Lowell D. Kispert [2]

1 Institute of Chemical Kinetics and Combustion, 630090 Novosibirsk, Russia
2 Department of Chemistry, University of Alabama, Tuscaloosa, AL 35401, USA
* Correspondence: polyakov@kinetics.nsc.ru

Carotenoids represent a ubiquitous and critically important class of natural isoprenoid pigments, synthesized de novo by plants, algae, and photosynthetic bacteria [1–3]. Their vibrant red, orange, and yellow hues are not merely for aesthetic purposes; they play fundamental roles in photoprotection and light harvesting within photosynthetic organisms. In the realm of human health, the biological significance of carotenoids, which must be acquired through the diet, has expanded far beyond their initial recognition. This Special Issue, entitled "The Role of Carotenoids in Health and Disease," features a collection of articles that describe cutting-edge research and emerging developments in this field. Topics covered include the features of biosynthesis and degradation of carotenoids; the health benefits and antioxidant functions of carotenoids; their application in cancer, and cardiovascular diseases; interaction of carotenoids with lipid membranes; and the use of various delivery systems to enhance the stability and bioavailability of carotenoids.

While the primary role of carotenoids is its potent antioxidant activity, encompassing the ability to quench singlet oxygen and scavenge peroxyl radicals, this is only one facet of a complex physiological profile. Individual carotenoids, such as β-carotene, α-carotene, and β-cryptoxanthin, function as crucial provitamin A precursors, which are metabolized into retinal and retinoic acid, compounds essential for vision, cell differentiation, and immune function. Beyond this, specific carotenoids like lutein and zeaxanthin accumulate in the macula lutea, protecting the retina from phototoxic blue light, while others demonstrate functions in cell membrane stabilization by influencing lipid packing and fluidity and in direct immune modulation by influencing lymphocyte proliferation and cytokine signaling. This multifaceted bioactivity has led to intensive research into their therapeutic potential, particularly concerning their interactions with free radicals. A growing body of evidence underscores their potential to mitigate the pathogenesis of major diseases, including myocardial infarction, cerebral thrombosis, and various neoplasms [3–5]. The common thread linking these conditions is the pervasive damage inflicted by reactive oxygen species (ROS) and oxidative stress on cellular lipids, proteins, and DNA. Carotenoids, acting as a first line of defense, offer a compelling nutritional strategy to modulate these pathways and potentially slow disease progression.

However, a profound paradox exists between the demonstrated therapeutic potential of carotenoids and their viable pharmaceutical application. This "carotenoid conundrum" is rooted in a suite of challenging inherent physicochemical properties:

- Extreme lipophilicity: Their highly hydrophobic, long conjugated polyene chain renders them virtually insoluble in aqueous biological fluids, leading to poor and variable oral bioavailability. This property also complicates parenteral administration, as they cannot be formulated in conventional intravenous solutions without risking precipitation or embolism.

- Profound chemical instability: The very feature that confers their antioxidant capacity, namely the conjugated electron system, makes them exceptionally susceptible to degradation. In the presence of oxygen, light, heat, and acids, they readily undergo oxidation and isomerization, losing their bioactivity and color.
- Pronounced photosensitivity: Exposure to light can rapidly degrade carotenoids, making storage, handling, and formulation into stable dosage forms a significant technical hurdle.

These challenges directly impede the effective use of carotenoids not only as therapeutic antioxidants but even as natural colorants in foods and supplements. The development of advanced, robust delivery systems is a critical and non-negotiable prerequisite for translating the preclinical promise of these low-molecular-weight, highly lipophilic compounds into clinical reality [6,7].

All articles included in this Special Issue, which are briefly summarized below, can be found via the following link: https://www.mdpi.com/journal/ijms/special_issues/YKFB4DDCZ1, accessed on 14 November 2025. Four review articles and four research articles are presented in this Special Issue.

The first review entitled "The endless world of carotenoids—structural, chemical and biological aspects of some rare carotenoids" focuses on various physicochemical properties of carotenoids, including its ability to form supramolecular "host–guest" complexes with water-soluble drug delivery systems, and their participation in various model redox processes [Contribution 1]. The application of electron paramagnetic resonance (EPR) spectroscopy and density functional theory (DFT) calculations has elucidated the structural characteristics of carotenoid paramagnetic intermediates—including radical cations and neutral radicals—and their functional roles in natural systems. These studies provide critical insights into the potential biological activity of carotenoids and their health implications. Notably, some rare carotenoids, such as crocin, siphonaxanthin, and sioxanthin, exhibit distinct structural features compared to conventional carotenoids, including additional functional groups or unique group positioning outside cyclic structures (e.g., sapronaxanthin, myxol, deinoxanthin, and sarcinaxanthin). The incorporation of carotenoids into host matrices via strategic molecular design or self-assembly enables stabilization through multiple hydrogen and coordination bonds. This supramolecular integration results in improved stability, increased oxidation potentials, and enhanced antioxidant activity, coupled with tunable photo-oxidation kinetics. Furthermore, embedding carotenoids in nonpolar environments can significantly improve their photostability. Additionally, the use of nanoscale supramolecular delivery systems has shown promise in enhancing both the stability and bioactivity of both conventional and rare carotenoids, offering potential advancements in their therapeutic and nutraceutical applications.

Wenjing Su with co-authors presented a review titled "Improving the treatment effect of carotenoids on Alzheimer's disease through various nano-delivery systems", which comprehensively summarizes how nanotechnology can overcome the inherent limitations of carotenoids to enhance their therapeutic potential against Alzheimer's disease [Contribution 2]. Various essential carotenoids, such as lutein, fucoxanthin, astaxanthin, lycopene and other, as well as rare carotenoids like crocin (natural pigments with potential antioxidant activity), demonstrate therapeutic potential in the management of various neurogenerative diseases, including Alzheimer's disease (AD). However, their lipophilic nature and highly unsaturated structures contribute to poor aqueous solubility, low stability, and limited bioavailability, restricting their clinical application. To overcome these challenges, the development of nanoscale drug delivery systems has emerged as a promising strategy to enhance carotenoid efficacy. These systems can significantly improve solubility, chemical stability, membrane permeability, and bioavailability, thereby optimizing their therapeutic effects. Recent advances suggest that tailored carotenoid delivery platforms may offer neuroprotective benefits, potentially mitigating AD progression through enhanced bioactivity and targeted delivery.

M. Pasenkiewicz-Gierula and colleagues presented the review on computational studies examining the interaction of carotenoids with model lipid membranes. By leveraging molecular modeling techniques, such as molecular dynamics (MD) simulations, the work provides atomic-level insights into the dynamic and energetic principles governing how carotenoids incorporate into and influence bilayer structure and properties [Contribution 3]. The absorption and transport of carotenoids in the human body have been extensively investigated. This review begins by summarizing recent experimental advances in these processes, covering key aspects such as carotenoid carriers, systemic transport, tissue delivery, and molecular-level interactions with membrane receptors. The core focus of this review, however, lies in the computational exploration of carotenoid intercalation and dynamic behavior within lipid bilayers. These computational studies are particularly significant because they bridge microscopic molecular behavior with macroscopic ensemble properties, providing critical insights into the structure–function relationships of carotenoids. By employing molecular dynamics simulations and other computational approaches, researchers can elucidate the atomic-level interactions between carotenoids, lipid bilayers, and proteins. Such detailed analyses not only aid in interpreting the macroscopic physicochemical properties of carotenoids but also advance our understanding of their biological roles and functional mechanisms.

The next review presented by Tomas Bas explores the innovative applications of carotenoids and their transformative potential for human health and medicine [Contribution 4]. The work comprehensively examines the therapeutic promise of these compounds in managing chronic diseases, such as cancer, cardiovascular disorders, and age-related macular degeneration. The review places particular emphasis on novel protective mechanisms—extending beyond conventional antioxidant activity—and highlights the resulting innovative pharmacological benefits that underscore their growing clinical relevance. The author discusses emerging evidence on carotenoid-mediated pathways that modulate oxidative stress, inflammation, and cellular signaling, offering new avenues for disease intervention. A key focus of this work is the examination of cutting-edge advancements in carotenoid extraction and bioavailability enhancement. Innovations such as supramolecular encapsulation systems and nanotechnology-driven delivery platforms are shown to significantly improve the solubility, stability, and targeted absorption of these bioactive compounds. These technological breakthroughs not only ensure reproducible product quality but also enable the customization of carotenoid-based therapies to meet individual patient profiles, thereby advancing the paradigm of precision nutrition and personalized medicine. By synthesizing contemporary research with pioneering methodologies, this review provides a forward-looking perspective on the translational potential of carotenoids. It establishes a robust framework for future investigations, setting a new standard for interdisciplinary research at the intersection of nutritional science, pharmacology, and nanotechnology.

Galina Brychkova and colleagues studied the features of biosynthesis and degradation of carotenoids in lettuce (*Lactuca sativa* L.) from seedling to harvest. Lettuce (*Lactuca sativa* L.) is a globally significant leafy vegetable, yet its cultivars exhibit substantial variability in carotenoid concentrations at harvest [Contribution 5]. To elucidate the regulatory mechanisms underlying this variation, authors conducted integrated transcriptomic and metabolomic analyses on inner and outer leaves from six cultivars at distinct developmental stages. This approach identified critical gene-to-metabolite networks governing the accumulation of two nutritionally essential carotenoids: β-carotene and lutein. These findings reveal that differential expression of carotenoid biosynthetic enzymes drives cultivar-specific disparities in lutein and β-carotene production. Notably, sustaining high carotenoid levels in leaves requires precise regulation of metabolic sinks, including the diversion of β-carotene and lutein toward zeaxanthin and subsequently abscisic acid (ABA) biosynthesis. A key conclusion is that lettuce harvested at the commercially dominant maturity stage—often coinciding with onset of senescence—shows marked declines in carotenoids and other vital metabolites. Authors propose that earlier harvest of less mature plants could enhance the nutritional value of lettuce, optimizing its content of bioactive compounds for human health.

B. Tatarowska and colleagues investigated the influence of genetic and agronomic factors on the concentration of key antioxidant compounds, total carotenoids (TC) and vitamin C (VC), by a comprehensive analysis of 65 potato cultivars from 10 countries, (Contribution 6). Their findings demonstrated a highly significant effect of cultivar, harvest year, and tuber flesh color on TC content. Quantitatively, TC levels were markedly higher in yellow-fleshed cultivars compared to white-fleshed ones and exhibited a weak correlation with annual variations. A similar trend was observed for VC, which was also present at higher concentrations in yellow-fleshed potatoes. Furthermore, statistical analysis confirmed that the concentration of TC was a significant determinant of the total antioxidant activity in the potato tubers.

The modulation of carotenoid and phospholipid content in staphylococcus aureus membrane was studied by Laura Zamudio-Chávez with co-authors [Contribution 7]. The membranes of *Staphylococcus aureus* contain carotenoids produced during staphyloxanthin biosynthesis, which function as virulence factors by scavenging reactive oxygen species (ROS) and inhibiting antimicrobial peptides (AMPs). In this study, authors demonstrate that oxygen-restricted growth conditions in *S. aureus* downregulate carotenoid biosynthesis and alter phospholipid composition in both biofilms and planktonic cells, as revealed by LC-MS analysis. Under oxygen limitation, the biophysical properties of *S. aureus* membranes undergo significant changes, including the following:

- Increased lipid headgroup spacing;
- Reduced bilayer core order;
- Elevated liquid crystalline to gel phase transition temperature.

Notably, carotenoid-deficient membranes exhibit a highly ordered gel phase at low temperatures, indicating that carotenoids play a key role in maintaining membrane fluidity. These findings suggest that *S. aureus* in hypoxic environments (e.g., abscesses) likely contains reduced carotenoid levels, leading to modified membrane biophysical properties that may influence bacterial survival and pathogenicity.

The study by M. Rozanowska et al. characterized the efficacy of lutein, zeaxanthin, taurine, and melanin as scavengers of cation radicals generated from visual cycle retinoids [Contribution 8]. In the retina, retinoids critical for vision are continuously exposed to oxidative stress, and their oxidation products can have detrimental effects. Using pulse radiolysis, authors demonstrated that lutein and zeaxanthin efficiently scavenge retinoid radical cations with bimolecular rate constants approaching diffusion-controlled limits. While lutein demonstrates superior scavenging ability for retinoid radical cations in vitro, the physiologically high concentration of ascorbate in the retina positions it as the predominant protector of all visual cycle retinoids against oxidative degradation. Notably, α-tocopherol contributes substantially to the protection of retinaldehyde but exhibits significantly lower efficacy in safeguarding retinol and retinyl palmitate. Although the protective role of lutein and zeaxanthin appears limited in the retinal periphery, their antioxidant activity is substantially more pronounced in the macula, where their concentrations are highest.

The results of these studies on carotenoids indicate their great potential for treating diseases and promoting health. By forming supramolecular host–guest complexes, insoluble carotenoids are useful for water-soluble drug delivery systems. EPR measurements have detected the structure of paramagnetic forms that are important for understanding the mechanisms of carotenoids activity. The bioavailability of carotenoids in human body has been advanced by improved carotenoid carriers, systematic transport, tissue delivery, and membrane receptors. In conclusion, while the pathophysiological rationale for using carotenoids in modern medicine is strong, the bridge between their inherent biological activity and their effective clinical application must be built upon the foundation of sophisticated delivery platforms. The future of carotenoid therapeutics lies as much in the innovation of formulation science as it does in the continued elucidation of their biological mechanisms. For instance, through various nanoscale drug delivery systems, the treatment of Alzheimer's diseases has been improved by carotenoids through carotenoid solubility, chemical stability, and membrane permeability. We expect that further discoveries in carotenoid research can eventually improve strategies for the treatment of a number of diseases.

Author Contributions: Conceptualization, N.E.P. and L.D.K.; writing—original draft preparation, N.E.P.; writing—review and editing, N.E.P. and L.D.K. All authors have read and agreed to the published version of the manuscript.

Acknowledgments: We thank the authors for submitting their interesting articles and reviews. Further appreciation is extended to the editorial staff of this journal.

Conflicts of Interest: The author declares no conflicts of interest.

List of Contributions:

1. Polyakov, N.; Focsan, A.; Gao, Y.; Kispert, L. The Endless World of Carotenoids—Structural, Chemical and Biological Aspects of Some Rare Carotenoids. *Int. J. Mol. Sci.* **2023**, 24, 9885. https://doi.org/10.3390/ijms24129885.
2. Su, W.; Xu, W.; Liu, E.; Su, W.; Polyakov, N. Improving the Treatment Effect of Carotenoids on Alzheimer's Disease through Various Nano-Delivery Systems. *Int. J. Mol. Sci.* **2023**, 24, 7652. https://doi.org/10.3390/ijms24087652.
3. Pasenkiewicz-Gierula, M.; Hryc, J.; Markiewicz, M. Dynamic and Energetic Aspects of Carotenoids In-and-Around Model Lipid Membranes Revealed in Molecular Modelling. *Int. J. Mol. Sci.* **2024**, 25, 8217. https://doi.org/10.3390/ijms25158217.
4. Bas, T. Bioactivity and Bioavailability of Carotenoids Applied in Human Health: Technological Advances and Innovation. *Int. J. Mol. Sci.* **2024**, 25, 7603. https://doi.org/10.3390/ijms25147603.
5. Brychkova, G.; de Oliveira, C.; Gomes, L.; de Souza Gomes, M.; Fort, A.; Esteves-Ferreira, A.; Sulpice, R.; McKeown, P.; Spillane, C. Regulation of Carotenoid Biosynthesis and Degradation in Lettuce (*Lactuca sativa* L.) from Seedlings to Harvest. *Int. J. Mol. Sci.* **2023**, 24, 10310. https://doi.org/10.3390/ijms241210310.
6. Tatarowska, B.; Milczarek, D.; Plich, J. The Content of Total Carotenoids, Vitamin C and Antioxidant Properties of 65 Potato Cultivars Characterised under the European Project ECOBREED. *Int. J. Mol. Sci.* **2023**, 24, 11716. https://doi.org/10.3390/ijms241411716.
7. Zamudio-Chávez, L.; Suesca, E.; López, G.; Carazzone, C.; Manrique-Moreno, M.; Leidy, C. Staphylococcus aureus Modulates Carotenoid and Phospholipid Content in Response to Oxygen-Restricted Growth Conditions, Triggering Changes in Membrane Biophysical Properties. *Int. J. Mol. Sci.* **2023**, 24, 14906. https://doi.org/10.3390/ijms241914906.
8. Rozanowska, M.; Edge, R.; Land, E.; Navaratnam, S.; Sarna, T.; Truscott, T. Scavenging of Cation Radicals of the Visual Cycle Retinoids by Lutein, Zeaxanthin, Taurine, and Melanin. *Int. J. Mol. Sci.* **2024**, 25, 506. https://doi.org/10.3390/ijms25010506.

References

1. González-Peña, M.A.; Ortega-Regules, A.E.; Anaya de Parrodi, C.; Lozada-Ramírez, J.D. Chemistry, Occurrence, Properties, Applications, and Encapsulation of Carotenoids—A Review. *Plants* **2023**, *12*, 313. [CrossRef]
2. Terao, J. Revisiting carotenoids as dietary antioxidants for human health and disease prevention. *Food Funct.* **2023**, *14*, 7799–7824. [CrossRef] [PubMed]
3. Farhana, A.; Saleem Khan, Y.; Alsrhani, A.; Manni, E.; Alameen, A.A.; Derafa, W.; Alhathlaul, N.; Atif, M.; Bashier Eltayeb, L. Antioxidant and Prooxidant Functions of Carotenoids in Human Health: Trigger Factors, Mechanism and Application. In *Recent Advances in Phytochemical Research*; IntechOpen: London, UK, 2025. [CrossRef]
4. Kispert, L.D.; Focsan, A.L. *Chemistry of Carotenoid Radicals and Complexes*; World Scientific Publishing Co. Pte. Ltd.: Singapore, 2024; ISBN 9789811278365.
5. Mukherjee, B.; Al Hoque, A.; Hota, S.H.; Gope, S.; Ray, M.; Barman, M.; Bhattacharya, S.; Chakraborty, S.; Das, L. Antioxidants and Their Physiological Role in Free Radical Scavenging. In *Dietary Supplements and Nutraceuticals*; Mukherjee, B., Ed.; Springer: Singapore, 2025. [CrossRef]
6. Arshad, M.T.; Maqsood, S.; Ikram, A.; Khan, A.A.; Raza, A.; Ahmad, A.; Gnedeka, K.T. Encapsulation Techniques of Carotenoids and Their Multifunctional Applications in Food and Health: An Overview. *Food. Sci. Nutr.* **2025**, *13*, e70310. [CrossRef] [PubMed]
7. Aanniz, T.; El Omari, N.; Elouafy, Y.; Benali, T.; Zengin, G.; Khalid, A.; Abdalla, A.N.; Sakran, A.M.; Bouyahya, A. Innovative Encapsulation Strategies for Food, Industrial, and Pharmaceutical Applications. *Chem. Biodivers.* **2024**, *21*, e202400116. [CrossRef] [PubMed]

International Journal of
Molecular Sciences

MDPI

Review

The Endless World of Carotenoids—Structural, Chemical and Biological Aspects of Some Rare Carotenoids

Nikolay E. Polyakov [1], A. Ligia Focsan [2], Yunlong Gao [3] and Lowell D. Kispert [4,*]

[1] Institute of Chemical Kinetics & Combustion, Institutskaya Str. 3, 630090 Novosibirsk, Russia; polyakov@kinetics.nsc.ru
[2] Department of Chemistry, Valdosta State University, Valdosta, GA 31698, USA; alfocsan@valdosta.edu
[3] College of Sciences, Nanjing Agricultural University, Nanjing 210095, China; yunlong@njau.edu.cn
[4] Department of Chemistry and Biochemistry, The University of Alabama, Tuscaloosa, AL 35487, USA
* Correspondence: lkispert@ua.edu

Abstract: Carotenoids are a large and diverse group of compounds that have been shown to have a wide range of potential health benefits. While some carotenoids have been extensively studied, many others have not received as much attention. Studying the physicochemical properties of carotenoids using electron paramagnetic resonance (EPR) and density functional theory (DFT) helped us understand their chemical structure and how they interact with other molecules in different environments. Ultimately, this can provide insights into their potential biological activity and how they might be used to promote health. In particular, some rare carotenoids, such as sioxanthin, siphonaxanthin and crocin, that are described here contain more functional groups than the conventional carotenoids, or have similar groups but with some situated outside of the rings, such as sapronaxanthin, myxol, deinoxanthin and sarcinaxanthin. By careful design or self-assembly, these rare carotenoids can form multiple H-bonds and coordination bonds in host molecules. The stability, oxidation potentials and antioxidant activity of the carotenoids can be improved in host molecules, and the photo-oxidation efficiency of the carotenoids can also be controlled. The photostability of the carotenoids can be increased if the carotenoids are embedded in a nonpolar environment when no bonds are formed. In addition, the application of nanosized supramolecular systems for carotenoid delivery can improve the stability and biological activity of rare carotenoids.

Keywords: rare carotenoids; xanthophylls; electron transfer; radical cation; carotenoid complexes; hydrogen bonding; nonpolar environment

1. Introduction

Carotenoids are a class of more than 1200 naturally occurring pigments synthesized by plants, algae and photosynthetic bacteria [1]. Carotenoids are important for human health because they act as antioxidants, helping to protect cells from damage caused by reactive oxygen species (ROS). They have been associated with a reduced risk of chronic diseases such as type 2 diabetes [2], cancer [3,4], heart disease or age-related macular degeneration [5]. Another important role is their provitamin A activity or the capability of some dietary carotenoids to form vitamin A by the action of dioxygenase enzymes [6]. For example, the dioxygenase enzyme β-Carotene 15-15′-oxygenase (BCO1) catalyzes the oxidative cleavage of dietary β-Carotene to retinal (vitamin A aldehyde) [7], which can be further reduced to retinol (vitamin A) or oxidized to retinoic acid (the biologically active form of vitamin A). Carotenoids such as β-carotene and α-carotene β-cryptoxanthin are considered provitamin A carotenoids, but any carotenoid with at least one unchanged β-ionone ring in its structure can have provitamin A activity. In addition to their health benefits, carotenoids in plants are also involved in photosynthesis. They are no longer considered just accessory pigments [8]; they have essential roles in photosynthesis [9], helping to capture light energy and transfer it to chlorophyll molecules and also protecting plants from damage caused by excess light and other environmental stresses [10]. Similarly, in the human body, carotenoids photoprotect against damage by intense light and harmful free radicals, and also maintain the structural and functional integrity of biological membranes. The mechanisms through which carotenoids may exert their health effects are complex, and further studies, including clinical studies, are needed to provide a comprehensive understanding. Carotenoids are indispensable for life. The world of carotenoids is endless. Their number and resourcefulness are immense, and only a tiny fraction of the 1200 carotenoids has been studied to date. Their complexity lies not only in their number, but in their different yet alike structures (given by the different functional groups and similar polyene chains) and their interactions with a multitude of different environments.

In the past 20 years, we have gained significant understanding about carotenoids' physicochemical properties and their interaction with other compounds in different environments [11]. We have mostly used advanced electron paramagnetic resonance (EPR) analysis techniques, such as continuous wave or pulsed electron nuclear double resonance (ENDOR), in combination with density functional theory (DFT) molecular orbital calculations, to elucidate the physisorption and electron and proton transfer processes that occur when carotenoids are adsorbed on solid artificial matrices [11]. Similar reactions were predicted in solution from electrochemistry studies performed in the 1990s, or in vivo in more recent studies performed in the 2000s. Additionally, EPR spin-trapping studies have been performed to characterize the inclusion complexes of carotenoids with different delivery systems [11]. In this paper, we review EPR and DFT studies relevant to understanding carotenoid chemistry, and we present the structural, chemical and biological aspects of several rare carotenoids that need to be considered when designing new systems for carotenoid delivery.

2. Studies of Electron and Proton Loss of Conventional Carotenoids by DFT and EPR

Carotenoids are prone to oxidation. We have studied the oxidation of carotenoids adsorbed on solid artificial matrices, such as silica alumina or imbedded in the pores of molecular sieves MCM-41, where their radical cation is formed by electron transfer from the carotenoid molecule to the matrix. Some of the most well-known carotenoids that we studied using EPR methods in combination with DFT include lycopene, β-carotene, zeaxanthin, canthaxanthin, lutein, astaxanthin and *cis*-bixin (Figure 1). With normal EPR at the X-band frequency (9 GHz), the carotenoid radical cation exhibits a single unresolved peak with g_{iso} = 2.0027, characteristic of organic π-radicals. In the year 1999, the EPR signal previously not resolved at the X-band frequency was resolved at a higher frequency (330 GHz) [12]. At higher frequencies, from 327 to 670 GHz, the unresolved line resolves into two peaks as a result of the g-anisotropy of $g_{\perp}$ = 2.0023 and $g_{||}$ = 2.0032, characteristic of a cylindrically symmetrical π-radical cation. Determining the g-tensors from high-field spectra is important for learning about molecular structure from its principal values. The difference $g_{xx} - g_{yy}$ decreases with increasing chain length. When $g_{xx} - g_{yy}$ approaches zero, the g-tensor becomes cylindrically symmetrical with $g_{xx} = g_{yy} = g_{\perp}$ and $g_{zz} = g_{||}$. This applies for the all-*trans* carotenoid radical cations and allows differentiation between carotenoid radical cations with cylindrical symmetry and other C-H organic radicals of different symmetry. The lack of temperature dependence of the EPR line widths over the range of 5–80 K at 327 GHz also suggests a rapid rotation of methyl groups even at 5 K, which averages out the proton couplings from three oriented β-protons. This results in isotropic β-proton couplings from rotating methyl groups [12].

Even though the number of hyperfine couplings is greatly reduced when considering this rapid rotation of the methyl groups, carotenoid radicals still contain a large number of anisotropic α-protons which give rise to numerous anisotropic coupling constants. These couplings cannot be determined by normal EPR. Instead, ENDOR techniques can be used to determine the hyperfine couplings of carotenoids adsorbed on silica alumina or in MCM-41. Continuous wave and pulsed ENDOR showed that for irradiated carotenoids on solid matrices, not only radical cations were formed, but also neutral radicals formed by the deprotonation of the radical cation, which is a weak acid. These neutral radicals formed by proton loss from the radical cations contain lots of similar hyperfine couplings to those of the radical cation, but ENDOR techniques helped to distinguish the two different radical species [11]. It is important to note that the presence of all radicals is enhanced by the irradiation of samples and the presence of metals, such as in metal-substituted MCM-41. The deprotonation of the radical cations to form neutral radicals determined by ENDOR on solid surfaces, which was also proven electrochemically in solution, needs to be considered in vivo where the radical cation is formed and is known to have a role in photoprotection mechanisms. We have also hypothesized that neutral radicals could have a role in an additional quenching mechanism to that of the radical cation [13].

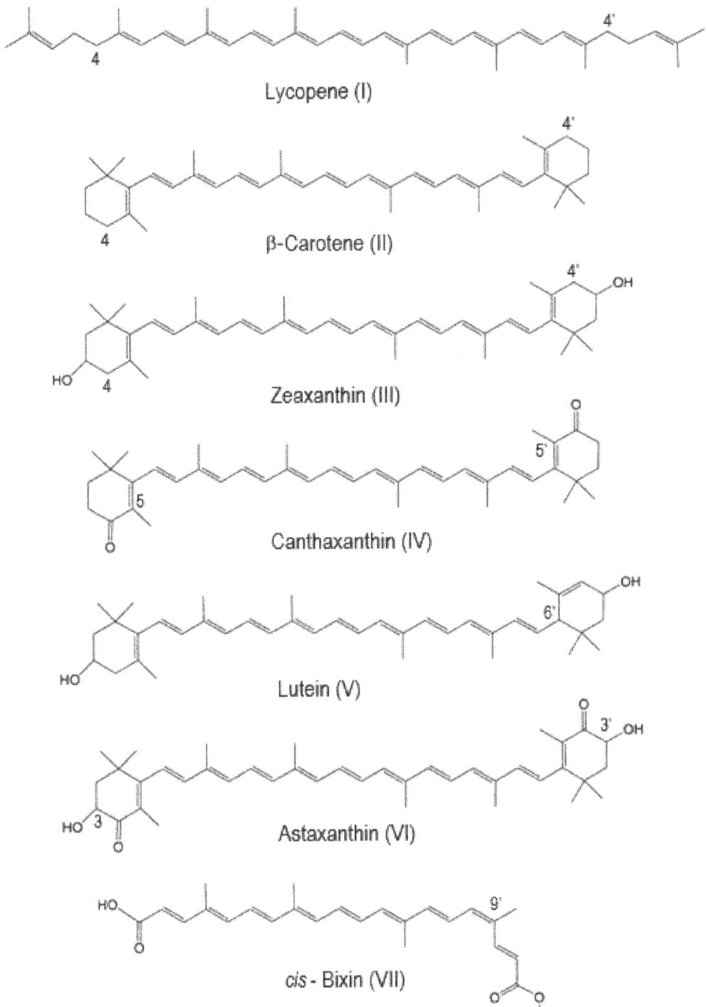

Figure 1. The structures of selected natural carotenoids.

DFT calculations were used starting in the early 2000s to determine the hyperfine coupling constants of radical cations and neutral radicals and to simulate spectra which matched the experimental ENDOR spectra and confirmed the identity of the radicals. DFT was also used to predict the most favorable positions for proton loss from the radical cation and establish the relative stability of the neutral radicals formed from the radical cation, as described next. Table 1 indicates the most favorable positions of proton loss from the radical cation for carotenoids listed in Figure 1.

Lycopene is a symmetric linear carotenoid that has ten methyl groups in four distinct positions: C1(1′), C5(5′), C9(9′) and C13(13′) (see Figure 1). The primed positions of this molecule are equivalent by symmetry to the unprimed positions. There is a smooth relationship between the relative energy ΔE(n) of a neutral radical formed by proton loss from the radical cation, and the conjugation or delocalization length, N, over which the unpaired spin density is distributed. The longer the conjugation length, the most stable the radical is. DFT has shown that the most stable neutral radicals for lycopene are formed by proton loss at the C4 or C4′ methylene positions, which extend conjugation. It is thus expected that proton loss occurs more favorably from the C4(4′) positions [14]. For β-carotene, which has two symmetric cyclohexene rings at the ends of the molecule, proton loss occurs also at the C4 methylene position, and symmetrically at C4′, rather than the methyl groups attached at C5(5′), C9(9′) and C13(13′) [15].

Zeaxanthin has the same structure as β-carotene with two additional hydroxyl groups on positions C3 and C3′, respectively. The two hydroxyl groups have no effect on the position of proton loss from a radical cation, so the most favorable neutral radicals form by proton loss at the C4(4′) methylene positions [16]. However, when the C4(4′) methylene positions contain carbonyl groups, for example in the case of canthaxanthin, proton loss occurs from the methyl groups attached at positions C5(5′). Lutein, an isomer of zeaxanthin which differs from it by one shifted double bond at C4′-C5′ (instead of C5′-C6′ in zeaxanthin), makes proton loss more favorable at the C6′ position instead of the C4′ position [15]. When both hydroxyl and carbonyl groups are present, such as in astaxanthin, it is possible for proton loss to occur at the C3 and C3′ positions [17]. Bixin is a C25 carotenoid, with only nine conjugated double bonds and thus a shorter-length carotenoid compared to the C40 structure of the others presented here. It is an asymmetric carotenoid with -COOH at one end and -$COOCH_3$ at the opposite end. Its IUPAC name is 6-methyl hydrogen (9′Z)-6,6′-diapocarotene-6,6′-dioate [1]. According to [18], bixin was the first *cis*-carotenoid to be isolated from natural sources and *trans*-bixin is a more stable isomer that the *cis* form. DFT calculations by Hernandez-Marin et al. [19] also show that, in most cases (with the exception of 13-cis auroxanthin), out of 11 carotenoids studied, the *trans* isomers are more stable than their corresponding 9- and 13-*cis* isomers, while the 15-*cis* isomers are the least stable isomers. However, upon oxidation of the neutral molecule to form the radical cation, our DFT calculations [20] show that the radical cation of *cis*-bixin becomes more stable than the *trans* radical cation of bixin. Furthermore, proton loss from *cis*-bixin occurs from the methyl group the C9′ position of the *cis*-bixin radical cation, on the side with the acetate group [20].

Proton loss from the radical cations of carotenoids has enabled us to predict the most acidic protons which are usually at the ends of the carotenoid to extend the conjugation length [14]. We have hypothesized that zeaxanthin and lutein radical cations' ability to deprotonate in light harvesting complex II (LHC II) and form neutral radicals could be linked to their quenching activity [21].

Table 1. The most favorable proton loss positions for selected carotenoids in our studies.

Carotenoid	The Most Favorable Proton Loss from the Radical Cation	Reference
Lycopene	C4(4′)	[14]
β-carotene	C4(4′)	[15]
Zeaxanthin	C4(4′)	[15,16]
Canthaxanthin	C5(5′)	[15]
Lutein	C6′	[15]
Astaxanthin	C3(C3′)	[17]
cis-Bixin	C9′	[20]

3. Improvement of Carotenoid Features by Nonpolar Environment and Absence of Hydrogen Bonding—Rare Carotenoids Discussion

The structure and type of a carotenoid (polar/nonpolar) and its position/orientation in the type of environment (polar/nonpolar) are also extremely important for the role that the carotenoid plays. The oxygen-containing groups (hydroxyl, carbonyl, carboxyl, acetate) situated at the ends of carotenoids can not only affect the position of proton loss as discussed above, but also play a role in anchoring the carotenoid in a certain environment.

DFT calculations, calorimetric experiments and EPR studies confirmed that carotenoids containing oxygen groups can form hydrogen bonds (H-bonds) on the surfaces of hosts [22]. This paper predicts the behavior of some rare carotenoids (Figure 2) containing oxygen groups.

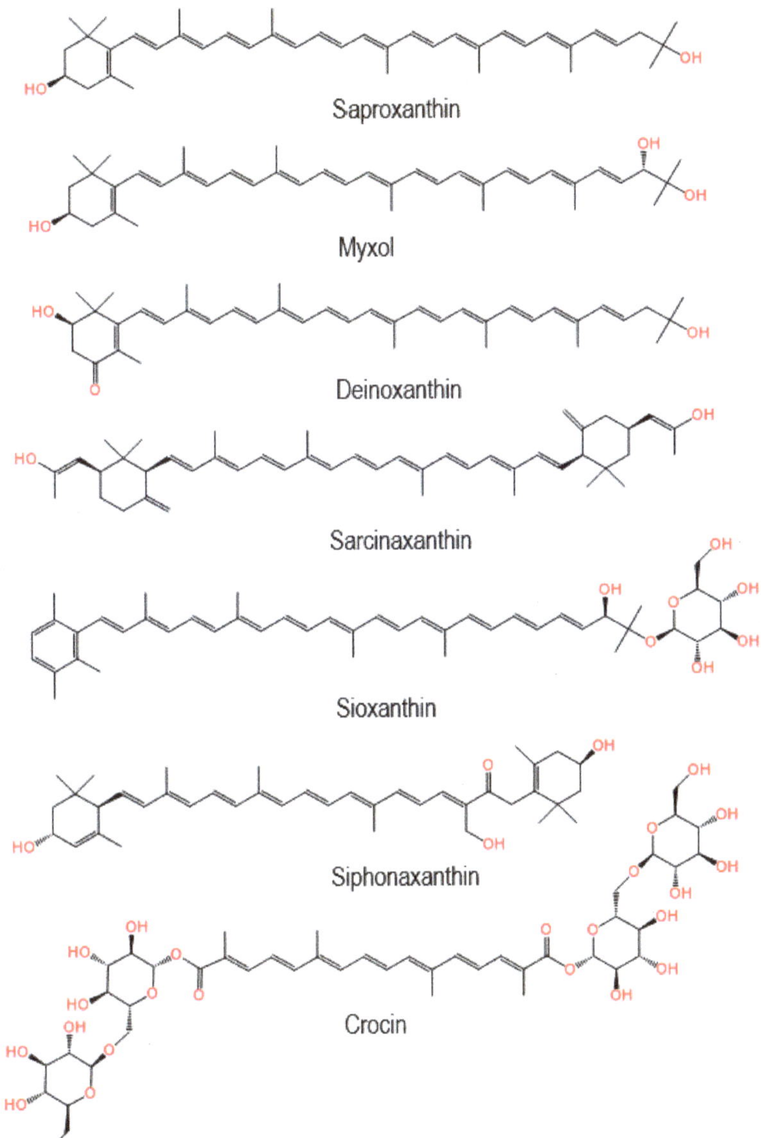

Figure 2. The structures of selected rare carotenoids.

Compared with conventional carotenoids, the rare carotenoids usually contain more oxygen-containing groups such as hydroxyl, keto, ester and ether groups, or a similar number of groups but situated somewhere else than the ring (see Figure 2). Both deinoxanthin and siphonaxanthin contain hydroxyl groups and a keto group. Deinoxanthin contains two hydroxyl groups and one keto group, while siphonaxanthin contains three hydroxyl groups and one keto group. Sioxanthin contains five hydroxyl groups and two ether groups; crocin contains fourteen hydroxyl groups, two ester groups, two keto groups and six ether groups.

It was confirmed that canthaxanthin can form hydrogen bonds (H-bonds) on surfaces of hosts such as MCM-41 [22]. For example, each canthaxanthin molecule can form two H-bonds on the surface of MCM-41 (see Figure 3). Hydrogen bonds are formed with the surface silanol groups in MCM-41.

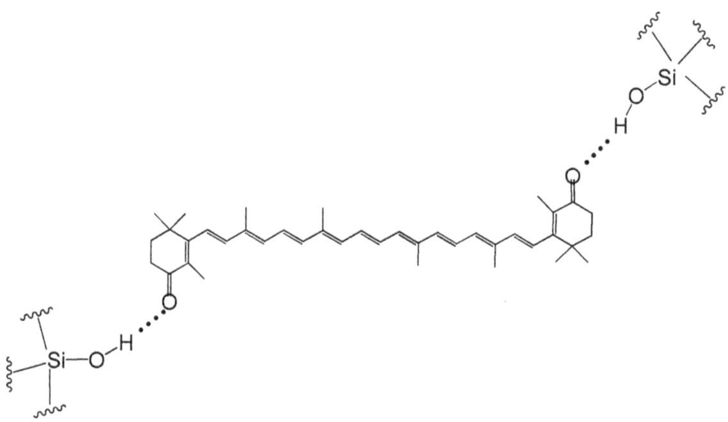

Figure 3. Formation of H-bonds of canthaxanthin with silanol groups on the surface of MCM-41.

If rare carotenoids with oxygen-containing groups like the ones described above are incorporated in host molecules, we hypothesize that multiple H-bonds can be formed (see Figure 4 for the rare carotenoid complexes). For example, deinoxanthin can form three H-bonds, and acts as both a H-bond donor and acceptor in the complex; myxol also forms three H-bonds, but can only act as a H-bond donor.

The carotenoids can be better stabilized when multiple bonds are formed. Our previous study [23] shows that both the HOMO and LUMO energies of carotenoids increase when the carotenoids act as H-bond donors, and the opposite is true if the carotenoids act as H-bond acceptors. Figure 5 shows the two modes for retinol adsorbed on the surface of MCM-41. DFT calculations demonstrated that both the HOMO and LUMO energies of retinol increase by ~0.2 eV compared with the free molecule when retinol acts as a H-bond donor, and the energies decrease by about ~0.1 eV when retinol acts as a H-bond acceptor [23]. This effect should be more pronounced if multiple H-bonds are formed (e.g., myxol acts as a H-bond donor in the 3 H-bonds and crocin acts as a H-bond donor in the 14 H-bonds, see Figure 4).

A study by Méndez-Hernández et al. [24] shows that there is a very strong linear correlation between DFT-calculated HOMO and LUMO energies (HLE) and the redox potentials of 51 polycyclic aromatic hydrocarbons (PAHs). The strong correlation obtained from the HLE and redox potentials of PAHs was independent of whether the solvent model was included in the calculations. This is consistent with our previous study [25], which shows that the oxidation potentials of carotenoids increase with decreasing the HOMO energies of the carotenoids. Carotenoids are more stable if their oxidation potentials are high because the carotenoids are less likely to be oxidized by metal ions, such as Cu^{2+} and Fe^{3+}. The antioxidant activity of a carotenoid's supramolecular complex is also high if the oxidation potential of the carotenoid is high because the scavenging ability of a carotenoid towards the free radicals increases nearly exponentially with increasing carotenoid oxidation potential, and the increase of 0.03–0.05 V causes the scavenging rate constant to increase about 30 times [26,27]. Our previous study also shows that the orientation of a carotenoid in the host molecule affects the antioxidant activity [25]. It is determined that free radicals abstract the most acidic hydrogen in the conjugated system of a carotenoid [27,28]. If access to the most acidic hydrogen is blocked for the free radicals, the antioxidant activity of the carotenoid decreases significantly. For instance, the most acidic hydrogen of deinoxanthin is at C3 (see Figure 4). If the hydrogen is close to the surface of the host molecule after the formation of the H-bonds, access to the hydrogen may be blocked, resulting in low antioxidant activity.

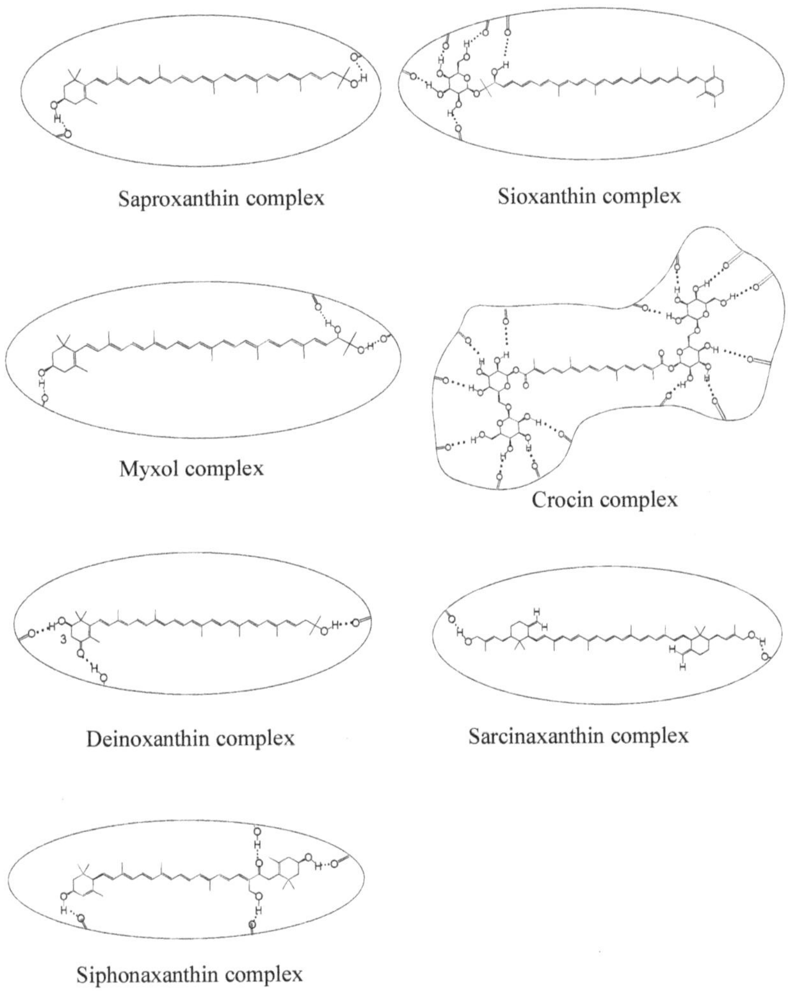

Figure 4. Complexes of the rare carotenoids indicating H bonding.

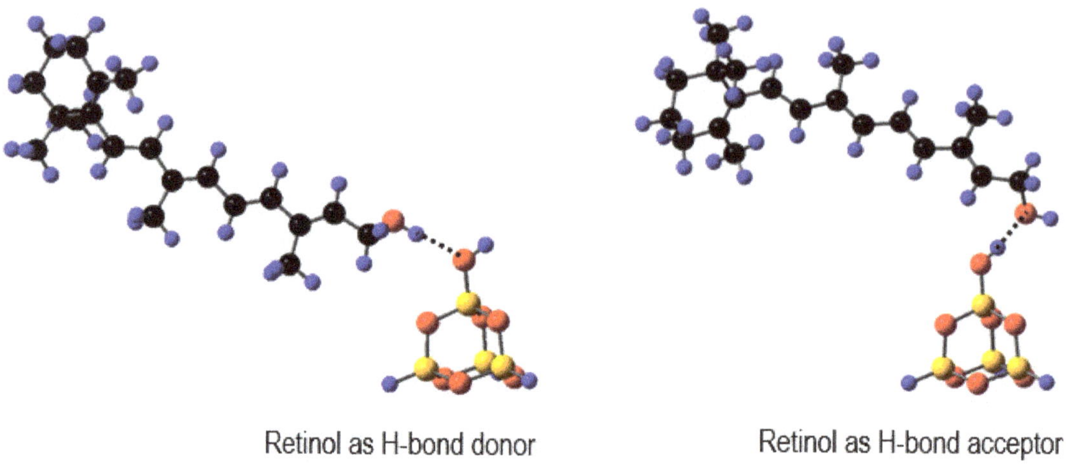

Figure 5. Two types of H-bonds for retinol adsorbed on the surface of MCM-41. H: blue, O: red, C: black and Si: yellow. Adapted from Ref. [23].

Our DFT calculations [23] also show that when a carotenoid acts as a H-bond acceptor, the formation of the H-bond decreases the energy of the LUMO of the carotenoid, which also stabilizes the neutral species more than the radical cation, disfavoring the photo-induced ET from the carotenoid to host molecules. This effect is more pronounced if the oxygen atom of the carotenoid is completely conjugated with the conjugated chain. However, if the carotenoid is a H-bond donor, the formation of the H-bond increases the energy of the LUMO of the carotenoid, which also stabilizes the radical cation more than the neutral species, thus significantly increasing the charge separation efficiency. The photoinduced electron transfer affects the photostability of carotenoids, which is related to the formation of carotenoid radicals [29]. Carotenoid radical cations are formed by photoinduced electron transfer to electron acceptors, and their deprotonation in the presence of proton acceptors creates carotenoid neutral radicals which are highly reactive species that can form other products as well as carotenoid dimmers [30]. It was also concluded in our study [31] that the photostability of a carotenoid is the highest if the environment of the carotenoid is nonpolar and no hydrogen bonds are formed because the electron transfer is intermolecular and thus is slow. There is an approximately exponential decrease in ET rate with increasing distance [32–35]. Another reason for the lower charge separation efficiency is that the nonpolar environment stabilizes the radical cation of the carotenoid less efficiently than the polar environment, according to our DFT calculations [31]. However, low photostability may be an advantage in some cases, such as in phototherapy, when radical cations or neutral radicals of carotenoids are needed. The rare carotenoids may contain several hydroxyl groups (e.g., crocin contains 14 hydroxyl groups) that can act as H-bond donors in the host molecules, and would be easily photo-oxidized. The rare carotenoids are good candidates for this purpose. The surface of a host can be modified so that a carotenoid can act as only a H-bond donor or as an acceptor. For example, when the surface functional group Ti-OH is replaced by Ti-F by ion exchange, 7-hydroxycoumarin (7-CN) acts as both a H-bond donor and acceptor on the surface of TiO_2, but acts as only a H-bond donor on F-TiO_2 (see Figure 6). EPR measurements showed five-fold increases in free radical yields for 7-CN on F-TiO_2 compared with TiO_2. DFT calculations for the 7-CN on TiO_2 and F-TiO_2 were performed to investigate these phenomena [36]. The calculations show that when 7-CN act as the H-bond donor, the driving force for photo-induced electron transfer from the dyes to TiO_2 is higher, and the dye's excited state mixes strongly with the TiO_2 conduction band states. This is attributed to the shorter distance between the coumarins and the surface of TiO_2 when coumarins act as the H-bond donors [36].

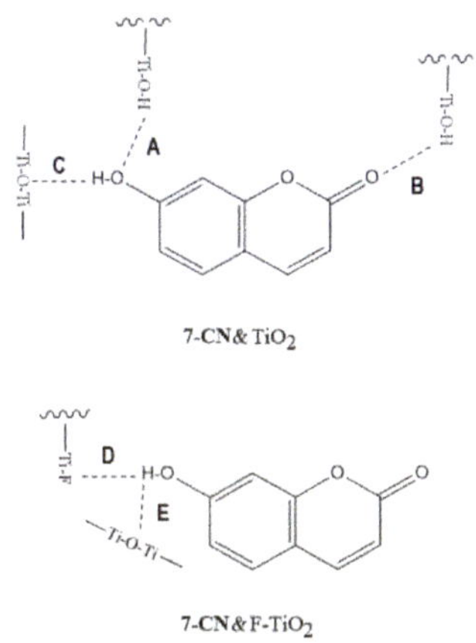

Figure 6. The formation of H-bonds of 7-CN on TiO_2 and F-TiO_2. 7-CN acts as both H-bond donor and acceptor on TiO_2, but acts as only H-bond donor on F-TiO_2. Adapted from Ref. [36].

Different anchoring modes of carotenoids on the surface of a semiconductor affect the bond length, degree of conjugation, energies of the HOMO and LUMO and the mixing of the carotenoids' excited states with the conduction band states of the semiconductor [37]. For example, three anchoring modes were observed for retinoic acid (RA) on the surface of TiO_2 (see Figure 7). Mode A: the carboxylic acid group loses a proton and forms two coordination bonds with two surface Ti atoms of TiO_2. Mode B: the carboxylic acid group loses a proton and forms one coordination bond with one surface Ti atom of TiO_2 and one H-bond with a Ti-OH group on the surface of TiO_2. Mode C: the carboxylic acid group does not lose a proton and forms one coordination bond with one surface Ti atom of TiO_2 and one H-bond due to the interaction between the OH group of the carboxylic acid group and the bridging oxygen atom on the surface of TiO_2. The degree of conjugation in RA increases from B to A to C. TD-DFT calculations [37] showed that compared with the free carotenoid, the maximum absorption wavelengths (λ_{maxs}) of A, B and C shift to the red because the degree of conjugation of RA anchored on TiO_2 increases, which is confirmed by the different colors of the samples. Mode B involving the surface Ti-OH group has the highest driving force for photo-induced ET. This is attributed to the lowest degree of conjugation of RA for this mode and the highest mixing of the RA's LUMO with the conduction band states of TiO_2.

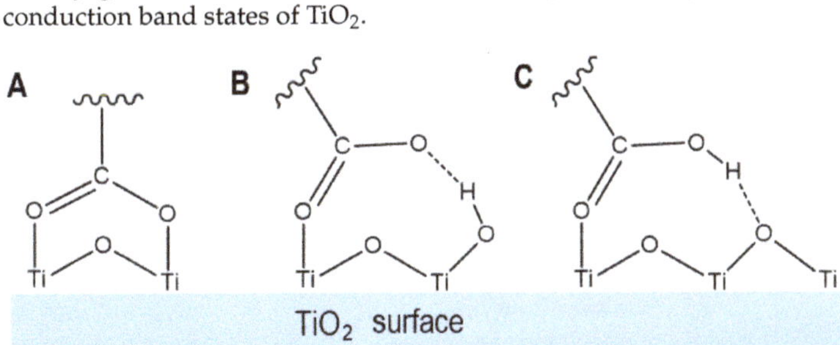

Figure 7. Three anchoring modes of a carboxylic acid group on the surface of TiO_2: (**A**) the carboxylic acid group loses a proton and forms two coordination bonds with two surface Ti atoms of TiO_2; (**B**) the carboxylic acid group loses a proton and forms one coordination bond with one surface Ti atom of TiO_2 and one hydrogen bond (H-bond) with a Ti-OH group on the surface of TiO_2; (**C**) the carboxylic acid group does not lose a proton and forms one coordination bond with one surface Ti atom of TiO_2 and one H-bond due to the interaction between the OH group of the carboxylic acid group and the bridging oxygen atom on the surface of TiO_2. Adapted from Ref. [37].

Carotene carotenoids can also form complexes with metal ions through metal–olefin interaction [38]. β-carotene can form a complex with Cu^{2+} on the surface of Cu-MCM-41, as examined by the EPR study [38]. Figure 8 shows the optimized complex from DFT calculations [39]. The EPR study showed that the formation of the complex favors the photoinduced electron transfer from β-carotene to Cu^{2+} and also permits thermal back electron transfer from Cu^{+} to β-carotene radical cation [38]. However, xanthophyll carotenoid canthaxanthin does not form a complex with Cu^{2+} on the surface of Cu-MCM-41. DFT calculations showed canthaxanthin prefers to form H-bonds with the surface silanol groups (−SiOH) because the interaction energy is higher than that between canthaxanthin and Cu^{2+} [22]. It is worth studying the rare carotenoids on the surfaces of semiconductors. For instance, deinoxanthin may form both coordination bonds and H-bonds on the surface of a semiconductor. It is important to know how the formation of those bonds affects the degree of conjugation, energies of the HOMO and LUMO, the mixing of the carotenoids' excited states with the conduction band states of the semiconductor, and thus the photoinduced charge separation of the carotenoids. There may be steric hindrance for the interactions, depending on the types of host molecules. Choosing appropriate host molecules for semiconductors can remove or mitigate the steric hindrance.

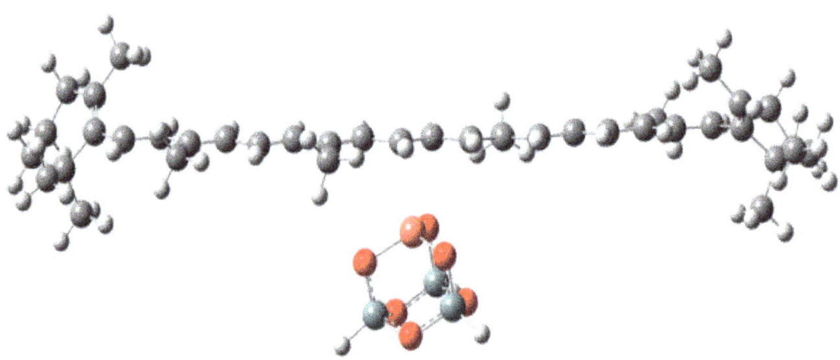

Figure 8. The optimized structures of Cu^{2+}-β-carotene complex by B3LYP/6-31G(d). H: light gray, O: red, Cu: orange and C: dark gray. Si: blue gray. Adapted from Ref. [39].

4. Improvement of Carotenoid Features by Supramolecular Delivery Systems

In addition to the previously studied carotenoids shown in Figure 1, the rare carotenoids shown in Figure 2 have similar limitations in practical applications. They are scarcely stable and lose most of their functionality after exposure to light, heat, oxygen and acids. Moreover, due to their low solubility in water, they have poor absorption and low bioavailability after oral administration. The bioavailability of carotenoids is affected not only by the type and amount of carotenoid but the molecular linkage, the matrix in which the carotenoid is incorporated, effectors of absorption and bioconversion, the nutrient status of the host, genetic factors, host-related factors and mathematical interactions [6].

Thus, despite their beneficial effects, chemical instability, low bioavailability and high susceptibility to process conditions drive researchers to find appropriate approaches to overcome the above-mentioned obstacles. From our own experience and published data of other authors [40–47], the use of supramolecular delivery systems can help to solve most of the above-mentioned problems of carotenoid applications in the medicine, cosmetics and food industries. In addition, insight from our EPR experiments has allowed us to find some unusual effects of supramolecular delivery systems on the physicochemical and photochemical properties of carotenoids, as discussed below. In this part, we will also briefly describe the biological activities of some rare carotenoids shown in Figure 2, the supramolecular systems used for carotenoids delivery and how their application can help to improve the stability and biological activity of rare carotenoids.

4.1. Biological Activities of Rare Carotenoids

Rare carotenoids, including siphonaxanthin, saproxanthin, myxol, sioxanthin and some others, contain oxygen groups and belong to the xanthophyll carotenoids. They are mainly present in some bacteria and marine algae. The most important feature of carotenoids is their antioxidant activity. Carotenoids are the most efficient natural quenchers of singlet oxygen due to their conjugated double bonds system, and this quenching ability increases with increasing the number of conjugated double bonds. Taking into account that singlet oxygen is the main reactive oxygen species (ROS) responsible for photo-oxidative damage to cell membranes and human tissues, carotenoids might have beneficial effects on skin protection [48]. Due to their electron-rich structure, carotenoids can also scavenge other free radicals, including hydroxyl and peroxyl radicals, as well as free radicals of xenobiotics [49–51]. The strong antioxidant activity of carotenoids reduces the risk of many types of diseases caused by free radicals, such as cardiovascular disease, cancer and other age-related diseases [52]. Assuming the antioxidant ability of carotenoids as an important factor against oxidative stress, the statistical data indicated a direct correlation between the use of carotenoids in diets and a decreased incidence of cancer types. Rare marine carotenoids show great chemical diversity, resulting in potential novel therapeutic properties, but many of these novel properties are still the subject of future studies.

Siphonaxanthin shows beneficial effects on health, including anticancer activity in the treatment of leukemia [53]. Additionally, it shows anti-angiogenic, antioxidant and anti-inflammatory activities [54,55]. Saproxanthin and myxol are potent antioxidants. They show high antioxidant activity against lipid peroxidation in the rat brain homogenate model and have a neuroprotective effect on L-glutamate-induced toxicity [56,57]. Myxol is also effective in strengthening biological membranes, reducing permeability to oxygen [58]. Deinoxanthin is a unique carotenoid synthesized by *Deinococcus radiodurans*, one of the most radioresistant organisms known for its high resistance to stresses including radiation and oxidants. Deinoxanthin exhibits significantly stronger ROS-scavenging activity in comparison with other carotenoids [59]. Using quantum chemical calculations, it was found that this carotenoid possesses lower triplet excitation energy than other carotenoids, such as β-carotene and zeaxanthin, which provides its strong potential in the energy-transfer-based ROS-scavenging process. In addition, authors show that the H-atom donating potential of deinoxanthin is also crucial for its strong antioxidant activity. On the other hand, an in vitro study demonstrates the novel functional property of deinoxanthin as a potent inducer of apoptosis in human cancer cells. Deinoxanthin treatment caused an increase in ROS in cancer cells, suggesting possible pro-oxidant activity [60]. The authors suggest that deinoxanthin could be potentially useful as a chemopreventive agent. When carotenoids are delivered with ROS-inducing cytotoxic drugs, they can minimize the adverse effects of these drugs on normal cells by acting as antioxidants and minimizing oxidative stress, without interfering with their cytotoxic effects on cancer cells as pro-oxidants which enhance oxidative stress in cancer cells [61].

Sioxanthin is a C_{40} carotenoid, glycosylated on one end of the molecule. Glycosylation is an unusual feature among carotenoids and sioxanthin represents a poorly studied group of carotenoids which are polar on one end and non-polar on the other [56,62]. Crocin, in contrast to all other carotenoids, is a water-soluble carotenoid glycosylated on both ends of the molecule [63]. Crocin is effective as an antioxidant and as a learning and memory enhancer, and it shows activity against brain neurodegenerative and Alzheimer's disorders. Crocin also shows an anti-angiogenesis effect and is likely to be involved in the regulation of molecules in the angiogenesis pathway [64]. However, similar to other highly bioactive carotenoids, it has limited use due to its instability at low pH and during heat and oxidative stress, and its low bioavailability, and is a unique C50 carotenoid isolated from marine bacteria, which shows high antioxidant activity in the singlet oxygen ($^1O_2$) quenching model [65].

Table 2 summarizes the biological roles of rare carotenoids described above. It can be concluded that although antioxidant activity is a common property of most carotenoids, its specific manifestation depends on the localization of the carotenoid in the cell and on its oxidative potential. As for the other properties of carotenoids, they significantly depend on the chemical structure.

Table 2. Biological roles of rare carotenoids.

Rare Carotenoid	Biological Role	Reference
Siphonaxanthin	• anticancer activity in the treatment of leukemia • anti-angiogenic, antioxidant and anti-inflammatory activities	[53] [54,55]
Saproxanthin	• potent antioxidant	[56,57]
Myxol	• potent antioxidant • strengthening of biological membranes, reducing permeability to oxygen	[56,57] [58]
Sioxanthin	• role in oxidative stress prevention/membrane fluidity	[56,62]
Deinoxanthin	• strong ROS-scavenging activity • potent inducer of apoptosis in human cancer cells, possible pro-oxidant activity	[59] [60]
Crocin	• antioxidant, learning and memory enhancer, activity against brain neurodegenerative and Alzheimer disorders • potent inducer of apoptosis and anti-angiogenesis effect	[63] [64]
Sarcinaxantin	• high antioxidant activity for singlet oxygen ($^1O_2$)	[65]

4.2. Supramolecular Delivery Systems of Carotenoids

The incorporation of carotenoids in nanosized supramolecular carriers can significantly change their physicochemical and biological properties, as well as their therapeutic potential. The nanoencapsulation of carotenoids is an effective strategy to improve their stability towards heat, light, oxygen, metal ions and processing conditions, as well as to increase water solubility [44–47]. In addition, nanoencapsulation can change the oxidation potentials of carotenoids [42], their absorption spectra [27] and their ability to form J- and H-type self-aggregates [27]. The latter property is especially important for xanthophyll carotenoids [66,67]. Additionally, nanoencapsulation enhances the bioavailability of carotenoids via modulating their release kinetics from the delivery system. The important property of supramolecular carriers is the possibility to modify their surface by attaching new functional groups, providing targeted delivery to the necessary organs or cell receptors [68,69]. The design of functional nanocarriers enables even on-demand drug release in specific microenvironments of various diseases. Nowadays, nanocarriers include lipid nanoparticles, polymers, micelles, inorganic nanoparticles, hybrid nanoparticles and others. Detailed descriptions of various nanocarriers can be found in numerous reviews [44–47]. Here, we will just briefly describe two delivery systems, polysaccharide arabinogalactan

(AG) and saponin glycyrrhizin or glycyrrhizic acid (GA) (see Figure 9), studied earlier by EPR and optical techniques [40–43].

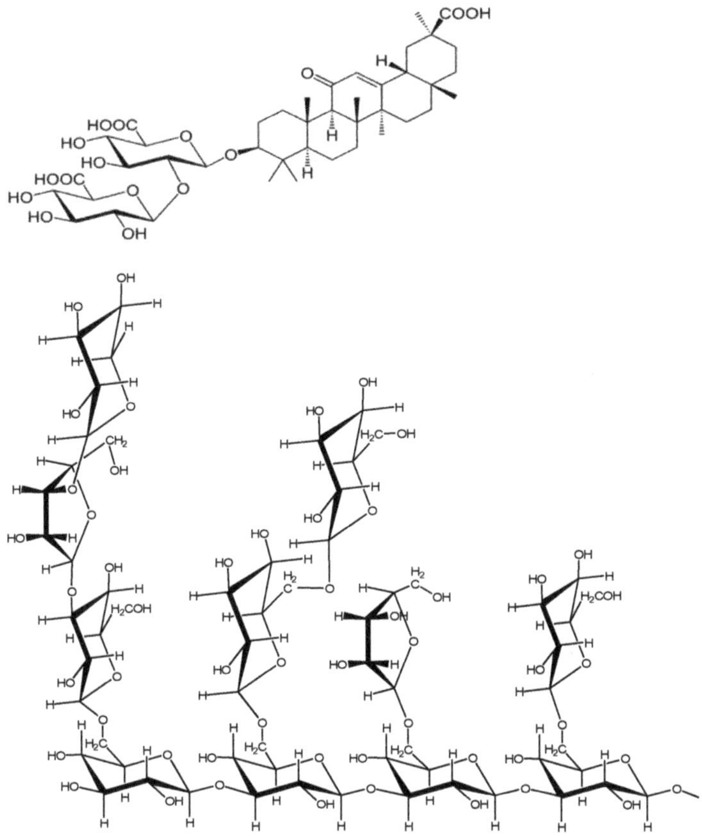

Figure 9. The structures of glycyrrhizin molecule (GA) (**top** image), and fragment of a branched polysaccharide arabinogalactan (AG) (**bottom** image).

Glycyrrhizin (GA) is the main active component of licorice root, and has frequently been used in traditional Persian, Chinese and Japanese medicine [70,71]. Recent studies have also shown its antioxidant [72], anticancer [73] and antivirus activities, including SARS-CoV-2 virus [74,75]. In addition, it was demonstrated that due to its amphiphilic nature, GA is able to form micelles in aqueous solutions, as well as water-soluble supramolecular complexes with hydrophobic drugs, including carotenoids [76]. Furthermore, the membrane-modifying ability of GA has been discovered as one of the key mechanisms of its activity [77,78].

The encapsulation of the xanthophyll carotenoids lutein and zeaxanthin into GA micelles protects these carotenoids from oxidation by ROS and metal ions [43]. It was demonstrated that GA forms supramolecular complexes with carotenoids, not only in water solutions with increasing carotenoid solubility by several orders, but also in non-aqueous organic solvents such as methanol, DMSO and acetonitrile [41]. This GA property is important for understanding the GA-assisted transport of carotenoid molecules through lipophilic cell membranes and their membrane protection properties. One of the most important biological properties of carotenoids is their antioxidant activity. Using the EPR spin-trapping technique, a synergetic effect of GA on the scavenging rates of hydroperoxyl radicals by carotenoids has been demonstrated [42]. The strongest increase in scavenging rates (20–30 times) was observed for carotenoids with high oxidation potentials. Since the scavenging rate is strongly potential-dependent, we suggested that GA affects the oxidation potential of the carotenoids. Electrochemical measurement of the oxidation potentials of the carotenoids zeaxanthin and canthaxanthin confirms this hypothesis. The increase in the oxidation potentials by 0.03–0.05 V was observed in the presence of GA [42].

Another characteristic feature of xanthophyll carotenoids is the ability to form J- and H-type aggregates in aqueous solution and in lipid membranes [51,52]. EPR spin-trapping and optical studies have shown that the formation of such aggregates significantly reduces the antioxidant and photoprotection abilities of these carotenoids [17]. The complexation of these carotenoids with GA reduces the aggregation ability of the xanthophyll carotenoids, and simultaneously increases their scavenging ability towards oxygen radicals. Taking into account the wide spectrum of biological activity of xanthophyll carotenoids, including eye and skin protection, GA could be considered a prospective carrier for rare carotenoids.

Arabinogalactan (AG) is a natural branched polysaccharide with a molecular mass near 16 kDa extracted from Siberian larch [79]. AG is a highly water-soluble polymer which produces low-viscosity water solutions. It is biodegradable, biocompatible and contains different functional groups, such as hydroxyl and carboxylic acid, that make it suitable for conjugation with functional groups for targeted delivery. The solubility of carotenoids complexes with AG prepared by the mechanochemical technique [40] were 2–5 mM in aqueous solution, which is several orders of magnitude higher than the solubility of free carotenoids in pure water (~1 nM) [80]. The mechanochemical method allows the preparation of inclusion complexes of drugs with natural or synthetic polymers in a solid state in one technological step without using any organic solvents [79]. It was shown that complexation with AG prevents the H-aggregate formation of xanthophyll carotenoids in the presence of water, similarly to what was detected for GA complexes [17]. Furthermore, the xanthophyll carotenoids lutein, astaxanthin and canthaxanthin have demonstrated an enhancement of photostability (5–10 times) and oxidation stability in AG complexes [40,43]. We assume that the main mechanism of the enhanced photostability and oxidation stability of these carotenoids in the polymer nanoparticles is their isolation from water molecules by incorporation into the hydrophobic polymer environment. As demonstrated by various EPR techniques, the oxidation of carotenoids results in the formation of carotenoid radical cations, which are unstable in aqueous solutions due to fast deprotonation and the formation of unstable neutral radicals [30,81]. The high effectiveness of AG as a delivery system for zeaxanthin and lutein was also confirmed in vivo using wild-type mice as a model [82]. Supplementation with these macular carotenoids can prevent and reduce the risk of age-related macular degeneration and other ocular diseases. Significant increases in lutein and zeaxanthin amounts in serum, liver and RPE/choroid of the mice were detected when they used the carotenoid-AG complex.

An important property of the carotenoid-AG inclusion complex, discovered by the EPR technique, is the stabilization of carotenoid radical cations. This feature has been demonstrated using EPR and ESEEM (electron spin echo envelope modulation) techniques under UV irradiation of the canthaxanthin-AG complex on the surface of titanium dioxide nanoparticles [40]. The carotenoid-AG complex, when irradiated on TiO_2, shows a significant increase in the EPR signal intensity of carotenoid radical cations compared to that of pure carotenoids. TiO_2 nanoparticles are widely used in photocatalysis and in artificial solar cells due to their ability to absorb light and to transfer negative or positive charge to a corresponding electron acceptor or donor [83]. The authors of [40] suggested that the significant increase in the yield of radical cations is due to the isolation of the carotenoid radical cation from the surface of TiO_2 by incorporation into the polysaccharide matrix. It allows more efficient charge separation and reduces the rate of back-electron transfer. Note that such a method of light energy transformation is similar to the mechanism used by plants for utilization of solar energy in photosynthesis. It is also very important that in contrast with the pure carotenoid-TiO_2 system, which allows carotenoid radical cations to be detected only at low temperatures (77 K), in the case of the carotenoid-AG complex, a significant increase in the stability of the radical cation was detected. Increasing the temperature up to room temperature does not lead to the disappearance of the spectrum. The lifetime of the canthaxanthin radical cation measured in [40] is approximately 10 days at room temperature. An increased stability of the carotenoid radical cation incorporated into a polymer host opens wide possibilities for the application of these complexes in the design of artificial light-harvesting, photoredox and catalytic devices. Additionally, the stabilization of carotenoid radical cations might be important for the regulation of the antioxidant/pro-oxidant activities of carotenoids. The stability of radical cations is a very important factor for the antioxidant activity of carotenoids. On the other hand, as mentioned above, the deprotonation of radical cations results in the formation of highly reactive neutral radicals. Yang, with co-authors, demonstrated that the generation of oxygen-irrelevant neutral radicals of carotenoids in the tumor microenvironment can offer novel opportunities to maximize the efficacy of chemodynamic therapy [84]. In this study, the authors demonstrated that proton-coupled electron transfer can promote the generation of neutral C-centered radicals of astaxanthin. The free radicals burst can significantly elevate free radical stress and induce cancer cell apoptosis.

4.3. Several Examples of Using Supramolecular Delivery Systems for Rare Carotenoids

In contrast to commonly used carotenoids, such as β-carotene, lutein, zeaxanthin, lycopene and astaxanthin, there are very few examples of using supramolecular delivery systems for rare carotenoids [43]. Esposito, with co-authors, developed lipid nanocarriers (ethosomes and organogels) for the cutaneous administration of crocin [85]. In vivo studies based on tape stripping and skin reflectance spectrophotometry have shown a more rapid anti-inflammatory effect and a rapid penetration of crocin from lipid nanocarriers, probably due to a strong interaction between phospholipids in the carrier and the lipids present in the stratum corneum. Since crocin has been shown to have the potential of inhibiting tumor genesis in a variety of cancers, there were attempts to use several drug delivery systems to improve its antitumor efficacy. For example, magnetite nanoparticles (MNPs) coated with natural polymers such as dextran have emerged as a promising technology for the targeted delivery of crocin to its site of action [86,87]. Coated MNPs can reduce the uptake by phagocytic cells, prolong the circulation time and passively accumulate in the tumor, leading to improvements in the therapeutic effect of crocin.

5. Conclusions

This paper predicts the behavior of some rare carotenoids that contain more oxygen groups than the conventional carotenoids or that are placed in positions other than on the rings. By careful design or self-assembly, these rare carotenoids can form multiple H-bonds and coordination bonds in host molecules. The stability, oxidation potentials and antioxidant activity of the carotenoids can be improved, and the photo-oxidation efficiency of the carotenoids can also be controlled. The photostability of the carotenoids can be increased if the carotenoids are embedded in a nonpolar environment when no bonds are formed.

Due to their lipophilicity, carotenoids also easily form "host-guest"-type supramolecular inclusion complexes with water-soluble compounds that have a suitable hydrophilic surface and hydrophobic interior. The incorporation of carotenoids results in noncovalent binding between the "guest" carotenoid and the "host" macromolecule and can significantly change the physical and chemical properties of carotenoids. Studies of the complexation of carotenoids with glycyrrhizin and arabinogalactan show that complexation affects the aggregation ability of some xanthophylls, their photostability and antioxidant activity. We have presented here our expertise on several conventional carotenoids studied by EPR and DFT and discussed the possible chemistry for some other several rare carotenoids, but endless possibilities remain for the approximately 1000 carotenoids left to be studied.

Author Contributions: Conceptualization, L.D.K.; writing—original draft preparation, N.E.P., A.L.F. and Y.G.; writing—review and editing, N.E.P., A.L.F., Y.G. and L.D.K.; project administration, L.D.K. All authors have read and agreed to the published version of the manuscript.

Funding: This research received no external funding.

Institutional Review Board Statement: Not applicable.

Informed Consent Statement: Not applicable.

Data Availability Statement: Not applicable.

Conflicts of Interest: The authors declare no conflict of interest.

References

1. Yabuzaki, J. Carotenoids database: Structures, chemical fingerprints and distribution among organisms. *Database J. Biol. Databases Curation* **2017**, *2017*, bax004. [CrossRef] [PubMed]
2. Wang, P.Y.; Fang, J.C.; Gao, Z.H.; Zhang, C.; Xie, S.Y. Higher intake of fruits, vegetables or their fiber reduces the risk of type 2 diabetes: A meta-analysis. *J. Diabetes Investig.* **2016**, *7*, 56–69. [CrossRef] [PubMed]
3. Aune, D.; Giovannucci, E.; Boffetta, P.; Fadnes, L.T.; Keum, N.; Norat, T.; Greenwood, D.C.; Riboli, E.; Vatten, L.J.; Tonstad, S. Fruit and vegetable intake and the risk of cardiovascular disease, total cancer and all-cause mortality-A systematic review and dose-response meta-analysis of prospective studies. *Int. J. Epidemiol.* **2017**, *46*, 1029–1056. [CrossRef] [PubMed]
4. Saini, R.K.; Rengasamy, K.R.; Mahomoodally, F.M.; Keum, Y.-S. Protective effects of lycopene in cancer, cardiovascular, and neurodegenerative diseases: An update on epidemiological and mechanistic perspectives. *Pharmacol. Res.* **2020**, *155*, 104730. [CrossRef] [PubMed]
5. Eggersdorfer, M.; Wyss, A. Carotenoids in human nutrition and health. *Arch. Biochem. Biophys.* **2018**, *652*, 18–26. [CrossRef]
6. Castenmiller, J.J.M.; West, C.E. Bioavailability and bioconversion of carotenoids. *Annu. Rev. Nutr.* **1998**, *18*, 19–38. [CrossRef] [PubMed]
7. dela Seña, C.; Riedl, K.M.; Narayanasamy, S.; Curley, R.W., Jr.; Schwartz, S.J.; Harrison, E.H. The human enzyme that converts dietary provitamin A carotenoids to vitamin A is a dioxygenase. *J. Biol. Chem.* **2014**, *289*, 13661–13666. [CrossRef]
8. Collini, E. Carotenoids in photosynthesis: The revenge of the "accessory" pigments. *Chem* **2019**, *5*, 494–495. [CrossRef]
9. Hashimoto, H.; Uragami, C.; Cogdell, R.J. Carotenoids and photosynthesis. In *Carotenoids in Nature*; Stange, C., Ed.; Subcellular Biochemistry; Springer: Cham, Switzerland, 2016; Volume 79.
10. Simkin, A.J.; Kapoor, L.; Doss, C.G.P.; Hofmann, T.A.; Lawson, T.; Ramamoorthy, S. The role of photosynthesis related pigments in light harvesting, photoprotection and enhancement of photosynthetic yield in planta. *Photosynth. Res.* **2022**, *152*, 23–42. [CrossRef]
11. Focsan, A.L.; Polyakov, N.E.; Kispert, L.D. Carotenoids: Importance in daily life—Insight gained from EPR and ENDOR. *Appl. Magn. Reson.* **2021**, *52*, 1093–1112. [CrossRef]
12. Konovalova, T.A.; Krzystek, J.; Bratt, P.J.; van Tol, J.; Brunel, L.-C.; Kispert, L.D. 95–670 GHz EPR studies of canthaxanthin radical cation stabilized on a silica−Alumina surface. *J. Phys. Chem. B* **1999**, *103*, 5782–5786. [CrossRef]
13. Focsan, A.L.; Kispert, L.D. Radicals formed from proton loss of carotenoid radical cations: A special form of carotenoid neutral radical occurring in photoprotection. *J. Photochem. Photobiol. B Biol.* **2017**, *166*, 148–157. [CrossRef] [PubMed]
14. Focsan, A.L.; Bowman, M.K.; Molnár, P.; Deli, J.; Kispert, L.D. Carotenoid radical formation: Dependence on conjugation length. *J. Phys. Chem. B* **2011**, *115*, 9495–9506. [CrossRef] [PubMed]
15. Lawrence, J.; Focsan, A.L.; Konovalova, T.A.; Molnar, P.; Deli, J.; Bowman, M.K.; Kispert, L.D. Pulsed electron nuclear double resonance studies of carotenoid oxidation in Cu(II)-substituted MCM-41 molecular sieves. *J. Phys. Chem. B* **2008**, *112*, 5449–5457. [CrossRef]
16. Focsan, A.L.; Bowman, M.K.; Konovalova, T.A.; Molnár, P.; Deli, J.; Dixon, D.A.; Kispert, L.D. Pulsed EPR and DFT characterization of radicals produced by photo-oxidation of zeaxanthin and violaxanthin on silica−alumina. *J. Phys. Chem. B* **2008**, *112*, 1806–1819. [CrossRef]
17. Focsan, A.L.; Bowman, M.K.; Shamshina, J.; Krzyaniak, M.D.; Magyar, A.; Polyakov, N.E.; Kispert, L.D. EPR study of the astaxanthin *n*-octanoic acid monoester and diester radicals on silica–alumina. *J. Phys. Chem. B* **2012**, *116*, 13200–13210. [CrossRef]
18. Rivera-Madrid, R.; Aguilar-Espinosa, M.; Cardenas-Conejo, Y.; Garza-Caligaris, L.E. Carotenoid derivates in achiote (*Bixa orellana*) seeds: Synthesis and health promoting properties. *Front. Plant Sci.* **2016**, *7*, 1406. [CrossRef]
19. Hernandez-Marin, E.; Galano, A.; Martínez, A. *Cis* carotenoids: Colorful molecules and free radical quenchers. *J. Phys. Chem. B* **2013**, *117*, 4050–4061. [CrossRef]
20. Tay-Agbozo, S.S.; Krzyaniak, M.D.; Bowman, M.K.; Street, S.; Kispert, L.D. DFT and ENDOR study of bixin radical cations and neutral radicals on silica–alumina. *J. Phys. Chem. B* **2015**, *119*, 7170–7179. [CrossRef]
21. Focsan, A.L.; Molnár, P.; Deli, J.; Kispert, L. Structure and properties of 9′-*cis* neoxanthin carotenoid radicals by electron paramagnetic resonance measurements and density functional theory calculations: Present in LHC II? *J. Phys. Chem.* **2009**, *113*, 6087–6096. [CrossRef]

22. Gao, Y.; Xu, D.; Kispert, L.D. Hydrogen bond formation between the carotenoid canthaxanthin and the silanol group on MCM-41 surface. *J. Phys. Chem. B* **2015**, *119*, 10488–10495. [CrossRef] [PubMed]
23. Gao, Y.; Chen, H.; Tay-Agbozo, S.; Kispert, L.D. Photo-induced electron transfer of carotenoids in mesoporous sieves (MCM-41) and surface modified MCM-41: The role of hydrogen bonds on the electron transfer. *J. Photochem. Photobiol. A Chem.* **2017**, *341*, 1–11. [CrossRef]
24. Méndez-Hernández, D.D.; Tarakeshwar, P.; Gust, D.; Moore, T.A.; Moore, A.L.; Mujica, V. Simple and accurate correlation of experimental redox potentials and DFT-calculated HOMO/LUMO energies of polycyclic aromatic hydrocarbons. *J. Mol. Model.* **2013**, *19*, 2845–2848. [CrossRef]
25. Gao, Y.; Focsan, A.L.; Kispert, L.D. Antioxidant activity in supramolecular carotenoid complexes favored by nonpolar environment and disfavored by hydrogen bonding. *Antioxidants* **2020**, *9*, 625. [CrossRef]
26. Polyakov, N.; Leshina, T.; Salakhutdinov, N.; Konovalova, T.; Kispert, L. Antioxidant and redox properties of supramolecular complexes of carotenoids with β-glycyrrhizic acid. *Free. Radic. Biol. Med.* **2006**, *40*, 1804–1809. [CrossRef] [PubMed]
27. Polyakov, N.E.; Magyar, A.; Kispert, L.D. Photochemical and optical properties of water-soluble xanthophyll antioxidants: Aggregation vs complexation. *J. Phys. Chem. B* **2013**, *117*, 10173–10182. [CrossRef]
28. Focsan, A.L.; Polyakov, N.E.; Kispert, L.D. Photo protection of *Haematococcus pluvialis* algae by astaxanthin: Unique properties of astaxanthin deduced by EPR, optical and electrochemical studies. *Antioxidants* **2017**, *6*, 80. [CrossRef] [PubMed]
29. Magyar, A.; Bowman, M.K.; Molnár, P.; Kispert, L. Neutral carotenoid radicals in photoprotection of wild-type *Arabidopsis thaliana*. *J. Phys. Chem. B* **2013**, *117*, 2239–2246. [CrossRef]
30. Gao, Y.; Webb, S.; Kispert, L.D. Deprotonation of carotenoid radical cation and formation of a didehydrodimer. *J. Phys. Chem. B* **2003**, *107*, 13237–13240. [CrossRef]
31. Gao, Y.; Focsan, A.L.; Kispert, L.D. The effect of polarity of environment on the antioxidant activity of carotenoids. *Chem. Phys. Lett.* **2020**, *761*, 138098. [CrossRef]
32. Closs, G.L.; Miller, J.R. Intramolecular long-distance electron transfer in organic molecules. *Science* **1988**, *240*, 440–447. [CrossRef] [PubMed]
33. Guarr, T.; McLendon, G. Quantum mechanical effects in inorganic and bioinorganic electron transfer. *Coord. Chem. Rev.* **1985**, *68*, 1–52. [CrossRef]
34. Hush, N. Distance dependence of electron transfer rates. *Coord. Chem. Rev.* **1985**, *64*, 135–157. [CrossRef]
35. Marcus, R.; Sutin, N. Electron transfers in chemistry and biology. *Biochim. Biophys. Acta* **1985**, *811*, 265–322. [CrossRef]
36. Gao, Y.; Lockart, M.; Kispert, L.D.; Bowman, M.K. Photo-induced charge separation in hydroxycoumarins on TiO_2 and F-TiO_2. *Dalton Trans.* **2019**, *48*, 10881–10891. [CrossRef]
37. Gao, Y.; Lockart, M.; Kispert, L.D.; Bowman, M.K. Photoinduced charge separation in retinoic acid on TiO_2: Comparison of three anchoring modes. *J. Phys. Chem. C* **2019**, *123*, 24634–24642. [CrossRef]
38. Gao, Y.; Konovalova, T.A.; Lawrence, J.N.; Smitha, M.A.; Nunley, J.; Schad, R.; Kispert, L.D. Interaction of carotenoids and Cu^{2+} in Cu-MCM-41: Distance-dependent reversible electron transfer. *J. Phys. Chem. B* **2003**, *107*, 2459–2465. [CrossRef]
39. Gao, Y.; Kispert, L.D. A DFT study of the interaction between olefins and Cu^{2+} on silica and MCM-41 model surfaces. *Dalton Trans.* **2014**, *43*, 6221–6228. [CrossRef]
40. Polyakov, N.E.; Leshina, T.V.; Meteleva, E.S.; Dushkin, A.V.; Konovalova, T.A.; Kispert, L.D. Water soluble complexes of carotenoids with arabinogalactan. *J. Phys. Chem. B* **2009**, *113*, 275–282. [CrossRef]
41. Polyakov, N.E.; Leshina, T.V.; Salakhutdinov, N.F.; Kispert, L.D. Host-guest complexes of carotenoids with β-glycyrrhizic acid. *J. Phys. Chem. B* **2006**, *110*, 6991–6998. [CrossRef]
42. Polyakov, N.E.; Leshina, T.V.; Konovalova, T.A.; Hand, E.O.; Kispert, L.D. Inclusion complexes of carotenoids with cyclodextrins: 1H NMR, EPR, and optical studies. *Free radic. Biol. Med.* **2004**, *36*, 872–880. [CrossRef]
43. Apanasenko, I.E.; Selyutina, O.Y.; Polyakov, N.E.; Suntsova, L.P.; Meteleva, E.S.; Dushkin, A.V.; Vachali, P.; Bernstein, P.S. Solubilization and stabilization of macular carotenoids by water soluble oligosaccharides and polysaccharides. *Arch. Biochem. Biophys.* **2015**, *572*, 58–65. [CrossRef] [PubMed]
44. Genç, Y.; Bardakci, H.; Yücel, Ç.; Karatoprak, G.; Akkol, E.K.; Barak, T.H.; Sobarzo-Sánchez, E. Oxidative stress and marine carotenoids: Application by using nanoformulations. *Mar. Drugs* **2020**, *18*, 423. [CrossRef] [PubMed]
45. Polyakov, N.E.; Kispert, L.D. Water soluble biocompatible vesicles based on polysaccharides and oligosaccharides inclusion complexes for carotenoid delivery. *Carbohydr. Polym.* **2015**, *128*, 207–219. [CrossRef] [PubMed]
46. Focsan, A.L.; Polyakov, N.E.; Kispert, L.D. Supramolecular carotenoid complexes of enhanced solubility and stability—The way of bioavailability improvement. *Molecules* **2019**, *24*, 3947. [CrossRef] [PubMed]
47. Soukoulis, C.; Bohn, T. A comprehensive overview on the micro- and nano-technological encapsulation advances for enhancing the chemical stability and bioavailability of carotenoids. *Crit. Rev. Food Sci. Nutr.* **2018**, *58*, 1–36. [CrossRef] [PubMed]
48. Terao, J.; Minami, Y.; Bando, N. Singlet molecular oxygen-quenching activity of carotenoids: Relevance to protection of the skin from photoaging. *J. Clin. Biochem. Nutr.* **2011**, *48*, 57–62. [CrossRef] [PubMed]
49. Polyakov, N.E.; Leshina, T.V.; Konovalova, T.A.; Kispert, L.D. Carotenoids as scavengers of free radicals in a Fenton reaction: Antioxidants or pro-oxidants? *Free. Radic. Biol. Med.* **2001**, *31*, 398–404. [CrossRef]
50. El-Agamey, A.; Lowe, G.M.; McGarvey, D.J.; Mortensen, A.; Phillip, D.M.; Truscott, T.; Young, A.J. Carotenoid radical chemistry and antioxidant/pro-oxidant properties. *Arch. Biochem. Biophys.* **2004**, *430*, 37–48. [CrossRef]

51. Young, A.J.; Lowe, G.L. Carotenoids—Antioxidant properties. *Antioxidants* **2018**, *7*, 28. [CrossRef]
52. Brotosudarmo, T.H.P.; Limantara, L.; Setiyono, E. Structures of astaxanthin and their consequences for therapeutic application. *Int. J. Food Sci.* **2020**, *2020*, 2156582. [CrossRef] [PubMed]
53. Ganesan, P.; Noda, K.; Manabe, Y.; Ohkubo, T.; Tanaka, Y.; Maoka, T.; Sugawara, T.; Hirata, T. Siphonaxanthin, a marine carotenoid from green algae, effectively induces apoptosis in human leukemia (HL-60) cells. *Biochim. Biophys. Acta—Gen. Subj.* **2011**, *1810*, 497–503. [CrossRef] [PubMed]
54. Sugawara, T.; Ganesan, P.; Li, Z.; Manabe, Y.; Hirata, T. Siphonaxanthin, a green algal carotenoid, as a novel functional compound. *Mar. Drugs* **2014**, *12*, 3660–3668. [CrossRef] [PubMed]
55. Manabe, Y.; Hirata, T.; Sugawara, T. Suppressive effects of carotenoids on the antigen-induced degranulation in RBL-2H3 rat basophilic leukemia cells. *J. Oleo Sci.* **2014**, *63*, 291–294. [CrossRef]
56. Shindo, K.; Misawa, N. New and rare carotenoids isolated from marine bacteria and their antioxidant activities. *Mar. Drugs* **2014**, *12*, 1690–1698. [CrossRef]
57. Shindo, K.; Kikuta, K.; Suzuki, A.; Katsuta, A.; Kasai, H.; Yasumoto-Hirose, M.; Matsuo, Y.; Misawa, N.; Takaichi, S. Rare carotenoids, (3R)-saproxanthin and (3R,2′S)-myxol, isolated from novel marine bacteria (*Flavobacteriaceae*) and their antioxidative activities. *Appl. Microbiol. Biotechnol.* **2007**, *74*, 1350–1357. [CrossRef]
58. Hamidi, M.; Safarzadeh Kozani, P.; Pierre, G.; Michaud, P.; Delattre, C. Marine bacteria versus microalgae: Who is the best for biotechnological production of bioactive compounds with antioxidant properties and other biological applications? *Mar. Drugs* **2020**, *18*, 28. [CrossRef]
59. Ji, H.-F. Insight into the strong antioxidant activity of deinoxanthin, a unique carotenoid in *Deinococcus radiodurans*. *Int. J. Mol. Sci.* **2010**, *11*, 4506–4510. [CrossRef]
60. Choi, Y.-J.; Hur, J.-M.; Lim, S.; Jo, M.; Kim, D.H.; Choi, J.-I. Induction of apoptosis by deinoxanthin in human cancer cells. *Anticancer. Res.* **2014**, *34*, 1829–1835.
61. Shin, J.; Song, M.-H.; Oh, J.-W.; Keum, Y.-S.; Saini, R.K. Pro-oxidant actions of carotenoids in triggering apoptosis of cancer cells: A review of emerging evidence. *Antioxidants* **2020**, *9*, 532. [CrossRef]
62. Richter, T.K.S.; Hughes, C.C.; Moore, B.S. Sioxanthin, a novel glycosylated carotenoid, reveals an unusual subclustered biosynthetic pathway. *Environ. Microbiol.* **2015**, *17*, 2158–2171. [CrossRef] [PubMed]
63. Singla, R.K.; Bhat, G.V. Crocin: An overview. *Indo Glob. J. Pharm. Sci.* **2011**, *1*, 281–286. [CrossRef]
64. Chen, S.-S.; Gu, Y.; Lu, F.; Qian, D.-P.; Dong, T.-T.; Ding, Z.-H.; Zhao, S.; Yu, Z.-H. Antiangiogenic effect of crocin on breast cancer cell MDA-MB-231. *J. Thorac. Dis.* **2019**, *11*, 4464–4473. [CrossRef] [PubMed]
65. Osawa, A.; Ishii, Y.; Sasamura, N.; Morita, M.; Kasai, H.; Maoka, T.; Shindo, K. Characterization and antioxidative activities of rare C-50 carotenoids-sarcinaxanthin, sarcinaxanthinmonoglucoside, and sarcinaxanthindiglucoside-obtained from *Micrococcus yunnanensis*. *J. Oleo Sci.* **2010**, *59*, 653–659. [CrossRef] [PubMed]
66. Wang, C.; Berg, C.J.; Hsu, C.-C.; Merrill, B.A.; Tauber, M.J. Characterization of carotenoid aggregates by steady-state optical spectroscopy. *J. Phys. Chem. B* **2012**, *116*, 10617–10630. [CrossRef]
67. Sujak, A.; Okulski, W.; Gruszecki, W.I. Organisation of xanthophyll pigments lutein and zeaxanthin in lipid membranes formed with dipalmitoylphosphatidylcholine. *Biochim. Biophys. Acta—Biomembr.* **2000**, *1509*, 255–263. [CrossRef] [PubMed]
68. Pinheiro, R.G.R.; Coutinho, A.J.; Pinheiro, M.; Neves, A.R. Nanoparticles for targeted brain drug delivery: What do we know? *Int. J. Mol. Sci.* **2021**, *22*, 11654. [CrossRef]
69. Martínez-Álvarez, O.; Calvo, M.M.; Gómez-Estaca, J. Recent advances in astaxanthin micro/nanoencapsulation to improve its stability and functionality as a food ingredient. *Mar. Drugs* **2020**, *18*, 406. [CrossRef]
70. Shibata, S. A Drug over the millennia: Pharmacognosy, chemistry, and pharmacology of licorice. *Yakugaku Zasshi* **2000**, *120*, 849–862. [CrossRef]
71. Hasan, K.; Ara, I.; Alam Mondal, M.S.; Kabir, Y. Phytochemistry, pharmacological activity, and potential health benefits of Glycyrrhiza glabra. *Heliyon* **2021**, *7*, e07240. [CrossRef]
72. Ageeva, A.A.; Kruppa, A.I.; Magin, I.M.; Babenko, S.V.; Leshina, T.V.; Polyakov, N.E. New aspects of the antioxidant activity of glycyrrhizin revealed by the CIDNP technique. *Antioxidants* **2022**, *11*, 1591. [CrossRef]
73. Su, X.; Wu, L.; Hu, M.; Dong, W.; Xu, M.; Zhang, P. Glycyrrhizic acid: A promising carrier material for anticancer therapy. *Biomed. Pharmacother.* **2017**, *95*, 670–678. [CrossRef]
74. Chrzanowski, J.; Chrzanowska, A.; Graboń, W. Glycyrrhizin: An old weapon against a novel coronavirus. *Phytother. Res.* **2021**, *35*, 629–636. [CrossRef] [PubMed]
75. Aqib, A.I.; Atta, K.; Muneer, A.; Arslan, M.; Shafeeq, M.; Rahim, K. Saponin and its derivatives (glycyrrhizin) and SARS-CoV-2. In *Application of Natural Products in SARS-CoV-2*; Niaz, K., Ed.; Academic Press: Cambridge, MA, USA, 2023; pp. 25–46.
76. Selyutina, O.; Polyakov, N. Glycyrrhizic acid as a multifunctional drug carrier—From physicochemical properties to biomedical applications: A modern insight on the ancient drug. *Int. J. Pharm.* **2019**, *559*, 271–279. [CrossRef] [PubMed]
77. Selyutina, O.Y.; Shelepova, E.A.; Paramonova, E.D.; Kichigina, L.A.; Khalikov, S.S.; Polyakov, N.E. Glycyrrhizin-induced changes in phospholipid dynamics studied by 1H NMR and MD simulation. *Arch. Biochem. Biophys.* **2020**, *686*, 108368. [CrossRef] [PubMed]
78. Kim, A.; Shelepova, E.; Evseenko, V.; Dushkin, A.; Medvedev, N.; Polyakov, N. Mechanism of the enhancing effect of glycyrrhizin on nifedipine penetration through a lipid membrane. *J. Mol. Liq.* **2021**, *344*, 117759. [CrossRef] [PubMed]

79. Dushkin, A.V.; Tolstikova, T.G.; Khvostov, M.V.; Tolstikov, G.A. Complexes of polysaccharides and glycyrrhizic acid with drug molecules-mechanochemical synthesis and pharmacological activity. In *The Complex World of Polysaccharides*; Karunaratne, D.N., Ed.; IntechOpen: Rijeka, Croatia, 2012; pp. 573–602.
80. Stancanelli, R.; Løjkner, L.D.; Larsen, K.L.; Guardo, M.; Cannavà, C.; Tommasini, S.; Ventura, C.A.; Calabrò, M.L.; Micali, N.; Villari, V.; et al. Structural and spectroscopic features of lutein/butanoyl-β-cyclodextrin nanoassemblies. *J. Pharm. Biomed.* **2012**, *71*, 214–218. [CrossRef] [PubMed]
81. Gao, Y.; Kispert, L.D. Reaction of carotenoids and ferric chloride: Equilibria, isomerization, and products. *J. Phys. Chem. B* **2003**, *107*, 5333–5338. [CrossRef]
82. Li, B.; Vachali, P.P.; Shen, Z.; Gorusupudi, A.; Nelson, K.; Besch, B.M.; Bartschi, A.; Longo, S.; Mattinson, T.; Shihab, S.; et al. Retinal accumulation of zeaxanthin, lutein, and β-carotene in mice deficient in carotenoid cleavage enzymes. *Exp. Eye Res.* **2017**, *159*, 123–131. [CrossRef]
83. Yucheng, L.; Yalin, L.; Zhifeng, R. Mini review on photocatalysis of titanium dioxide nanoparticles and their solar applications. *Nano Energy* **2013**, *2*, 1031–1045.
84. Yang, X.; Guo, S.; Wang, L.; Guan, S.; Zhou, S.; Lu, J. Function toggle of tumor microenvironment responsive nanoagent for highly efficient free radical stress enhanced chemodynamic therapy. *Nano Res.* **2022**, *15*, 8228–8236. [CrossRef]
85. Esposito, E.; Drechsler, M.; Huang, N.; Pavoni, G.; Cortesi, R.; Santonocito, D.; Puglia, C. Ethosomes and organogels for cutaneous administration of crocin. *Biomed. Microdevices* **2016**, *18*, 108. [CrossRef] [PubMed]
86. Muthiah, M.; Park, I.-K.; Cho, C.-S. Surface modification of iron oxide nanoparticles by biocompatible polymers for tissue imaging and targeting. *Biotechnol. Adv.* **2013**, *31*, 1224–1236. [CrossRef] [PubMed]
87. El-Kharrag, R. Development of a Therapeutic Model of Early Liver Cancer Using Crocin-Coated Magnetite Nanoparticles. Ph.D. Dissertation, United Arab Emirates University, Al Ain, Abu Dhabi, 2015.

International Journal of
Molecular Sciences

MDPI

Review

Improving the Treatment Effect of Carotenoids on Alzheimer's Disease through Various Nano-Delivery Systems

Wenjing Su [1], Wenhao Xu [1], Enshuo Liu [1], Weike Su [1] and Nikolay E. Polyakov [2,3,*]

1 Collaborative Innovation Center of Yangtze River Delta Region Green Pharmaceuticals, Zhejiang University of Technology, Hangzhou 310014, China
2 Institute of Solid State Chemistry and Mechanochemistry, 630128 Novosibirsk, Russia
3 Institute of Chemical Kinetics and Combustion, 630090 Novosibirsk, Russia
* Correspondence: polyakov@kinetics.nsc.ru

Abstract: Natural bioactive compounds have recently emerged as a current strategy for Alzheimer's disease treatment. Carotenoids, including astaxanthin, lycopene, lutein, fucoxanthin, crocin and others are natural pigments and antioxidants, and can be used to treat a variety of diseases, including Alzheimer's disease. However, carotenoids, as oil-soluble substances with additional unsaturated groups, suffer from low solubility, poor stability and poor bioavailability. Therefore, the preparation of various nano-drug delivery systems from carotenoids is a current measure to achieve efficient application of carotenoids. Different carotenoid delivery systems can improve the solubility, stability, permeability and bioavailability of carotenoids to a certain extent to achieve Alzheimer's disease efficacy. This review summarizes recent data on different carotenoid nano-drug delivery systems for the treatment of Alzheimer's disease, including polymer, lipid, inorganic and hybrid nano-drug delivery systems. These drug delivery systems have been shown to have a beneficial therapeutic effect on Alzheimer's disease to a certain extent.

Keywords: carotenoids; Alzheimer's disease; nanosized drug delivery systems

1. Introduction

Alzheimer's disease (AD) is the most common form of dementia, and has become the main health problem of the elderly [1]. According to a recent report, over 46 million people suffer from Alzheimer's disease, and that number is expected to grow to 130 million by 2050. AD is a recessive disease, one which generally occurs in people more than 55 years old and worsens with age [2,3]. AD is accompanied by progressive and irreversible dementia, memory loss, pessimism and other behavioral changes, loss of social status, speech loss without changing sensorimotor function, and additional symptoms [4,5]. Although the exact biological facts of Alzheimer's disease are still unclear, current research shows that the etiology of Alzheimer's disease is multifactorial; such factors as oxidative stress, loss of synaptic neurons, low levels of acetylcholine, β-Amyloid aggregation and metal accumulation all contribute to the development of Alzheimer's disease [6]. Amyloid plaques can aggregate into senile plaques on the outer surface of blood vessels and brain neurons, and into intracellular aggregates of neurofibrillary tangles arising from hyperphosphorylated tau [7–10]. The formation of reactive oxygen species (ROS) is a key factor in the development of AD. ROS can attack and destroy proteins, DNA, lipids and other macromolecules of living cells [11].

Carotenoids are naturally occurring pigments with powerful antioxidant properties [12–14], which are secondary metabolites produced by various enzymatic reactions [15,16]. Carotenoids have the ability to quench reactive oxygen species, free radicals and singlet oxygen. Their strong antioxidant power is primarily due to the presence of long-chain conjugated olefins in their structure, which makes them ideal candidates for scavenging free radicals [17,18]. Dietary intake of carotenoids has been reported to be associated with the prevention and treatment of many diseases, such as cardiovascular disease, cancer and age-related macular degeneration [19,20], with particular contributions to the prevention of brain diseases [21]. It plays a crucial role in the human brain and has a variety of roles in animals and plants [22–24]. Despite this, bioavailability and stability are major challenges for these natural compounds. Fortunately, structural modifications of these compounds can improve their biological function [25]. One of the approaches that makes it possible to overcome these disadvantages of carotenoids and significantly improve their bioavailability and stability is the use of drug delivery systems [26–31]. Recently, the development of nano-drug delivery systems has created exciting opportunities for the prevention and treatment of AD. Heretofore-poorly-distributed drugs are now prepared in nano-drug delivery systems. The drug delivery system has a favorable interaction with endothelial microvascular cells at the blood–brain barrier and can produce elevated concentrations of the drug in the brain parenchyma. These nano-drug delivery systems can achieve better efficacy and safety [32–35].

2. Carotenoids in Alzheimer's Disease Treatment

2.1. Pathogenesis of Alzheimer's Disease

The etiology of Alzheimer's disease remains unclear due to the multi-factorial nature of the disease's process. The formation of amyloid plaques is a key factor in the development of AD. These plaques can aggregate into senile plaques on the outer surface of blood vessels and brain neurons, and in intracellular aggregations of neurofibrillary tangles generated from hyperphosphorylated tau [7–10,36]. Naturally, oxidative stress (OS) is a major feature of AD. The reasons why neurons are extremely sensitive to OS include the following: (1) the energy generated by oxidative phosphorylation in neuronal mitochondria is extremely important [37,38]; (2) about 20% of the oxygen generated by respiration is used by neurons, 1–2% of which is converted into reactive oxygen species to cause OS [39–42]; (3) metal ions in neurons accumulate and catalyze ROS production in the brain as the aging process progresses [43]; (4) polyunsaturated fatty acids in neurons are susceptible to oxidation [44]; and (5) neurons have relatively low levels of antioxidants and related enzymes [45,46]. In general, OS level increases with age and is an important factor in inducing AD. ROS has the ability to attack and destroy a variety of macromolecules, including proteins, DNA and lipids of living cells [11,47–50].

2.2. Overview of Carotenoids

Carotenoids (Figure 1) are the most prevalent class of isoprenoid yellow-orange pigments that can be synthesized by photosynthetic organisms and fungal microorganisms and bacteria [51,52]. It can be categorized into two main groups: (1) nonpolar carotenes, such as β-carotene and lycopene, which are the hydrocarbon compound that carry no functional groups [53,54]; (2) polar xanthophylls, for instance, astaxanthin, lutein, and canthaxanthin, the structures of which contain hydrogen, carbon, and oxygen [17]. Carotenoids can also be divided into pro- and non-pro vitamin A, which cannot be converted into retinoids [53]. Carotenoids are known to be efficient compounds due to their antioxidant properties and nontoxic nature, which can minimize the risk of age-related muscular disorders [55].

2.3. Therapeutic Mechanisms of Carotenoids on Alzheimer's Disease

The interest in carotenoids has increased dramatically over the last decade due to their newly discovered activities, in particular, their neuroprotective properties. Neuroprotective mechanisms of carotenoids include antioxidant, anti-inflammatory, and anti-apoptotic activities, as well as the potential to promote neural plasticity. Although the exact molecular mechanisms of neurodegenerative diseases are still being elucidated, aging is considered as a primary risk factor for their development, including development of Alzheimer's disease [56]. It is usually accepted that increased inflammation and oxidative stress within the brain contribute to neurodegeneration. The brain is usually susceptible to higher oxidative stress due to its high metabolic activity and the presence of various oxidized compounds.

Oxidative stress can harm biomolecules (peptides, lipids and so on) and lead to neuronal dysfunction over time. The antioxidant activity of carotenoids is perhaps the known property responsible for its health benefit in prevention of neurodegenerative diseases.

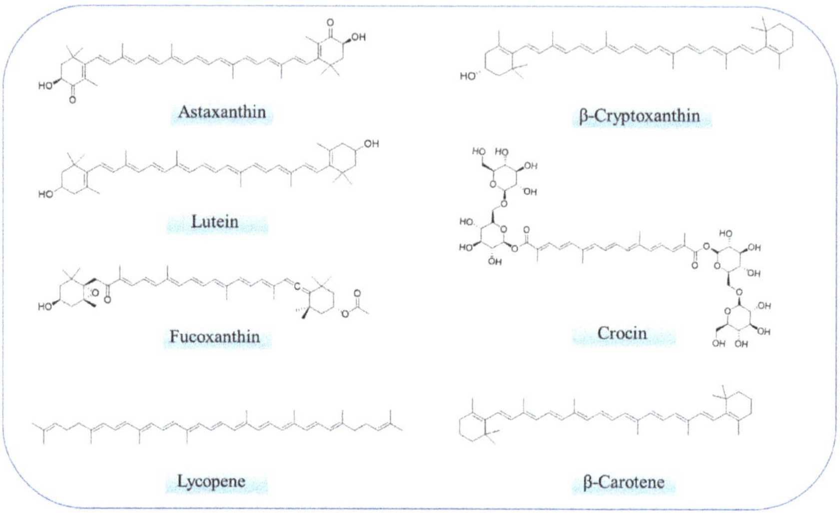

Figure 1. Chemical structures of some carotenoids.

Carotenoids are known to act as singlet oxygen quenchers and free radical scavengers and are used to combat oxidative stress in organisms. Singlet oxygen quenching, important in photosynthesis, relies on the energy transfer between electrophilic singlet oxygen and the carotenoid skeleton. The scavenging rate increases with the conjugation length. There are three main types of carotenoid radical scavenging reactions: (1) electron transfer between free radicals and carotenoids resulting in the formation of carotenoid radical cations or carotenoid radical anions; (2) formation of free radical adducts; and (3) hydrogen atom transfer to form neutral carotenoid group [19]. The structure of carotenoids gives them a strong antioxidant capacity, which protects cells against OS mediated by a variety of stressors by reducing DNA damage and activating the endogenous antioxidant enzymes, including superoxide dismutase and catalase. Promoting the efficacy of endogenous antioxidant enzymes is considered to be an important mechanism of some carotenoids' action in neuroprotection in the brain [57].

As an example, some authors reported successful application of carotenoid astaxanthin in the reducing of neurotoxicity induced by the amyloid-β fragments in cell culture models of Alzheimer's disease [58–60]. The authors report that astaxanthin protected primary hippocampal neurons from amyloid-β induced ROS generation and calcium dysregulation. Another example of effective anti-AD carotenoid is lycopene; see, for example, review [61]. In vitro studies have shown that pretreatment with this carotenoid reduces amyloid-β induced cellular damage and prevented amyloid-β 1–42-stimulated cellular apoptosis via the inhibition of ROS production, reducing mitochondrial dysfunction and expression of pro-apoptotic factors, and thus inhibiting pro-inflammatory response in microglia. In vivo studies on animal models have replicated the results observed in cell cultures. Humans with AD have significantly lower plasma concentrations of lycopene than do of healthy subjects [62]. Administration of the carotenoid-enriched products in humans is useful in increasing serum carotenoid levels and reducing OS, and thereby preventing neurodegeneration and delaying the onset of dementia.

3. Carotenoid-Loaded Nanocarriers for Alzheimer's Disease Therapy

The brain is a special organ that is protected by two major barriers, the blood–brain barrier (BBB) with its 20 m^2 surface area, and the blood–cerebrospinal-fluid barrier (BCSFB) [63]. Because of the large molecular weight of carotenoids, they do not penetrate easily to the brain, which presents one of the most important challenges in the development of drugs for the central nervous system [64]. Recently, the development of nano-based drug delivery systems has created exciting opportunities for the prevention and treatment of AD. Heretofore-poorly-distributed drugs are now prepared using nano-drug delivery systems. The drug delivery system has an excellent interaction with endothelial microvascular cells at the blood–brain barrier and is capable of producing elevated drug concentrations in the cerebral parenchyma. First, nanocarriers can pass through the blood–brain barrier passively (through the direct plasma membrane) or actively (endocytosis, pinocytosis, etc.) through transmembrane channels. Secondly, functional groups on the surface of nanocarriers (polysorbate surfactant layer or covalent binding of apolipoprotein, etc.) can enhance the efficiency of the carrier system in penetrating through the blood–brain barrier. The nanocarrier system is easily phagocytized by mononuclear phagocyte system, and then degraded or metabolized by lysosomes. The chemical groups on the surface of inorganic nanomaterials can be metabolized by enzymes or non-enzymes. Organic nanomaterials may first decompose and then metabolize into smaller particles. In the liver, if too large to pass through the pores between the transcellular hepatic sinusoidal endothelial cells, small-particle-sized materials can pass through the pores into the perisinusoidal space and then into the hepatocytes, where they are subsequently passed by monooxygenases, transferases, esters metabolism by enzymes, and epoxide hydrolases [65]. Therefore, these nano-drug delivery systems can achieve better efficacy and safety. Nano-delivery systems such as polymer/biopolymer nano-carriers, lipid-based nano-carriers, inorganic nano-carriers, and hybrid nano-carriers have been used (Figure 2). In the following subsections, a brief overview of nanocarriers will be made.

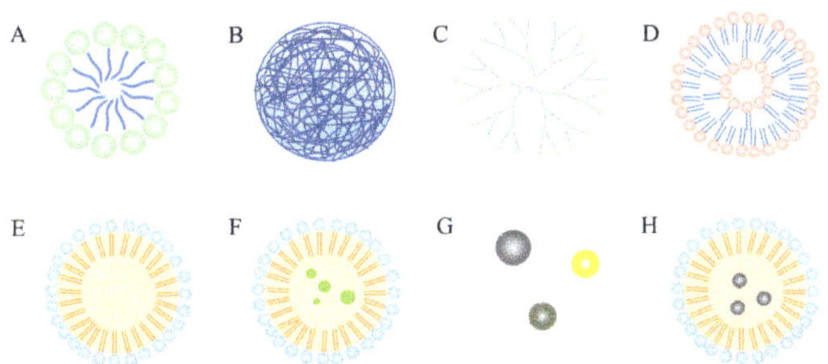

Figure 2. Different types of carotenoids nano-delivery systems for Alzheimer's disease treatment ((**A**) Polymeric micelles; (**B**) Polymeric nanoparticles; (**C**) Dendrimers; (**D**) Liposomes; (**E**) Solid lipid nanoparticles; (**F**) Nanostructured lipid carriers; (**G**) Inorganic nanocarriers; and (**H**) Hybrid nanocarriers).

3.1. *Polymeric Nanocarriers*

Polymer nanocarriers based on biodegradable and biocompatible properties include polymer micelles, polymer nanoparticles, and dendrimers. It is an ideal drug delivery system for carotenoids for the prevention and treatment of AD. Nanoparticles made of arabinogalactan, whey proteins, poly (butyl cyanoacrylate), casein, starch, and so on have been widely studied for drug delivery in the AD [33,35].

3.1.1. Polymeric Micelles

Micelles are core-shell structures formed spontaneously by amphiphilic molecules in water [66]. Proper micelles can be obtained by adjusting the fraction of monomers in the bulk copolymer so that most hydrophobic drugs can be easily incorporated into the core of the micelles. The function of the micellar shell is to protect the drug from interactions with serum proteins and non-target cells. In addition, targeted drug delivery can be achieved by end-functionalization of micelles with block copolymers of peptides, sugars, and additional components. Nanoscale micelles minimize the clearance of micelles from the body, prolong the action time of the drug, and improve the bioavailability of the drug. For instance, micelles prepared from Pluronic block copolymers have been the most studied. Both in vivo and in vitro experiments have shown that the micelle has the ability to enhance the drug's ability to penetrate the blood–brain barrier [67]. In summary, micelles can respond to external or internal stimuli and thus play a role in stabilizing, targeting or controlling the release of drugs [68,69], which are ideal drug delivery systems for the treatment of central nervous system diseases, particularly AD, due to their excellent ability to penetrate the blood–brain barrier [70–73].

3.1.2. Polymeric Nanoparticles

Polymer nanoparticles and micelles have some similar characteristics, such as loading efficiency, versatility, stimulus response (including light, temperature, enzyme, pH, and other biological and chemical agents) and so on [74]. Amphiphilic polymers with different structures, lengths and charges can be used to prepare polymer nanoparticles. They vary in size, shape and stability and can be used to encapsulate hydrophilic and hydrophobic drug molecules, including macromolecules such as carotenoids [28,31,75]. Polymer nanoparticles are widely used as biodegradable materials in the medical field. Commonly used polymers include polylactides, polyglycolides, poly-ε-caprolactone, and polyethylene glycol. Although these materials have been approved by the FDA for use in the medical field, they are not considered ideal for the treatment of central nervous system disorders due to their poor solubility and degradation in acidic byproducts. Acrylic polymer nanoparticles, especially poly (butyl cyanoacrylate) (PBCA) nanoparticles, have been widely used in the delivery of drugs in the central nervous system [76]. PBCA can be rapidly degraded in vivo to reduce toxicity due to polymer accumulation in the central nervous system. Drugs used to treat diseases of the central nervous system through the PBCA nanoparticle delivery system include doxorubicin, temozolomide, methotrexate, etc. [77,78]. Compared to acrylic polymers, polyester nanoparticles may be a safer choice for brain drug delivery because the degradation products are mainly water and carbon dioxide [79]. The drug delivery systems of polymer nanoparticles prepared by arabinogalactan [28], polylactic acid and poly (lactide co glycolide) also have been used for drug delivery in the central nervous system [80,81].

3.1.3. Dendrimers

Dendrimers are highly branched molecules with a 3D structure consisting of repetitive monomeric units with highly branched structures [82]. The modifiability of its surface structure gives dendrimers versatility, and the presence of a hydrophobic core enables encapsulation of genes, nucleic acids, and other drug molecules through electrostatic interactions or conjugation for the treatment of central nervous system disorders such as AD [64,83,84]. Polymeric dendrites have been developed for the treatment of Alzheimer's disease, Parkinson's disease, multiple sclerosis, ischemic stroke and other central nervous system disorders. Although dendrimers have a well-defined structure, their deeply branched and cross-linked nature makes it difficult to predict their degenerate outcome. Therefore, their inherent toxicity needs to be confirmed by other in vivo studies [85–87].

3.2. Lipid-Based Nanocarriers

3.2.1. Liposomes

One of the drug delivery systems that has received increasing attention is the liposome, which is a spherical vesicle composed of a unilamellar or multilamellar phospholipid bilayer [88]. Liposomes have excellent biocompatibility and biodegradability, low toxicity, and enable the targeted delivery of lipophilic and hydrophilic drugs. The greatest advantage of liposomes for central nervous system (CNS) delivery is that they can be easily surface-modified to prepare advanced liposomes such as immunoliposomes for targeted delivery [88]. For example, immunoliposomes carrying anti-epidermal growth factor receptor (anti-EGFR) antibodies can enhance the ability of vinorelbine and doxorubicin to target brain tumor cells [89]. Although liposomes have the problems of rapid clearance rate in vivo, low stability and inability to achieve long-term drug release, these problems have been solved step by step. For instance, modification of liposomes with polyethylene glycol (PEG) can reduce the possibility of liposomes being engulfed by macrophages and prolong their time in blood circulation, and the surface-functionalization of liposomes can improve their stability and efficacy. It is expected that additional liposomes will be used for the delivery of drugs in the central nervous system [90,91].

3.2.2. Solid Lipid Nanoparticles (SLNs)

SLN is normally composed of a lipid matrix in the solid state at room temperature and dispersed in water or a solution composed of surfactants for stabilization at body temperature. Fatty acids, cholesterol, monostearin, etc. are commonly used lipid matrices for the preparation of SLNs [92,93]. Solid lipid nanoparticles are produced from natural materials or natural lipids, are biocompatible, and do not affect the internal and external environment of cells after degradation, which makes them less immunogenic. At the same time, solid lipid nanoparticles are modest in size, flow rapidly in the blood, and are not easily absorbed by macrophages, thus facilitating the continuous release of therapeutic drugs in the body. Because solid lipid nanoparticle delivery systems are able to bypass P-glycoprotein by cell-by-cell percolation, they facilitate the penetration of lipophilic drugs across the blood–brain barrier. In fact, SLN can also bind apolipoprotein to target brain tissues [94,95]. Bhatt et al. prepared astaxanthin solid lipid nanoparticles by a double lotion solvent displacement method and administered the drug through the nose to increase the efficiency of brain targeting. The results of biological distribution experiments indicate relatively high concentrations of the drug in the brain and that intranasal administration of astaxanthin maximizes nerve protection against oxidative stress [96]. Therefore, solid lipid nanoparticles have become the best candidate drug delivery system for CNS drugs. Their low encapsulation efficiency due to the lipid core, which does not allow for the generation of empty space to encapsulate the underlying material in the solid phase again during crystallization, limits the large-scale production and development of solid-state lipid nanoparticles [70,97–99].

3.2.3. Nanostructured Lipid Carriers (NLCs)

To improve the inherent shortcomings of SLNs for high load efficiency, nanostructured lipid carriers came into being [100]. Nanostructured lipid carriers are composed of liquid and solid lipids (inner layer) and water emulsifiers (outer layer). They are transformed forms of SLN. The difference between SLN and NLC is that in NLC, 5–40% of the solid phase is exchanged with the liquid phase [101], liquid lipids or oils including fatty acid esters or alcohols such as 2-octanol are mixed with solid lipids [102,103], and lipids with different chain lengths, such as mono-, di-, and triglycerides, are also used to increase the space of the delivery system [104]; note that some hydrophobic drugs have better solubility in liquid lipids, which makes nanostructured lipid carriers more loading-efficient [105,106]. Rodriguez Ruiz et al. used sunflower seed oil as liquid lipid to synthesize astaxanthin loaded nanostructured lipid carriers by a green method. The results showed that nanostructured lipid carriers could stabilize astaxanthin molecules and maintain and enhance its antioxidant activity [107].

3.3. *Inorganic Nanocarriers*

Inorganic nanocarriers include cerium dioxide, iron oxide, gold, inorganic quantum dots and so on. Inorganic nano-delivery systems have been used for drug delivery due to their excellent physical and chemical properties, including size, shape, surface functionality, chemical structure, and high specific surface area, as well as imaging capabilities, so they can play their therapeutic and diagnostic roles at the same time. However, their biocompatibility, biodegradability and safety become the difficulties that restrict their wide application [108–110].

3.4. Hybrid Nanocarriers

Hybrid nanocarriers are composed of lipid, organic, and inorganic polymers [97,111,112]. In general, the organic polymer or inorganic substance acts as the core, and the lipid layer acts as the shell. The outer lipid can inhibit the diffusion of water to the inner layer, thus delaying the degradation of the inner polymer and ensuring that the loaded drug molecules are slowly and continuously released. Hybrid nanocarriers have controllable drug release capacity, high loading efficiency, biocompatibility and biodegradability, and are good carriers for the treatment of central nervous system diseases [97]. For instance, gold nanoparticles have been coated with PEG through pH-sensitive hydrazine bonds and modified with gadolinium-chelate and lipoprotein receptor-related protein-1 (LRP-1) recognition peptides. After BBB penetration and PEG cleavage, the mixed nanoparticles aggregated in the acidic tumor environment, resulting in an increased magnetic resonance imaging (MRI) signal [113].

4. Different Nano-Encapsulated Carotenoids in Alzheimer's Disease Therapy

4.1. Crocin and Crocetin

Crocin (Figure 1) and crocetin are mainly derived from saffron and the fruits of gardenia. Crocin, a water-soluble carotenoid, is composed of a conjugated polyene skeleton and sugar substituents at both ends. When the substituents at both ends are hydrogen atoms, it is a crocetin, which is insoluble in water. Numerous studies have shown that crocin and crocetin have disparate pharmacological effects, such as anti-Alzheimer, antioxidant, anti-tumor, anti-inflammatory, memory enhancing, antidepressant, etc. [114,115].

Sonali [116] and his colleagues obtained Crocus sativus extract from Crocus sativus (stigma) by using the cold-maceration method. The main ingredient in the extract is crocin. They combined extract with various excipients (hydroxypropyl methyl cellulose, ethyl cellulose, and Eudragit), which were prepared by physical mixing into particles, and then filled into hard gelatin capsules. The results showed that the capsule formulation had better dissolution properties and the in vitro pharmacokinetics showed more plasma exposure volume. The plasma concentration of crocin was increased by a factor of 3.3 compared to the extract. Their study has shown that crocin enhances the clearance of the amyloid-β in an AD brain by inducing the expression of P-gp. Hence, it has therapeutic and protective effects against Alzheimer's disease.

Another study of crocin application against AD has recently been published by Song et al. [33]. The authors reported that they had successfully made crocin-loaded hollow dextran sulfate/chitosan-coated zein nanoparticles using a layer-by-layer self-assembly technique. Compared with uncoated crocin-loaded zein nanoparticles, coated nanoparticles showed better controlled release effects in vitro (simulated gastrointestinal digestion) and stronger antioxidant activity. Experimental results in AD cell models showed that these nanoparticles can reduce the amyloid-β concentration in differentiated SH-SY5Y cells by 43%. These results prove that dextran sulfate/chitosan-coated crocin-loaded zein nanoparticles could be a hopeful delivery system for treating AD.

To overcome the poor water solubility and bioavailability of crocetin, Wong et al. [117] prepared an innovative water-soluble crocetin-γ-cyclodextrin inclusion complex for intravenous injection. The inclusion complex was shown to be safe in cell experiments. Pharmacokinetic and biodistribution experiments have shown that the inclusion complex can improve the bioavailability of crocetin and achieve the effect of facilitating the crocetin delivery system through the blood–brain barrier to the brain. The pharmacological mechanism of crocetin is through reducing the production of amyloid-β protein.

4.2. Astaxanthin

Astaxanthin (AST, 3,3′-dihydroxy-β, β-carotene-4,4′-dione, Figure 1) is a xanthophyll-type carotenoid [118]. It is synthesized by algae, bacteria or yeasts and can also accumulate in birds, fish and crustaceans through the food chain [119]. Unlike other carotenoids, astaxanthin contains two ionone rings at each end of the carbon chain with long-chain conjugated double bonds. Their unique molecular configuration and size allow astaxanthin molecules to be inserted vertically into the phospholipid bilayer of the cell membrane, allowing astaxanthin to prevent lipid peroxidation and protect the integrity of the cell membrane [120]. It is by far the most powerful natural antioxidant found in nature. Its strong antioxidant effect is due to the effective stability of the ionone rings and polyene skeletons against free radicals. The mechanism of antioxidant activity in AST involves the absorption of free radicals into the polyene chain, providing electrons or forming chemical bonds with the active material [121]. In addition to this, astaxanthin has anti-inflammatory, anti-diabetic and anti-cancer activity and cardiovascular disease prevention properties [122], as well as efficacy against nervous system diseases [123]. Because of astaxanthin's beneficial effects on biological systems, the FDA has approved the use of the pigment as a food colorant in animal and fish feed [124]. It has been assumed that the anti-cancer activity of AST might be related to its ability to chelate metal ions and produce neutral radicals [125–127].

Debora et al. formulated stealth lipid nanoparticles loaded with AST as a potential strategy for evading the defense lines represented by macrophages and improving the stability of the drug for the achievement of good bioavailability in the brain [34]. Prakash et al. reported the preparation of astaxanthin solid lipid nanoparticles by solvent displacement using citric acid, lecithin and poloxamer. Astaxanthin solid lipid nanoparticles have shown a strong neuroprotective effect against oxidative stress in neuronal cell lines. Published γ scintigraphy and radiological data showed that 99mTc AST solid lipid nanoparticles were directly transported from nose to brain. These findings confirm that these nanoparticles can be effectively used for brain targeting and can provide protection against a variety of neurological diseases, firmly suggesting that astaxanthin can provide neuroprotection against oxidative stress-induced cell damage [96].

4.3. Lycopene

Lycopene, a natural carotenoid, is widely found in fruits as diverse as pink guavas, tomatoes and red-skinned watermelons [128,129]. In the clinic, lycopene is the subject of numerous studies investigating its anti-cancer, cardiovascular disease prevention, liver protection, and other alternative effects. In 2003, Zhao et al. [130] used mice as test subjects and compared lycopene to aspirin, which they found to have similar anti-inflammatory activity. It is precisely because lycopene has neuroprotective activities, such as antioxidant and anti-inflammatory properties, that it is expected to be used to prevent Alzheimer's disease [131]. However, lycopene has a poor bioavailability and is easily affected by the presence of cis-isomers and other carotenoids [132]. Therefore, an appropriate administrative approach is required. Nano-delivery systems have great potential and advantages in improving bioavailability and drug stability [61].

4.4. Lutein

Lutein, a dietary carotenoid, comes from foods such as egg yolk, corn, kiwi, persimmon, and green vegetables. Its structure is conjugated polyene skeleton and ketone at both ends. Due to the particularity of its structure, it is easily affected by environmental factors such as temperature, light, oxygen and so on [133]. Similar to other carotenoids, it has antioxidant, anti-inflammatory and anti-tumor activity, and the potential to protect against several diseases, including Alzheimer's disease and age-related macular degeneration. However, its poor absorption and bioavailability due to its low solubility limits its use in the food and pharmaceutical industries [134]. Therefore, it is necessary to develop appropriate dosage forms in order to successfully deliver it.

Through electrostatic interaction and nanoprecipitation method, Dhas and Mehta [35] prepared spherical nanoparticles with chitosan as shell and PLGA as core with the particle size less than 150 nm. The nanoparticles were administered through the nasal cavity. This is a relatively short route of drug delivery, allowing the target drug to reach the brain more quickly. In vitro results from the co-culture BBB model show that the nanoparticles are able to penetrate the blood–brain barrier more effectively than pure lutein suspension and PLGA nanoparticles. In addition, in vitro release experiments, in vitro toxicity experiments, and cell uptake experiments have demonstrated, respectively, the effect of sustained release, the safety of nanoparticles, and the ability of vesicle-mediated endocytosis to achieve more efficient delivery. Antioxidant experiments have demonstrated the excellent ROS scavenging activity of nanoparticles. The delivery of lutein from the nasal cavity to the brain can be effective in reducing oxidative stress and thus achieving efficacy in treating AD.

4.5. Fucoxanthin

Fucoxanthin, an orange pigment and one of the most abundant carotenoids in nature [135], is found in the chloroplasts of brown algae [136]. Recently, fucoxanthin has been reported to have a variety of biological activities, including anti-cancer, antioxidant, anti-angiogenesis, anti-diabetes, anti-obesity, anti-inflammatory, and anti-malaria activities [137]. Fucoxanthin has previously been reported to have neuroprotective effects against Alzheimer's disease [32,138–141]. Fucoxanthin has also been shown to be safe in preclinical and small-population clinical studies, but its low bioavailability in the central nervous system limits its clinical use. To overcome this problem, nanoparticles with a diameter of about 200 nm and negative charge have been synthesized, and their penetration into the central nervous system is recommended [32]. Fucoxanthin nanoparticles can continuously release fucoxanthin in physiological environments. Fucoxanthin nanoparticles showed significant inhibition of amyloid-β formation of fibrils and oligomers. Most importantly, intravenous injection of fucoxanthin nanoparticles can prevent amyloid-β oligomer induced cognitive impairment in AD mice more effectively than fucoxanthin. These results indicate that nanoparticles can improve the bioavailability of drug delivery and enhance its efficacy in the treatment of AD, which may make it possible to use them in the treatment of AD in the future [32].

5. Conclusions and Perspectives

Nowadays, it is clear that carotenoids have many benefits for health and positive nutritional effects and can reduce the risk of many diseases. However, there are some critical points to be considered: (1) Most carotenoids play a synergistic role when combined with other compounds, and the single form of carotenoids may not be effective, but from another perspective if two or more carotenoids are put together inside the nanocarrier there may be competition for absorption, which leads to lower bioavailability. (2) Carotenoids are unstable, and easy to transform into different compounds; therefore, the safety of carotenoids needs additional research. (3) The therapeutic effect varies from person to person, thus the effective dose is an unknown problem.

Carotenoid-based nano-drug delivery systems are feasible for effective disease prevention and treatment. This review presents a series of examples of carotenoid nano-delivery systems for Alzheimer's disease. Each technology has its own strengths and limitations. To some extent, nano-delivery systems can improve the loading capacity, bioavailability, bioactivity, stability and solubility of carotenoids. In the author's opinion, polymeric micelles are more suitable for the delivery of carotenoids. First, the polymer micelles can be adjusted to a suitable size to accommodate carotenoids of different sizes. Second, the modifiability and ease of modification of the polymer surface increases the functional properties of carotenoids. Finally, carotenoid polymers can make it easier to pass through the blood–brain barrier by adjusting the hydrophilic–lipophilic balance. However, industrial production of nanomedicines is still in its early stages. Safety and health concerns need to be explored in depth before widespread consumption. First, each process or material must be formally approved by regulatory authorities. However, the regulatory framework for the inclusion of nano-carriers in pharmaceutical products is still in flux. State agencies are expected to add initiatives and some legislation to regulate and monitor the proper development and application of nanoparticles in food and drug formulations.

In modern medicine, the idea of "synergy" between drug and carrier has attracted increased attention, seeking to preserve and improve the health benefits of various drugs in prevention and the treatment of many diseases. In the case of carotenoids, drug delivery systems can assist these bioactive compounds in exerting greater biological activity and stability. On the other hand, some nano-delivery systems can also play more functional roles, including targeted delivery to the brain or other organs, or overcome the blood–brain-barrier. Thirdly, since most of the nanocarriers are natural protein or polysaccharide components, they can provide the body with some needed nutrients to a certain extent and improve the efficiency of disease prevention and treatment.

Author Contributions: Conceptualization, N.E.P. and W.S. (Wenjing Su); methodology, W.X.; validation, W.S. (Weike Su); investigation, E.L.; resources, N.E.P.; writing—original draft preparation, W.S. (Wenjing Su) and E.L.; writing—review and editing, N.E.P. and W.S. (Wenjing Su); project administration, W.S. (Weike Su). All authors have read and agreed to the published version of the manuscript.

Funding: Authors acknowledge the core funding from the National Key R&D Program of China (Grant No. 2021YFC2101005) and the core funding from the Russian Federal Ministry of Science and Higher Education (projects Nos. 0304-2017-0009 and 0301-2019-0005).

Institutional Review Board Statement: Not applicable.

Informed Consent Statement: Not applicable.

Data Availability Statement: Not applicable.

Conflicts of Interest: The authors declare no conflict of interest.

References

1. Paris, D.; Beaulieu-Abdelahad, D.; Bachmeier, C.; Reed, J.; Ait-Ghezala, G.; Bishop, A.; Chao, J.; Mathura, V.; Crawford, F.; Mullan, M. Anatabine lowers Alzheimer's A beta production in vitro and in vivo. *Eur. J. Pharmacol.* **2011**, *670*, 384–391. [CrossRef] [PubMed]
2. Fratiglioni, L.; Launer, L.J.; Andersen, K.; Breteler, M.M.B.; Copeland, J.R.M.; Dartigues, J.F.; Lobo, A.; Martinez-Lage, J.; Soininen, H.; Hofman, A.; et al. Incidence of dementia and major subtypes in Europe: A collaborative study of population-based cohorts. *Neurology* **2000**, *54*, S10–S15.
3. Lobo, A.; Launer, L.J.; Fratiglioni, L.; Andersen, K.; Di Carlo, A.; Breteler, M.M.B.; Copeland, J.R.M.; Dartigues, J.F.; Jagger, C.; Martinez-Lage, J.; et al. Prevalence of dementia and major subtypes in Europe: A collaborative study of population-based cohorts. *Neurology* **2000**, *54*, S4–S9. [PubMed]
4. Jonsson, L.; Jonhagen, M.E.; Kilander, L.; Soininen, H.; Hallikainen, M.; Waldemar, G.; Nygaard, H.; Andreasen, N.; Winblad, B.; Wimo, A. Determinants of costs of care for patients with Alzheimer's disease. *Int. J. Geriatr. Psychiatry* **2006**, *21*, 449–459. [CrossRef] [PubMed]
5. Citron, M. Strategies for disease modification in Alzheimer's disease. *Nat. Rev. Neurosci.* **2004**, *5*, 677–685. [CrossRef] [PubMed]
6. Chon, S.-H.; Yang, E.-J.; Lee, T.; Song, K.-S. Beta-Secretase (BACE1) inhibitory and neuroprotective effects of p-terphenyls from Polyozellus multiplex. *Food Funct.* **2016**, *7*, 3834–3842. [CrossRef]
7. Golde, T.E. The A beta hypothesis: Leading us to rationally-designed therapeutic strategies for the treatment or prevention of Alzheimer disease. *Brain Pathol.* **2005**, *15*, 84–87. [CrossRef]
8. Hardy, J.; Selkoe, D.J. Medicine-The amyloid hypothesis of Alzheimer's disease: Progress and problems on the road to therapeutics. *Science* **2002**, *297*, 353–356. [CrossRef]
9. Kabir, M.T.; Uddin, M.S.; Setu, J.R.; Ashraf, G.M.; Bin-Jumah, M.N.; Abdel-Daim, M.M. Exploring the Role ofPSENMutations in the Pathogenesis of Alzheimer's Disease. *Neurotox. Res.* **2020**, *38*, 833–849. [CrossRef]
10. Kabir, M.T.; Uddin, M.S.; Zaman, S.; Begum, Y.; Ashraf, G.M.; Bin-Jumah, M.N.; Bungau, S.G.; Mousa, S.A.; Abdel-Daim, M.M. Molecular Mechanisms of Metal Toxicity in the Pathogenesis of Alzheimer's Disease. *Mol. Neurobiol.* **2021**, *58*, 1–20. [CrossRef]
11. Cores, A.; Abril, S.; Michalska, P.; Duarte, P.; Olives, A.I.; Martin, M.A.; Villacampa, M.; Leon, R.; Menendez, J.C. Bisavenathramide Analogues as Nrf2 Inductors and Neuroprotectors in In Vitro Models of Oxidative Stress and Hyperphosphorylation. *Antioxidants* **2021**, *10*, 941. [CrossRef] [PubMed]

12. Bucciantini, M.; Leri, M.; Nardiello, P.; Casamenti, F.; Stefani, M. Olive Polyphenols: Antioxidant and Anti-Inflammatory Properties. *Antioxidants* **2021**, *10*, 1044. [CrossRef] [PubMed]
13. Loi, M.; Paciolla, C. Plant Antioxidants for Food Safety and Quality: Exploring New Trends of Research. *Antioxidants* **2021**, *10*, 972. [CrossRef] [PubMed]
14. Rodriguez-Yoldi, M.J. Anti-Inflammatory and Antioxidant Properties of Plant Extracts. *Antioxidants* **2021**, *10*, 921. [CrossRef] [PubMed]
15. Bolhassani, A.; Khavari, A.; Bathaie, S.Z. Saffron and natural carotenoids: Biochemical activities and anti-tumor effects. *Biochim. Biophys. Acta-Rev. Cancer* **2014**, *1845*, 20–30. [CrossRef] [PubMed]
16. Cong, L.; Wang, C.; Chen, L.; Liu, H.; Yang, G.; He, G. Expression of phytoene synthase1 and Carotene Desaturase crtl Genes Result in an Increase in the Total Carotenoids Content in Transgenic Elite Wheat (*Triticum aestivum* L.). *J. Agric. Food Chem.* **2009**, *57*, 8652–8660. [CrossRef] [PubMed]
17. Prakash, D.; Gupta, C. Carotenoids: Chemistry and Health Benefits. *Phytochem. Nutraceut. Importance* **2014**, 181–195.
18. Focsan, A.L.; Polyakov, N.E.; Kispert, L.D. Carotenoids: Importance in Daily Life-Insight Gained from EPR and ENDOR. *Appl. Magn. Reson.* **2021**, *52*, 1093–1112. [CrossRef]
19. Fiedor, J.; Burda, K. Potential Role of Carotenoids as Antioxidants in Human Health and Disease. *Nutrients* **2014**, *6*, 466–488. [CrossRef]
20. Widomska, J.; Zareba, M.; Subczynski, W.K. Can Xanthophyll-Membrane Interactions Explain Their Selective Presence in the Retina and Brain? *Foods* **2016**, *5*, 7. [CrossRef]
21. Lakey-Beitia, J.; Doens, D.; Kumar, D.J.; Murillo, E.; Fernandez, P.L.; Rao, K.S.; Durant-Archibold, A.A. Anti-amyloid aggregation activity of novel carotenoids: Implications for Alzheimer's drug discovery. *Clin. Interv. Aging* **2017**, *12*, 815–822. [CrossRef] [PubMed]
22. Bohn, T. Carotenoids and Markers of Oxidative Stress in Human Observational Studies and Intervention Trials: Implications for Chronic Diseases. *Antioxidants* **2019**, *8*, 179. [CrossRef] [PubMed]
23. Iddir, M.; Yaruro, J.F.P.; Cocco, E.; Hardy, E.M.; Appenzeller, B.M.R.; Guignard, C.; Larondelle, Y.; Bohn, T. Impact of Protein-Enriched Plant Food Items on the Bioaccessibility and Cellular Uptake of Carotenoids. *Antioxidants* **2021**, *10*, 1005. [CrossRef]
24. Young, A.J.; Lowe, G.L. Carotenoids-Antioxidant Properties. *Antioxidants* **2018**, *7*, 28. [CrossRef]
25. Kim, S.H.; Kim, M.S.; Lee, B.Y.; Lee, P.C. Generation of structurally novel short carotenoids and study of their biological activity. *Sci. Rep.* **2016**, *6*, 21987. [CrossRef]
26. Polyakov, N.E.; Leshina, T.V.; Salakhutdinov, N.F.; Konovalova, T.A.; Kispert, L.D. Antioxidant and redox properties of supramolecular complexes of carotenoids with beta-glycyrrhizic acid. *Free Radic. Biol. Med.* **2006**, *40*, 1804–1809. [CrossRef] [PubMed]
27. Polyakov, N.E.; Leshina, T.V.; Salakhutdinov, N.F.; Kispert, L.D. Host-guest complexes of carotenoids with beta-glycyrrhizic acid. *J. Phys. Chem. B* **2006**, *110*, 6991–6998. [CrossRef]
28. Polyakov, N.E.; Leshina, T.V.; Meteleva, E.S.; Dushkin, A.V.; Konovalova, T.A.; Kispert, L.D. Water Soluble Complexes of Carotenoids with Arabinogalactan. *J. Phys. Chem. B* **2009**, *113*, 275–282. [CrossRef]
29. Polyakov, N.E.; Kispert, L.D. Water soluble biocompatible vesicles based on polysaccharides and oligosaccharides inclusion complexes for carotenoid delivery. *Carbohydr. Polym.* **2015**, *128*, 207–219. [CrossRef]
30. Focsan, A.L.; Polyakov, N.E.; Kispert, L.D. Supramolecular Carotenoid Complexes of Enhanced Solubility and Stability-The Way of Bioavailability Improvement. *Molecules* **2019**, *24*, 3947. [CrossRef]
31. Apanasenko, I.E.; Selyutina, O.Y.; Polyakov, N.E.; Suntsova, L.P.; Meteleva, E.S.; Dushkin, A.V.; Vachali, P.; Bernstein, P.S. Solubilization and stabilization of macular carotenoids by water soluble oligosaccharides and polysaccharides. *Archives Biochem. Biophys.* **2015**, *572*, 58–65. [CrossRef] [PubMed]
32. Yang, M.; Jin, L.; Wu, Z.; Xie, Y.; Zhang, P.; Wang, Q.; Yan, S.; Chen, B.; Liang, H.; Naman, C.B.; et al. PLGA-PEG Nanoparticles Facilitate In Vivo Anti-Alzheimer's Effects of Fucoxanthin, a Marine Carotenoid Derived from Edible Brown Algae. *J. Agric. Food Chem.* **2021**, *69*, 9764–9777. [CrossRef] [PubMed]
33. Song, G.; Liu, J.; Wang, Q.; Wang, D.; Chu, B.; Li, L.; Xiao, G.; Gong, J.; Zheng, F. Layer-by-layer self-assembly of hollow dextran sulfate/chitosan-coated zein nanoparticles loaded with crocin: Fabrication, structural characterization and potential biological fate. *Food Hydrocoll.* **2022**, *125*, 107420. [CrossRef]
34. Santonocito, D.; Raciti, G.; Campisi, A.; Sposito, G.; Panico, A.; Siciliano, E.A.; Sarpietro, M.G.; Damiani, E.; Puglia, C. Astaxanthin-Loaded Stealth Lipid Nanoparticles (AST-SSLN) as Potential Carriers for the Treatment of Alzheimer's Disease: Formulation Development and Optimization. *Nanomaterials* **2021**, *11*, 391. [CrossRef] [PubMed]
35. Dhas, N.; Mehta, T. Cationic biopolymer functionalized nanoparticles encapsulating lutein to attenuate oxidative stress in effective treatment of Alzheimer's disease: A non-invasive approach. *Int. J. Pharm.* **2020**, *586*, 119553. [CrossRef]
36. Neve, R.L.; Robakis, N.K. Alzheimer's disease: A re-examination of the amyloid hypothesis. *Trends Neurosci.* **1998**, *21*, 15–19. [CrossRef]
37. Dominiak, K.; Jarmuszkiewicz, W. The Relationship between Mitochondrial Reactive Oxygen Species Production and Mitochondrial Energetics in Rat Tissues with Different Contents of Reduced Coenzyme Q. *Antioxidants* **2021**, *10*, 533. [CrossRef] [PubMed]

38. Kolodziej, F.; O'Halloran, K.D. Re-Evaluating the Oxidative Phenotype: Can Endurance Exercise Save the Western World? *Antioxidants* **2021**, *10*, 609. [CrossRef]
39. Olowe, R.; Sandouka, S.; Saadi, A.; Shekh-Ahmad, T. Approaches for Reactive Oxygen Species and Oxidative Stress Quantification in Epilepsy. *Antioxidants* **2020**, *9*, 990. [CrossRef]
40. Picazo, C.; Molin, M. Impact of Hydrogen Peroxide on Protein Synthesis in Yeast. *Antioxidants* **2021**, *10*, 952. [CrossRef]
41. Ullah, H.; Di Minno, A.; Santarcangelo, C.; Khan, H.; Daglia, M. Improvement of Oxidative Stress and Mitochondrial Dysfunction by beta-Caryophyllene: A Focus on the Nervous System. *Antioxidants* **2021**, *10*, 546. [CrossRef]
42. Woo, J.; Cho, H.; Seol, Y.; Kim, S.H.; Park, C.; Yousefian-Jazi, A.; Hyeon, S.J.; Lee, J.; Ryu, H. Power Failure of Mitochondria and Oxidative Stress in Neurodegeneration and Its Computational Models. *Antioxidants* **2021**, *10*, 229. [CrossRef]
43. Kaur, I.; Behl, T.; Aleya, L.; Rahman, M.H.; Kumar, A.; Arora, S.; Akter, R. Role of metallic pollutants in neurodegeneration: Effects of aluminum, lead, mercury, and arsenic in mediating brain impairment events and autism spectrum disorder. *Environ. Sci. Pollut. Res.* **2021**, *28*, 8989–9001. [CrossRef]
44. Shichiri, M. The role of lipid peroxidation in neurological disorders. *J. Clin. Biochem. Nutr.* **2014**, *54*, 151–160. [CrossRef] [PubMed]
45. Lima, L.W.; Nardi, S.; Santoro, V.; Schiavon, M. The Relevance of Plant-Derived Se Compounds to Human Health in the SARS-CoV-2 (COVID-19) Pandemic Era. *Antioxidants* **2021**, *10*, 1031. [CrossRef]
46. Ofosu, F.K.; Mensah, D.-J.F.; Daliri, E.B.-M.; Oh, D.-H. Exploring Molecular Insights of Cereal Peptidic Antioxidants in Metabolic Syndrome Prevention. *Antioxidants* **2021**, *10*, 518. [CrossRef] [PubMed]
47. Barnham, K.J.; Masters, C.L.; Bush, A.I. Neurodegenerative diseases and oxidative stress. *Nat. Rev. Drug Discov.* **2004**, *3*, 205–214. [CrossRef] [PubMed]
48. Halliwell, B. Reactive Oxygen Species in Living Systems-Source, Biochemistry, and Role in Human-Disease. *Am. J. Med.* **1991**, *91*, S14–S22. [CrossRef] [PubMed]
49. Kabir, M.T.; Tabassum, N.; Uddin, M.S.; Aziz, F.; Behl, T.; Mathew, B.; Rahman, M.H.; Akter, R.; Rauf, A.; Aleya, L. Therapeutic Potential of Polyphenols in the Management of Diabetic Neuropathy. *Evid.-Based Complement. Altern. Med.* **2021**, *2021*, 9940169. [CrossRef] [PubMed]
50. Park, H.; Kim, J.; Shin, C.; Lee, S. Intersection between Redox Homeostasis and Autophagy: Valuable Insights into Neurodegeneration. *Antioxidants* **2021**, *10*, 694. [CrossRef]
51. Milani, A.; Basirnejad, M.; Shahbazi, S.; Bolhassani, A. Carotenoids: Biochemistry, pharmacology and treatment. *Br. J. Pharmacol.* **2017**, *174*, 1290–1324. [CrossRef] [PubMed]
52. Huang, J.J.; Lin, S.; Xu, W.; Cheung, P.C.K. Occurrence and biosynthesis of carotenoids in phytoplankton. *Biotechnol. Adv.* **2017**, *35*, 597–618. [CrossRef] [PubMed]
53. Ribeiro, D.; Freitas, M.; Silva, A.M.S.; Carvalho, F.; Fernandes, E. Antioxidant and pro-oxidant activities of carotenoids and their oxidation products. *Food Chem. Toxicol.* **2018**, *120*, 681–699. [CrossRef]
54. González-Peña, M.A.; Ortega-Regules, A.E.; Anaya de Parrodi, C.; Lozada-Ramírez, J.D. Chemistry, Occurrence, Properties, Applications, and Encapsulation of Carotenoids—A Review. *Plants* **2023**, *12*, 313. [CrossRef] [PubMed]
55. Meyers, K.J.; Mares, J.A.; Igo, R.P., Jr.; Truitt, B.; Liu, Z.; Millen, A.E.; Klein, M.; Johnson, E.J.; Engelman, C.D.; Karki, C.K.; et al. Genetic Evidence for Role of Carotenoids in Age-Related Macular Degeneration in the Carotenoids in Age-Related Eye Disease Study (CAREDS). *Investig. Ophthalmol. Vis. Sci.* **2014**, *55*, 587–599. [CrossRef]
56. Arvanitakis, Z.; Fleischman, D.A.; Arfanakis, K.; Leurgans, S.E.; Barnes, L.L.; Bennett, D.A. Association of white matter hyperintensities and gray matter volume with cognition in older individuals without cognitive impairment. *Brain Struct. Funct.* **2016**, *221*, 2135–2146. [CrossRef]
57. Stranahan, A.M.; Mattson, M.P. Recruiting adaptive cellular stress responses for successful brain ageing. *Nat. Rev. Neurosci.* **2012**, *13*, 209–216. [CrossRef]
58. Lobos, P.; Bruna, B.; Cordova, A.; Barattini, P.; Galaz, J.L.; Adasme, T.; Hidalgo, C.; Munoz, P.; Paula-Lima, A. Astaxanthin Protects Primary Hippocampal Neurons against Noxious Effects of Abeta-Oligomers. *Neural Plast.* **2016**, *2016*, 3456783. [CrossRef]
59. Chang, C.-H.; Chen, C.-Y.; Chiou, J.-Y.; Peng, R.Y.; Peng, C.-H. Astaxanthine Secured Apoptotic Death of PC12 Cells Induced by beta-Amyloid Peptide 25-35: Its Molecular Action Targets. *J. Med. Food* **2010**, *13*, 548–556. [CrossRef]
60. Wang, H.-Q.; Sun, X.-B.; Xu, Y.-X.; Zhao, H.; Zhu, Q.-Y.; Zhu, C.-Q. Astaxanthin upregulates heme oxygenase-1 expression through ERK1/2 pathway and its protective effect against beta-amyloid-induced cytotoxicity in SH-SY5Y cells. *Brain Res.* **2010**, *1360*, 159–167. [CrossRef]
61. Ratto, F.; Franchini, F.; Musicco, M.; Caruso, G.; Di Santo, S.G. A narrative review on the potential of tomato and lycopene for the prevention of Alzheimer's disease and other dementias. *Crit. Rev. Food Sci. Nutr.* **2022**, *62*, 4970–4981. [CrossRef] [PubMed]
62. Dias, I.H.K.; Polidori, M.C.; Li, L.; Weber, D.; Stahl, W.; Nelles, G.; Grune, T.; Griffiths, H.R. Plasma Levels of HDL and Carotenoids are Lower in Dementia Patients with Vascular Comorbidities. *J. Alzheimers Dis.* **2014**, *40*, 399–408. [CrossRef] [PubMed]
63. Neuwelt, E.A.; Bauer, B.; Fahlke, C.; Fricker, G.; Iadecola, C.; Janigro, D.; Leybaert, L.; Molnar, Z.; O'Donnell, M.E.; Povlishock, J.T.; et al. Engaging neuroscience to advance translational research in brain barrier biology. *Nat. Rev. Neurosci.* **2011**, *12*, 169–182. [CrossRef] [PubMed]
64. Pardridge, W.M. The blood-brain barrier: Bottleneck in brain drug development. *NeuroRx J. Am. Soc. Exp. NeuroTherapeutics* **2005**, *2*, 3–14. [CrossRef]

65. Guiot, C.; Zullino, S.; Priano, L.; Cavalli, R. The physics of drug-delivery across the blood-brain barrier. *Ther. Deliv.* **2016**, *7*, 153–156. [CrossRef]
66. Kedar, U.; Phutane, P.; Shidhaye, S.; Kadam, V. Advances in polymeric micelles for drug delivery and tumor targeting. *Nanomed.-Nanotechnol. Biol. Med.* **2010**, *6*, 714–729. [CrossRef] [PubMed]
67. Kabanov, A.V.; Alakhov, V.Y. Pluronic (R) block copolymers in drug delivery: From micellar nanocontainers to biological response modifiers. *Crit. Rev. Ther. Drug Carr. Syst.* **2002**, *19*, 1–72. [CrossRef]
68. Ahmad, Z.; Shah, A.; Siddiq, M.; Kraatz, H.B. Polymeric micelles as drug delivery vehicles. *RSC Adv.* **2014**, *4*, 17028–17038. [CrossRef]
69. Xu, W.; Ling, P.X.; Zhang, T.M. Polymeric Micelles, a Promising Drug Delivery System to Enhance Bioavailability of Poorly Water-Soluble Drugs. *J. Drug Deliv.* **2013**, *2013*, 340315. [CrossRef]
70. Craparo, E.F.; Bondi, M.L.; Pitarresi, G.; Cavallaro, G. Nanoparticulate Systems for Drug Delivery and Targeting to the Central Nervous System. *Cns Neurosci. Ther.* **2011**, *17*, 670–677. [CrossRef]
71. Masserini, M. Nanoparticles for brain drug delivery. *ISRN Biochem.* **2013**, *2013*, 238428. [CrossRef] [PubMed]
72. Morshed, R.A.; Cheng, Y.; Auffinger, B.; Wegscheid, M.L.; Lesniak, M.S. The potential of polymeric micelles in the context of glioblastoma therapy. *Front. Pharmacol.* **2013**, *4*, 157. [CrossRef] [PubMed]
73. Yang, H. Nanoparticle-Mediated Brain-Specific Drug Delivery, Imaging, and Diagnosis. *Pharm. Res.* **2010**, *27*, 1759–1771. [CrossRef] [PubMed]
74. Wang, Y.C.; Shim, M.S.; Levinson, N.S.; Sung, H.W.; Xia, Y.N. Stimuli-Responsive Materials for Controlled Release of Theranostic Agents. *Adv. Funct. Mater.* **2014**, *24*, 4206–4220. [CrossRef]
75. Elsabahy, M.; Wooley, K.L. Design of polymeric nanoparticles for biomedical delivery applications. *Chem. Soc. Rev.* **2012**, *41*, 2545–2561. [CrossRef] [PubMed]
76. Rempe, R.; Cramer, S.; Huewel, S.; Galla, H.-J. Transport of Poly(n-butylcyano-acrylate) nanoparticles across the blood-brain barrier in vitro and their influence on barrier integrity. *Biochem. Biophys. Res. Commun.* **2011**, *406*, 64–69. [CrossRef] [PubMed]
77. Kim, H.R.; Andrieux, K.; Gil, S.; Taverna, M.; Chacun, H.; Desmaele, D.; Taran, F.; Georgin, D.; Couvreur, P. Translocation of poly(ethylene glycol-co-hexadecyl)cyanoacrylate nanoparticles into rat brain endothelial cells: Role of apolipoproteins in receptor-mediated endocytosis. *Biomacromolecules* **2007**, *8*, 793–799. [CrossRef] [PubMed]
78. Tian, X.-H.; Lin, X.-N.; Wei, F.; Feng, W.; Huang, Z.-C.; Wang, P.; Ren, L.; Diao, Y. Enhanced brain targeting of temozolomide in polysorbate-80 coated polybutylcyanoacrylate nanoparticles. *Int. J. Nanomed.* **2011**, *6*, 445–452.
79. Dechy-Cabaret, O.; Martin-Vaca, B.; Bourissou, D. Controlled ring-opening polymerization of lactide and glycolide. *Chem. Rev.* **2004**, *104*, 6147–6176. [CrossRef]
80. Titov, E.A.; Novikov, M.A.; Sosedova, L.M. Effect of silver nanoparticles encapsulated in a polymer matrix on the structure of nervous tissue and expression of caspase-3. *Nanotechnol. Russ.* **2015**, *10*, 640–644. [CrossRef]
81. Gao, X.; Wu, B.; Zhang, Q.; Chen, J.; Zhu, J.; Zhang, W.; Rong, Z.; Chen, H.; Jiang, X. Brain delivery of vasoactive intestinal peptide enhanced with the nanoparticles conjugated with wheat germ agglutinin following intranasal administration. *J. Control. Release* **2007**, *121*, 156–167. [CrossRef] [PubMed]
82. Svenson, S. Dendrimers as versatile platform in drug delivery applications. *Eur. J. Pharm. Biopharm.* **2009**, *71*, 445–462. [CrossRef] [PubMed]
83. He, H.; Li, Y.; Jia, X.-R.; Du, J.; Ying, X.; Lu, W.-L.; Lou, J.-N.; Wei, Y. PEGylated Poly(amidoamine) dendrimer-based dual-targeting carrier for treating brain tumors. *Biomaterials* **2011**, *32*, 478–487. [CrossRef] [PubMed]
84. Menjoge, A.R.; Kannan, R.M.; Tomalia, D.A. Dendrimer-based drug and imaging conjugates: Design considerations for nanomedical applications. *Drug Discov. Today* **2010**, *15*, 171–185. [CrossRef]
85. Madaan, K.; Kumar, S.; Poonia, N.; Lather, V.; Pandita, D. Dendrimers in drug delivery and targeting: Drug-dendrimer interactions and toxicity issues. *J. Pharm. Bioallied Sci.* **2014**, *6*, 139–150. [PubMed]
86. Somani, S.; Dufes, C. Applications of dendrimers for brain delivery and cancer therapy. *Nanomedicine* **2014**, *9*, 2403–2414. [CrossRef] [PubMed]
87. Zhao, J.; Zhang, B.; Shen, S.; Chen, J.; Zhang, Q.; Jiang, X.; Pang, Z. CREKA peptide-conjugated dendrimer nanoparticles for glioblastoma multiforme delivery. *J. Colloid Interface Sci.* **2015**, *450*, 396–403. [CrossRef]
88. Webb, M.S.; Rebstein, P.; Lamson, W.; Bally, M.B. Liposomal drug delivery: Recent patents and emerging opportunities. *Recent Pat. Drug Deliv. Formul.* **2007**, *1*, 185–194. [CrossRef]
89. Wu, D.F.; Yang, J.; Pardridge, W.M. Drug targeting of a peptide radiopharmaceutical through the primate blood-brain barrier in vivo with a monoclonal antibody to the human insulin receptor. *J. Clin. Investig.* **1997**, *100*, 1804–1812. [CrossRef]
90. Akbarzadeh, A.; Rezaei-Sadabady, R.; Davaran, S.; Joo, S.W.; Zarghami, N.; Hanifehpour, Y.; Samiei, M.; Kouhi, M.; Nejati-Koshki, K. Liposome: Classification, preparation, and applications. *Nanoscale Res. Lett.* **2013**, *8*, 1–9. [CrossRef]
91. Roux, E.; Passirani, C.; Scheffold, S.; Benoit, J.P.; Leroux, J.C. Serum-stable and long-circulating, PEGylated, pH-sensitive liposomes. *J. Control. Release* **2004**, *94*, 447–451. [CrossRef] [PubMed]
92. Zhang, Q.-Z.; Zha, L.-S.; Zhang, Y.; Jiang, W.-M.; Lu, W.; Shi, Z.-Q.; Jiang, X.-G.; Fu, S.-K. The brain targeting efficiency following nasally applied MPEG-PLA nanoparticles in rats. *J. Drug Target.* **2006**, *14*, 281–290. [CrossRef] [PubMed]
93. Wong, H.L.; Bendayan, R.; Rauth, A.M.; Li, Y.Q.; Wu, X.Y. Chemotherapy with anticancer drugs encapsulated in solid lipid nanoparticles. *Adv. Drug Deliv. Rev.* **2007**, *59*, 491–504. [CrossRef] [PubMed]

94. Goppert, T.M.; Muller, R.H. Polysorbate-stabilized solid lipid nanoparticles as colloidal carriers for intravenous targeting of drugs to the brain: Comparison of plasma protein adsorption patterns. *J. Drug Target.* **2005**, *13*, 179–187. [CrossRef] [PubMed]
95. Kerwin, B.A. Polysorbates 20 and 80 used in the formulation of protein biotherapeutics: Structure and degradation pathways. *J. Pharm. Sci.* **2008**, *97*, 2924–2935. [CrossRef]
96. Bhatt, P.C.; Srivastava, P.; Pandey, P.; Khan, W.; Panda, B.P. Nose to brain delivery of astaxanthin-loaded solid lipid nanoparticles: Fabrication, radio labeling, optimization and biological studies. *RSC Adv.* **2016**, *6*, 10001–10010. [CrossRef]
97. Weber, S.; Zimmer, A.; Pardeike, J. Solid Lipid Nanoparticles (SLN) and Nanostructured Lipid Carriers (NLC) for pulmonary application: A review of the state of the art. *Eur. J. Pharm. Biopharm.* **2014**, *86*, 7–22. [CrossRef]
98. Saneja, A.; Dubey, R.D.; Alam, N.; Khare, V.; Gupta, P.N. Co-formulation of P-glycoprotein Substrate and Inhibitor in Nanocarriers: An Emerging Strategy for Cancer Chemotherapy. *Curr. Cancer Drug Targets* **2014**, *14*, 419–433. [CrossRef]
99. Feng, L.; Mumper, R.J. A critical review of lipid-based nanoparticles for taxane delivery. *Cancer Lett.* **2013**, *334*, 157–175. [CrossRef]
100. Muller, R.H.; Radtke, M.; Wissing, S.A. Nanostructured lipid matrices for improved microencapsulation of drugs. *Int. J. Pharm.* **2002**, *242*, 121–128. [CrossRef]
101. Katouzian, I.; Esfanjani, A.F.; Jafari, S.M.; Akhavan, S. Formulation and application of a new generation of lipid nano-carriers for the food bioactive ingredients. *Trends Food Sci. Technol.* **2017**, *68*, 14–25. [CrossRef]
102. Iqbal, M.A.; Md, S.; Sahni, J.K.; Baboota, S.; Dang, S.; Ali, J. Nanostructured lipid carriers system: Recent advances in drug delivery. *J. Drug Target.* **2012**, *20*, 813–830. [CrossRef] [PubMed]
103. Rizwanullah, M.; Ahmad, J.; Amin, S. Nanostructured Lipid Carriers: A Novel Platform for Chemotherapeutics. *Curr. Drug Deliv.* **2016**, *13*, 4–26. [CrossRef]
104. Naseri, N.; Valizadeh, H.; Zakeri-Milani, P. Solid Lipid Nanoparticles and Nanostructured Lipid Carriers: Structure, Preparation and Application. *Adv. Pharm. Bull.* **2015**, *5*, 305–313. [CrossRef] [PubMed]
105. Alam, T.; Pandit, J.; Vohora, D.; Aqil, M.; Ali, A.; Sultana, Y. Optimization of nanostructured lipid carriers of lamotrigine for brain delivery: In vitro characterization and in vivo efficacy in epilepsy. *Expert Opin. Drug Deliv.* **2015**, *12*, 181–194. [CrossRef]
106. Li, F.; Wang, Y.; Liu, Z.; Lin, X.; He, H.; Tang, X. Formulation and characterization of bufadienolides-loaded nanostructured lipid carriers. *Drug Dev. Ind. Pharm.* **2010**, *36*, 508–517. [CrossRef]
107. Rodriguez-Ruiz, V.; Salatti-Dorado, J.A.; Barzegari, A.; Nicolas-Boluda, A.; Houaoui, A.; Caballo, C.; Caballero-Casero, N.; Sicilia, D.; Bastias Venegas, J.; Pauthe, E.; et al. Astaxanthin-Loaded Nanostructured Lipid Carriers for Preservation of Antioxidant Activity. *Molecules* **2018**, *23*, 2601. [CrossRef]
108. Glaser, T.; Han, I.; Wu, L.; Zeng, X. Targeted Nanotechnology in Glioblastoma Multiforme. *Front. Pharmacol.* **2017**, *8*, 166. [CrossRef]
109. Cherukula, K.; Lekshmi, K.M.; Uthaman, S.; Cho, K.; Cho, C.-S.; Park, I.-K. Multifunctional Inorganic Nanoparticles: Recent Progress in Thermal Therapy and Imaging. *Nanomaterials* **2016**, *6*, 76. [CrossRef]
110. Mahmoudi, K.; Hadjipanayis, C.G. The application of magnetic nanoparticles for the treatment of brain tumors. *Front. Chem.* **2014**, *2*, 109. [CrossRef]
111. Burton, G.W.; Ingold, K.U. Beta-Carotene: An unusual type of lipid antioxidant. *Science* **1984**, *224*, 569–573. [CrossRef] [PubMed]
112. Palozza, P.; Serini, S.; Maggiano, N.; Angelini, M.; Boninsegna, A.; Di Nicuolo, F.; Ranelletti, F.O.; Calviello, G. Induction of cell cycle arrest and apoptosis in human colon adenocarcinoma cell lines by beta-carotene through down-regulation of cyclin A and Bcl-2 family proteins. *Carcinogenesis* **2002**, *23*, 11–18. [CrossRef] [PubMed]
113. Gao, X.H.; Yue, Q.; Liu, Z.N.; Ke, M.J.; Zhou, X.Y.; Li, S.H.; Zhang, J.P.; Zhang, R.; Chen, L.; Mao, Y.; et al. Guiding Brain-Tumor Surgery via Blood-Brain-Barrier-Permeable Gold Nanoprobes with Acid-Triggered MRI/SERRS Signals. *Adv. Mater.* **2017**, *29*, 1603917. [CrossRef] [PubMed]
114. Moratalla-Lopez, N.; Jose Bagur, M.; Lorenzo, C.; Martinez-Navarro, M.E.; Rosario Salinas, M.; Alonso, G.L. Bioactivity and Bioavailability of the Major Metabolites of *Crocus sativus* L. Flower. *Molecules* **2019**, *24*, 2827. [CrossRef] [PubMed]
115. Batarseh, Y.S.; Bharate, S.S.; Kumar, V.; Kumar, A.; Vishwakarma, R.A.; Bharate, S.B.; Kaddoumi, A. Crocus sativus Extract Tightens the Blood-Brain Barrier, Reduces Amyloid beta Load and Related Toxicity in 5XFAD Mice. *Acs Chem. Neurosci.* **2017**, *8*, 1756–1766. [CrossRef] [PubMed]
116. Bharate, S.S.; Kumar, V.; Singh, G.; Singh, A.; Gupta, M.; Singh, D.; Kumar, A.; Vishwakarma, R.A.; Bharate, S.B. Preclinical Development of Crocus sativus-Based Botanical Lead IIIM-141 for Alzheimer's Disease: Chemical Standardization, Efficacy, Formulation Development, Pharmacokinetics, and Safety Pharmacology. *Acs Omega* **2018**, *3*, 9572–9585. [CrossRef] [PubMed]
117. Wong, K.H.; Xie, Y.; Huang, X.; Kadota, K.; Yao, X.-S.; Yu, Y.; Chen, X.; Lu, A.; Yang, Z. Delivering Crocetin across the Blood-Brain Barrier by Using gamma-Cyclodextrin to Treat Alzheimer's Disease. *Sci. Rep.* **2020**, *10*, 3654. [CrossRef]
118. Donoso, A.; Gonzalez-Duran, J.; Agurto Munoz, A.; Gonzalez, P.A.; Agurto-Munoz, C. Therapeutic uses of natural astaxanthin: An evidence-based review focused on human clinical trials. *Pharmacol. Res.* **2021**, *166*, 105479. [CrossRef]
119. Routray, W.; Dave, D.; Cheema, S.K.; Ramakrishnan, V.V.; Pohling, J. Biorefinery approach and environment-friendly extraction for sustainable production of astaxanthin from marine wastes. *Crit. Rev. Biotechnol.* **2019**, *39*, 469–488. [CrossRef]
120. Kidd, P. Astaxanthin, Cell Membrane Nutrient with Diverse Clinical Benefits and Anti-Aging Potential. *Altern. Med. Rev.* **2011**, *16*, 355–364.
121. Liu, X.; Shibata, T.; Hisaka, S.; Osawa, T. Astaxanthin inhibits reactive oxygen species-mediated cellular toxicity in dopaminergic SH-SY5Y cells via mitochondria-targeted protective mechanism. *Brain Res.* **2009**, *1254*, 18–27. [CrossRef]

122. Bauer, A.; Minceva, M. Direct extraction of astaxanthin from the microalgae Haematococcus pluvialis using liquid-liquid chromatography. *RSC Adv.* **2019**, *9*, 22779–22789. [CrossRef] [PubMed]
123. Che, H.; Li, Q.; Zhang, T.; Wang, D.; Yang, L.; Xu, J.; Yanagita, T.; Xue, C.; Chang, Y.; Wang, Y. Effects of Astaxanthin and Docosahexaenoic-Acid-Acylated Astaxanthin on Alzheimer's Disease in APP/PS1 Double-Transgenic Mice. *J. Agric. Food Chem.* **2018**, *66*, 4948–4957. [CrossRef] [PubMed]
124. Pashkow, F.J.; Watumull, D.G.; Campbell, C.L. Astaxanthin: A novel potential treatment for oxidative stress and inflammation in cardiovascular disease. *Am. J. Cardiol.* **2008**, *101*, 58D–68D. [CrossRef] [PubMed]
125. Yang, X.; Guo, S.; Wang, L.; Guan, S.; Zhou, S.; Lu, J. Function toggle of tumor microenvironment responsive nanoagent for highly efficient free radical stress enhanced chemodynamic therapy. *Nano Res.* **2022**, *15*, 8228–8236. [CrossRef]
126. Polyakov, N.E.; Focsan, A.L.; Bowman, M.K.; Kispert, L.D. Free Radical Formation in Novel Carotenoid Metal Ion Complexes of Astaxanthin. *J. Phys. Chem. B* **2010**, *114*, 16968–16977. [CrossRef]
127. Focsan, A.L.; Bowman, M.K.; Shamshina, J.; Krzyaniak, M.D.; Magyar, A.; Polyakov, N.E.; Kispert, L.D. EPR Study of the Astaxanthin n-Octanoic Acid Monoester and Diester Radicals on Silica-Alumina. *J. Phys. Chem. B* **2012**, *116*, 13200–13210. [CrossRef]
128. Kong, K.W.; Ismail, A. Lycopene content and lipophilic antioxidant capacity of by-products from Psidium guajava fruits produced during puree production industry. *Food Bioprod. Process.* **2011**, *89*, 53–61. [CrossRef]
129. Perkins-Veazie, P.; Collins, J.K.; Pair, S.D.; Roberts, W. Lycopene content differs among red-fleshed watermelon cultivars. *J. Sci. Food Agric.* **2001**, *81*, 983–987. [CrossRef]
130. Zhao, Y.P.; Yu, W.L.; Hu, W.L.; Ying, Y. Anti-inflammatory and anticoagulant activities of lycopene in mice. *Nutr. Res.* **2003**, *23*, 1591–1595.
131. Przybylska, S. Lycopene-a bioactive carotenoid offering multiple health benefits: A review. *Int. J. Food Sci. Technol.* **2020**, *55*, 11–32. [CrossRef]
132. Shi, J.; Le Maguer, M. Lycopene in tomatoes: Chemical and physical properties affected by food processing. *Crit. Rev. Food Sci. Nutr.* **2000**, *40*, 1–42. [CrossRef] [PubMed]
133. Chuacharoen, T.; Sabliov, C.M. Stability and controlled release of lutein loaded in zein nanoparticles with and without lecithin and pluronic F127 surfactants. *Colloids Surf. A-Physicochem. Eng. Asp.* **2016**, *503*, 11–18. [CrossRef]
134. do Prado Silva, J.T.; Tonin Geiss, J.M.; Oliveira, S.M.; Brum, E.d.S.; Sagae, S.C.; Becker, D.; Leimann, F.V.; Ineu, R.P.; Guerra, G.P.; Goncalves, O.H. Nanoencapsulation of lutein and its effect on mice's declarative memory. *Mater. Sci. Eng. C-Mater. Biol. Appl.* **2017**, *76*, 1005–1011. [CrossRef] [PubMed]
135. Dembitsky, V.M.; Maoka, T. Allenic and cumulenic lipids. *Prog. Lipid Res.* **2007**, *46*, 328–375. [CrossRef]
136. Heo, S.-J.; Jeon, Y.-J. Protective effect of fucoxanthin isolated from Sargassum siliquastrum on UV-B induced cell damage. *J. Photochem. Photobiol. B-Biol.* **2009**, *95*, 101–107. [CrossRef]
137. Peng, J.; Yuan, J.-P.; Wu, C.-F.; Wang, J.-H. Fucoxanthin, a Marine Carotenoid Present in Brown Seaweeds and Diatoms: Metabolism and Bioactivities Relevant to Human Health. *Mar. Drugs* **2011**, *9*, 1806–1828. [CrossRef]
138. Xiang, S.; Liu, F.; Lin, J.; Chen, H.; Huang, C.; Chen, L.; Zhou, Y.; Ye, L.; Zhang, K.; Jin, J.; et al. Fucoxanthin Inhibits beta-Amyloid Assembly and Attenuates beta-Amyloid Oligomer-Induced Cognitive Impairments. *J. Agric. Food Chem.* **2017**, *65*, 4092–4102. [CrossRef]
139. Oliyaei, N.; Moosavi-Nasab, M.; Tanideh, N.; Iraji, A. Multiple roles of fucoxanthin and astaxanthin against Alzheimer's disease: Their pharmacological potential and therapeutic insights. *Brain Res. Bull.* **2022**, *193*, 11–21. [CrossRef]
140. Li, N.; Gao, X.; Zheng, L.; Huang, Q.; Zeng, F.; Chen, H.; Farag, M.A.; Zhao, C. Advances in fucoxanthin chemistry and management of neurodegenerative diseases. *Phytomedicine* **2022**, *105*, 154352. [CrossRef]
141. Lee, A.H.; Hong, S.-C.; Park, I.; Yoon, S.; Kim, Y.; Kim, J.; Yang, S.-H. Validation of Fucoxanthin from Microalgae *Phaeodactylum tricornutum* for the Detection of Amyloid Burden in Transgenic Mouse Models of Alzheimer's Disease. *Appl. Sci.* **2021**, *11*, 5878. [CrossRef]

International Journal of
Molecular Sciences

MDPI

Review

Dynamic and Energetic Aspects of Carotenoids In-and-Around Model Lipid Membranes Revealed in Molecular Modelling

Marta Pasenkiewicz-Gierula *, Jakub Hryc and Michal Markiewicz

Department of Computational Biophysics and Bioinformatics, Faculty of Biochemistry, Biophysics and Biotechnology, Jagiellonian University, 30-387 Krakow, Poland; jakub.hryc@doctoral.uj.edu.pl (J.H.); m.markiewicz@uj.edu.pl (M.M.)

* Correspondence: marta.pasenkiewicz-gierula@uj.edu.pl; Tel.: +4812-664-6518; Fax: +4812-664-6902

Abstract: In contrast to plants, humans are unable to synthesise carotenoids and have to obtain them from diet. Carotenoids fulfil several crucial biological functions in the organism; however, due to poor solubility in water, their bioavailability from plant-based food is low. The processes of carotenoid absorption and availability in the human body have been intensively studied. The recent experimental findings concerning these processes are briefly presented in the introductory part of this review, together with a summary of such topics as carotenoid carriers, body transport and tissue delivery, to finally report on molecular-level studies of carotenoid binding by membrane receptors. The main message of the review is contained in the section describing computational investigations of carotenoid intercalation and dynamic behaviour in lipid bilayers. The relevance of these computational studies lies in showing the direct link between the microscopic behaviour of molecules and the characteristics of their macroscopic ensembles. Furthermore, studying the interactions between carotenoids and lipid bilayers, and certainly proteins, on the molecular- and atomic-level using computational methods facilitates the interpretation and explanation of their macroscopic properties and, hopefully, helps to better understand the biological functions of carotenoids.

Keywords: spontaneous membrane intercalation; preferred membrane orientation; free energy gain; timescales; lipid-soluble antioxidant; *trans-cis* photoisomerisation

1. Carotenoids: Basic Information

Carotenoids are pigments synthesised predominantly by photosynthetic organisms, i.e., plants, algae and cyanobacteria, but some non-photosynthetic bacteria and fungi can also do so, e.g., [1,2]. In contrast, animals and humans are unable to synthesise carotenoids de novo even though they are critically important to their nutrition and health. In plants, carotenoids play fundamental roles in photosynthesis and photoprotection, e.g., [3–5]. In humans, carotenoids are vital in maintaining good health, e.g., as antioxidants, they reduce the risk of various chronic diseases such as cancer and cardiovascular diseases; β- and α-carotenes are precursors of vitamin A, whose deficiency can cause blindness; xanthophylls lutein and zeaxanthin, as macular pigments, decrease the onset of age-related macular degeneration and cataract, e.g., [2,6–9]. Lutein, as well as zeaxanthin, accumulate in the human brain improve its cognitive functions [10–12].

At present, 1204 natural carotenoids have been identified from 722 source organisms. Their chemical structures and other information are collected in the Carotenoids Database (http://carotenoiddb.jp, (accessed on 1 July 2024)) established by Yabuzaki [13]. Among the known carotenoids, lycopene, β-carotene (β,β-carotene), α-carotene ((6′R)-β,ε-carotene), lutein and zeaxanthin (Figure 1) are indicated as the most beneficial to humans, e.g., [6,14,15]. The key structural element of all of them is a nearly linear [16] nonpolar polyene chain consisting of eight isoprene units (40-carbon carotenoids) with conjugated double bonds in the *trans* conformation. With the exception of lycopene, these carotenoids have two terminal ionone rings in the C6 and C6′ positions of the polyene chain (Figure 1). In the β- and α-carotene, the rings are unsubstituted, and in lutein and zeaxanthin, they are monohydroxylated at positions C3 and C3′. Thus, lycopene, β- and α-carotene are non-polar and belong to the group of carotenes, whereas lutein and zeaxanthin are oxygenated derivatives of α- and β-carotene, respectively, and belong to the group of xanthophylls. It is worth noting that both halves (C6–C15 and C15′–C6′, Figure 1) of the polyene chain of the carotenoids are related by 180° rotation about a vertical axis placed at the chain centre (C_{2h} symmetry). Apart from the polyene chain symmetry, lycopene, β-carotene and zeaxanthin are symmetric molecules (C_{2h} symmetry), as both ionone rings of β-carotene and zeaxanthin are the same (β) and lycopene has no terminal rings. In contrast, both α-carotene and lutein have two different rings (β and ε); thus, the molecules are asymmetric. Each of the small differences in the chemical structures of the molecules has its biological relevance and determines carotenoid-protein and carotenoid–lipid bilayer interactions.

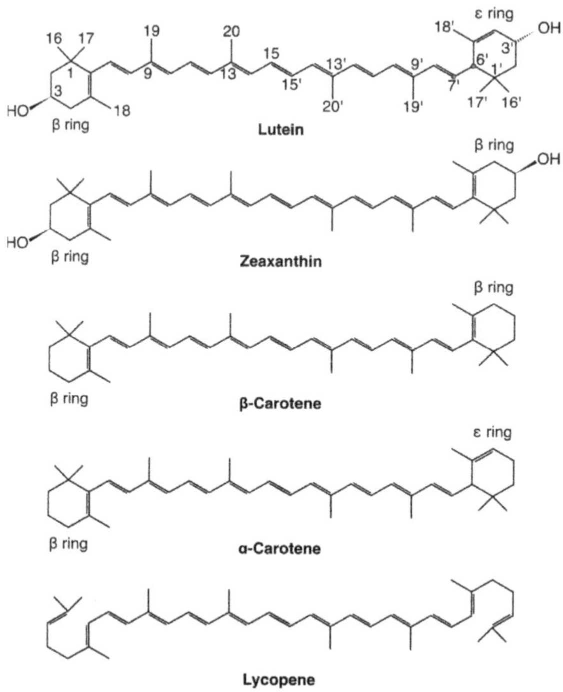

Figure 1. The chemical structures of lutein, zeaxanthin, β-carotene, α-carotene and lycopene. The ionone ring and polyene chain atoms that are used in the text are indicated in lutein only. The atoms are numbered according to the IUPAC convention. The C6′ and C7′ atoms of lutein are indicated with red circles. The chemical symbols for carbon atoms, C and the hydrogen atoms have been omitted except for the OH groups of lutein and zeaxanthin.

2. The Journey of Carotenoids through the Human Body

As was mentioned above, humans are unable to synthesise carotenoids de novo, so they have to obtain them from their diet. The main source of dietary carotenoids are coloured fruits and green leafy vegetables [6]. To become available to the human organism, carotenoids must first be released from the cellular matrix of fruit- and vegetable-based food products. This takes place in the gastrointestinal tract during digestion. As carotenoids are highly nonpolar, to facilitate their absorption in the intestine, they are dispersed into the lipid droplets in the stomach. Then, they are incorporated into bile salt and other lipid mixed micelles in the small intestine [17–20]. Next, free carotenoids are absorbed by intestinal absorptive cells via scavenger receptors or passive diffusion, incorporated into ultra-low-density lipoproteins (chylomicrons) and secreted into the lymph, e.g., [17–19,21–24]. In general, chylomicrons are taken up by the liver where carotenoids are integrated into either very low-density lipoproteins (VLDL), low-density lipoproteins (LDL) or high-density lipoproteins (HDL) [25–28] or stored. Roughly, xanthophylls are associated mainly with HDL, and carotenes mainly with LDL and VLDL [23,25,26,28,29], although the exact distribution of carotenoids among lipoproteins is not known for certain [27]. Lipoproteins transport carotenoids in the blood and deliver them to specific tissues for more or less selective uptake.

During their digestion, secretion and delivery, carotenoids exchange, in the free form, between tissues and supramolecular ensembles, which they are temporarily associated with, several times before they reach the sites of their main biological activity. There is a large number of such sites in the human organism; thus, carotenoids are widely distributed among various organs and tissues. However, different tissues accumulate different amounts of carotenoids and for different purposes. For example, the liver collects virtually all carotenoids mainly to transfer them to different lipoproteins and release them into the bloodstream. Adipose tissue accumulates carotenoids rather indiscriminately mainly to store them. In contrast, lutein and zeaxanthin are the only carotenoids that incorporate into the macula lutea, a specialised tissue at the centre of the retina of the human eyes. Additionally, lutein and zeaxanthin are predominant carotenoids in human brain tissue. Carotenoids also accumulate in other tissues, but those mentioned above are the most representative. There are details concerning the tissue distribution of carotenoids and their functions in Refs. [10,11,23,30–32], but as pointed out by Landrum [33], carotenoids are rather non-specifically accumulated in tissues, with important exceptions for lycopene in the prostate and lutein and zeaxanthin in the retina.

From a practical perspective, information on how carotenoids are transferred from the plasma lipoproteins to specific tissues is, undoubtedly, the most useful. It would be natural to expect that there are several transfer mechanisms specific to the tissue and the type of carrier. Extensive molecular biology research has revealed that the selective transfer of carotenoid molecules from their carriers to tissues occurs via cell membrane receptors [34]. In the case of macular pigments, intake of zeaxanthin into the retinal pigment epithelia (RPE) cells proceeds via scavenger receptor class B type 1 (abbreviated as SR-B1, SRB1 or SCARB1), which is a multifunctional receptor that binds a broad range of lipoproteins, with a high affinity for HDL [35–37]. The structure and the detailed structure-function analysis of the full-length homology model of human SR-B1 predicted by the molecular modelling software package trRosetta [38] are presented in Ref. [39]. However, intake of lutein into the RPE cells proceeds via the LDL receptor (LDLR) [35,36]. The receptor for LDL-bound β-carotene, let alone that for LDL-bound α-carotene, has not yet been identified conclusively, although it is likely that it is LDLR [32,35,40], but SR-B1 has also been considered [32,41].

β- and α-carotene are provitamin A carotenoids. In the human intestine, about half of them are converted to vitamin A, and the other half are absorbed intact. In β-carotene conversion, the molecule is cleaved by β-carotene 15,15′-oxygenase 1 at the C15,C15′ double bond into two retinal molecules that are subsequently converted into two retinol (vitamin A) molecules [42]. Cleavage of α-carotene eventually gives one retinol and one α-retinol molecule. In the liver, retinol associates with soluble retinol binding protein 4 (RBP4). The RBP4-retinol complex is released into the bloodstream and delivered to tissues via stimulated by retinoic acid 6 (STRA6) receptor, which transports retinol across the cell membrane [43–45]. In peripheral cells, all-*trans*-retinol (vitamin A) is converted into 11-*cis*-retinal, a light-sensitive molecule, and into all-*trans* retinoic acid, a signalling molecule [44]. In the eye, 11-*cis*-retinal is the chromophore of rod and cone opsins. After binding retinal, apo-protein opsin becomes a rod (rhodopsin) or cone (cone opsin) photoreceptor protein. As a signalling molecule, all-*trans* retinoic acid binds to the ligand-binding domains of two retinoic acid nuclear receptors, RAR and RXR, while RAR/RXR heterodimers bind specifically to DNA [44,46].

RBP4 that binds and delivers retinol to specific tissue does not bind α-retinol. Nevertheless, animals fed with α-retinol maintained about 50% of the growth rate of those fed with retinol [47]. This indicates that even though α-retinol has no vitamin A activity, it probably shares the ability of retinoic acid to act as a transcription factor [47]. This also indicates that α-retinol is transported as α-retinyl ester in lipoproteins, as does retinyl ester, which is a storage form of retinol.

3. Carotenoid–Receptor Binding: Molecular-Level Studies

The transfer of carotenoids from plasma lipoproteins to their receptors on the cells of the targeted tissues is a molecular-level process. The methods employed in studying such a process are both experimental and computational. Naturally, molecular-level experiments can provide information with different resolution. Thus, experiments carried out at the macroscopic molecular level have provided physicochemical and biochemical data on carotenoid–protein interactions [27,28,45,48–53]. Experiments carried out at the microscopic molecular level, in combination with a computational approach, have provided high-resolution data of the atomic structure of some carotenoid-binding proteins and the location and structure of the carotenoid molecules bound by them [34,50,54–60]. The results of these and earlier studies have greatly enhanced the knowledge and deepened the understanding of carotenoid absorption, transport and tissue delivery. However, there are still many topics concerning carotenoid binding, transfer and insertion to and between proteins and lipoproteins that have yet to be revealed. It is thus important to study the processes and elucidate their molecular and atomic-level details, as a deeper understanding of them may help in improving the bioavailability of carotenoids in the body [23,61].

4. Carotenoids: Transfer from Receptor to Membrane

In the body, carotenoids are transported in lipoproteins, and from them, they are transferred to the cell through more or less specific membrane-bound receptors, but the ultimate location of biologically active carotenoids in the cell is predominantly the cell membrane. Unfortunately, there is not much well-confirmed information regarding how they arrive there. This is mainly because determining the three-dimensional structures of large multi-domain membrane proteins is a complex and difficult process and not always successful; consequently, none of the structures of human carotenoid receptors have been solved so far. Knowledge of the structure will certainly make predicting how a carotenoid molecule inserts into the membrane feasible. Fortunately, important progress has been made recently as the full-length human SR-B1 receptor has been expressed and purified [53] and, moreover, its 3D structure predicted by homology modelling has been published [39]. The homology model structure of SR-B1 has also been predicted with SWISS-MODEL [62] in Ref. [34] and with MODELLER [63] in Ref. [60]. All three predicted structures have shown that the extracellular domain of SR-B1 has a large, predominantly hydrophobic [60] cavity that serves as a tunnel through which a lipid (carotenoid) passes from the lipoprotein to the plasma membrane. The putative sites where the lipid (carotenoid) enters and exits the tunnel [34] are marked in Figure 2.

The situation concerning the membrane receptor, STRA6, for retinol binding protein RBP is more fortunate as its dimeric structure has been solved by cryo-electron microscopy [45]. Knowledge of the STRA6 structure has made it possible to indicate the probable routes of retinol transfer from RBP through STRA6 to the membrane [45]. The routes are schematically shown in Figure 3B.

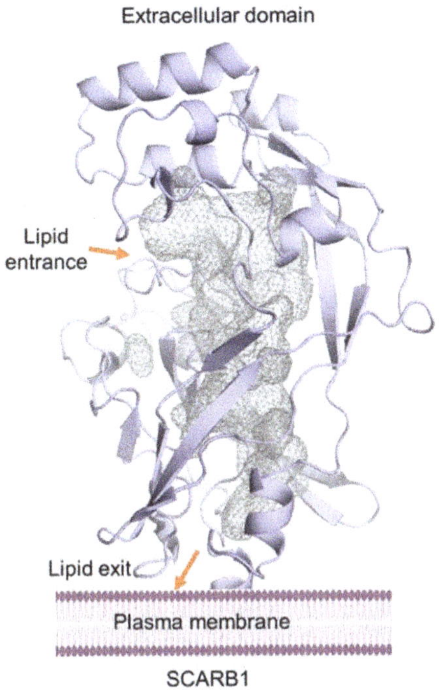

Figure 2. Homology model of class B scavenger receptor. Extracellular domain model of the human SCARB1 (SR-B1) receptor with a proposed lipid entering and exiting tunnel highlighted with a grey colour mesh. Figure. From Bandara, S., & von Lintig, J. (2022). Aster la vista: Unraveling the biochemical basis of carotenoid homeostasis in the human retina. BioEssays, 44, e2200133 [34]. https://doi.org/10.1002/bies.202200133. CC BY 4.0.

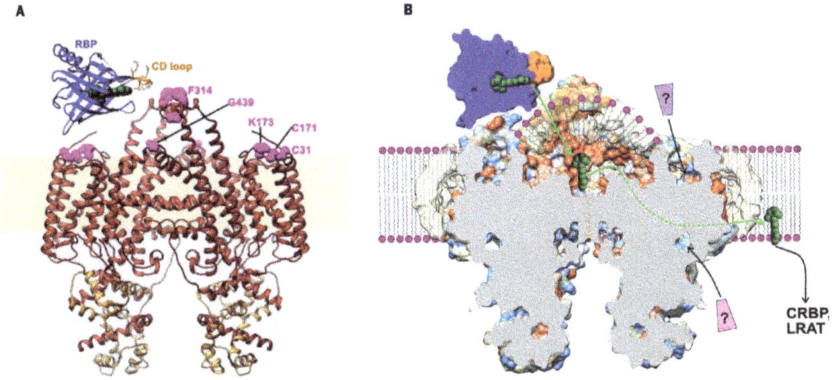

Figure 3. Possible mechanism for STRA6-mediated retinol uptake. (**A**) Ribbon representation of STRA6 (*red* OK) and RBP (*blue*). Retinol is shown as *green* spheres. The lipid bilayer is shaded light yellow. (**B**) Schematic of STRA6-mediated retinol (*green* spheres) release from RBP into the outer cleft and translocation to the lipid bilayer (shown as *purple* spheres and wavy lines) through the lateral window. Question marks indicate putative ligand binding sites; *green* and *orange* arrows show two potential retinol exit pathways. Figure (modified). From: Yunting Chen et al., Structure of the STRA6 receptor for retinol uptake. *Science* **353**, aad8266 (2016). https://doi.org/10.1126/science.aad8266 [45]. Reprinted with permission from AAAS.

Experimental studies with atomic resolution provide detailed structures of biomolecules but are not able to record their motion simultaneously. This is because they do not possess sufficiently high concurrent spatial and temporal resolution. At present, only computational molecular modelling methods possess such resolution. Unfortunately, computer simulations of the lipid translocation from the donor lipoprotein to the lipid bilayer through the receptor tunnel have not yet been published, so this issue still remains elusive.

5. Carotenoids in the Membrane: Atomic-Level Motional Studies

Classical molecular dynamics (MD) simulation is one of the computational molecular modelling methods. This widely used method has an atomic spatial resolution and sub-picosecond (ps) temporal resolution but has its limitations too. Its main limitations concern the size of the simulated system and the time of its "observation". Nevertheless, at present, the all-atom MD simulation allows systems to be studied that contain over a million atoms and processes on the μs timescale sampled with 2–4 fs time steps [64]. However, it should be remembered that each model predicted in MD simulations has to be validated through experiments.

There are several computational and experimental papers where the effect of carotenoids on lipid bilayers have been investigated, predicted and reviewed, e.g., [65–70]. In general, they have focused on the effect of carotenoids on the lipids in the bilayer and analysed changes in their ordering, orientation and dynamics, as well as on the bilayer physical, mechanical and phase properties. Conversely, here, the main focus is put on a carotenoid molecule in the lipid bilayer and, in particular, on its membrane intercalation and dynamic behaviour. The studies that are reviewed below were performed using computational methods, but the significance of computer predictions is discussed in the light of experimental findings.

The first MD simulation study on the behaviour of carotenoids in lipid bilayers was carried out on a simple model where four β-carotene molecules were inserted into a hydrated palmitoyl-oleoyl-phosphatidylcholine (POPC) bilayer, parallel to the bilayer normal (vertically), in Ref. [71]. During 4 ns MD simulation, the molecules tilted by ~20° and located their rings near the positions of the phospholipid carbonyl groups; the simulation was, though, too short to allow for larger configurational changes in the system. The next MD simulation study focused on spontaneous intercalation into and subsequent orientation of lutein molecules in the POPC bilayer [72]. In each of two bilayer systems, six lutein molecules were placed horizontally on the bilayer surfaces on the side of the hydrating water. In each of them, one molecule intercalated into the bilayer within the first two ns of MD simulation from the β-ring side, and five did not. These ten molecules (five in each system) aggregated in the water, and during 200 ns simulation, they remained there as aggregates. The remaining two lutein molecules intercalated into the upper leaflet of the bilayer, one vertically and the other horizontally. At ~70–80 ns of MD simulations, the molecules changed their positions to horizontal and vertical (transmembrane), respectively, and remained in these positions until the end of the 200 ns simulations. These orientations were stabilised mainly via hydrogen bonds (H-bond) between the lutein OH groups, and, in the case of the vertical orientation, phosphate and carbonyl groups of POPC and water, and in the case of the horizontal orientation, carbonyl groups of POPC and water [72].

The study of lutein intercalation into a POPC bilayer was pursued further in Ref. [73]. In this study, four bilayer systems were built, each containing six lutein molecules. In the initial structures, the luteins were situated almost vertically to the bilayer plane with the "entering ring" of each molecule placed within the regions of the PC phosphate groups and the rest of the molecule immersed in the hydrating water. In two systems, the "entering ring" was the lutein β ring, and in another two, it was the ε ring. The systems were MD simulated for 20–100 ns depending on the time it took for lutein to intercalate fully into the bilayer. Of the 24 lutein molecules present in the systems, 12 either fully or partially spontaneously intercalated in the bilayer, 10 intercalated from the β ring side (~80%) and 2 intercalated from the ε ring side (~17%). The remaining 12 molecules aggregated in the water phase and stayed there as aggregates during the whole simulation time. Five of the intercalated molecules were analysed in detail. Four of them intercalated vertically, and one intercalated horizontally. The time needed by a vertically translocating molecule to reach the interface of the other bilayer side ranged between ~7.5–15 and ~95 ns, both for the β and ε "entering ring". It is interesting to note that both the β and the ε ring of the lutein that intercalated into the bilayer horizontally were H-bonded with 1–3 water molecules during the whole simulation time. Irrespective of the final orientation of the lutein in the bilayer, both its rings were H-bonded with water and PC polar groups; these interactions stabilised the orientations [73].

To understand the reason for the discrepancy between the number of lutein molecules intercalated with the "entering β ring" and with the "entering ε ring", the barriers to the β and ε ring lutein intercalation into the bilayer were evaluated from the free energy profiles of transfer of a lutein molecule from the water phase into the bilayer [73]. The profiles were calculated using umbrella sampling simulations [74]. The positions of the barriers to lutein intercalation from the β and the ε ring end coincided with the region of the rigid POPC glycerol groups. The barriers were low, although that encountered by the ε ring was ~2 kcal/mol higher than that encountered by the β ring. The difference in the barrier heights stemmed from different orientations of the β and ε rings relative to the flat polyene chain, which resulted from *sp2* hybridisation of the C6 carbon atom and *sp3* hybridisation of the C6′ carbon atom. When lutein intercalated into the bilayer from the β ring end, the system gained 19.5 ± 1.0 kcal/mol, and from the ε ring end, the gain was 17.9 ± 1.1 kcal/mol [73].

The first MD simulation study where reorientation of carotenoid molecules in the lipid bilayer was analysed in detail was described in Ref. [75]. In this study, either a single β-carotene or zeaxanthin molecule was inserted into a dimyristoyl-PC (DMPC) bilayer either vertically or horizontally, with the latter in the central bilayer region. During 100 ns and 200 ns simulations, the orientation of the β-carotenes varied more than that of the zeaxanthins, but the molecules eventually assumed relatively stable orientations, and the β-carotenes were more tilted than the zeaxanthins. These results were confirmed in umbrella sampling simulations [74], which provided free energy profiles for the rotation of β-carotene and zeaxanthin from the horizontal to the vertical position in the bilayer. The relatively flat profile for β-carotene between 90 and 35° implied that in this range of angles, β-carotene had no preferential orientation, whereas the preferential orientation of zeaxanthin in the bilayer was ~35°. Moreover, the system gained over 3 kcal/mol when zeaxanthin rotated from 90 to 35°. For both carotenoids, orientations between 35 and 0° were energetically unfavoured. Moreover, on the basis of the model that utilised a detailed conformational analysis of the single-bond torsion angles along the conjugated polyene chain of β-carotene and zeaxanthin, the authors estimated the effective length of the chain of the molecules in the bilayer and in water. The transmembrane position of zeaxanthin made its chain more extended, which might have resulted in its enhanced antioxidant capability. However, this was not the case with β-carotene [75].

Orientations of xanthophylls in the phospholipid bilayer were also analysed in Ref. [76]. In this study, which was carried out both experimentally and computationally, free energy profiles for rotating zeaxanthin and lutein from the horizontal (90°) to the vertical (10°) position in the DMPC bilayer were calculated using umbrella sampling simulations [74]. The profile for the zeaxanthin was similar to that in Ref. [75]. The minor differences concerned the free energy minimum, which was at ~25° in Ref. [76], and the energy gain due to rotation from 90 to 25°, which was ~4 kcal/mol in Ref. [76]. The free energy profile for rotating lutein in the bilayer showed that the energy gain due to rotation was 3 kcal/mol, thus smaller than that for zeaxanthin, but the free energy minimum was also at ~25°. As their experiment had no means of determining the molecule's orientation at the free energy minimum, the authors compared the average values of the xanthophyll tilt obtained from unbiased MD simulations and from experimental measurements; the values turned out to be close to each other, which mutually validated the model and the measurements [76]. The values of the average tilt of lutein derived from the experiment and computation were virtually the same as those, respectively, of zeaxanthin. This result was surprising because, as could be anticipated from the difference in the free energy profiles, the horizontal orientation of lutein in the bilayer was ~10 times more probable than that of zeaxanthin [76].

The study carried out in Ref. [77] was a cutting-edge piece of research performed both experimentally and computationally. Its aim was to identify a regulatory mechanism of controlling the intensity of light reaching photoreceptors in the human eye. This molecular-level mechanism, which took place in the retinal cell membranes, was based on the rotation of xanthophyll molecules induced by *trans-cis* photoisomerisation. Mutually perpendicular orientations of *trans* and *cis* xanthophyll isomers acted as modulators of light absorption. In the computational part of the research, unbiased MD and umbrella sampling [74] simulations were employed to predict the orientation of zeaxanthin and its two isomers, 9-*cis* and 13-*cis*, in two bilayers. One of them was composed of dipalmitoyl-PC (DPPC), and the other was composed of lipids typical for the retinal cell membrane, distearoyl-PC (DSPC), stearoyl-docosahexaenoyl-phosphatidylethanolamine (SDPE) and SD-phosphatidylserine (SDPS). The simulations confirmed the previous results [76] that *trans* zeaxanthin in the bilayer was oriented vertically, whereas 9-*cis* and 13-*cis* zeaxanthin were in 2 and 45%, respectively, of cases oriented horizontally (Figure 4). Both the simulations and the experimental measurements identified the same molecular mechanisms of controlling the intensity of light reaching photoreceptors in the retina [77].

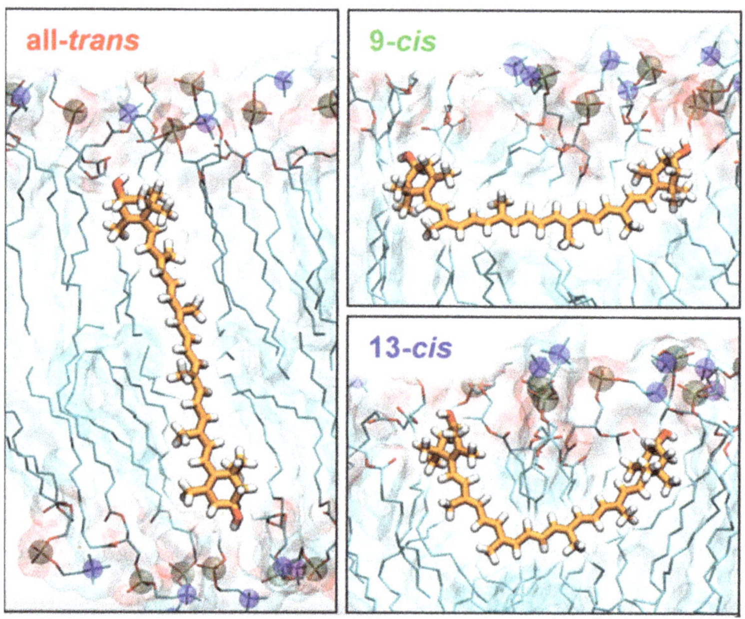

Figure 4. Representative structures for the perpendicular orientation of all-trans zeaxanthin (**left**) and the horizontal orientations of its 9-*cis* and 13-*cis* isomers (**right**) in the DPPC membrane. Figure. Form *J. Phys. Chem. B* 2021, 125, 23, 6090–6102. https://doi.org/10.1021/acs.jpcb.1c01198. [77] CC BY 4.0.

Orientations of echinenone (β,β-caroten-4-one, 4-keto-β-carotene) and β-carotene in the lipid bilayer were investigated in Ref. [78]. In this study, a mixed-lipid bilayer made of PC and PE with 18- and 16-carbon atom acyl chains of varying unsaturation that represented lipid composition of natural egg yolk, was used. The bilayer containing one molecule of echinenone or β-carotene was MD simulated for 1 μs. The results showed that although β-carotene could reorient in the bilayer rather freely, it resided most often (78% of the simulation time) in the middle of one of the bilayer leaflets oriented almost horizontally, at 80–85°, with the bilayer normal. Less often (22% of the simulation time), β-carotene was oriented 40–45° relative to the bilayer normal. Echinenone in the bilayer was in a transmembrane position with a preferential tilt at ~30°. The results for β-carotene were in general agreement with those in Ref. [75], and the orientation and position of echinenone in the bilayer were similar to those of zeaxanthin in Ref. [76]. However, detailed comparison is not fully justified, as in both studies, related but not the same quantities were analysed, and the bilayers used had different lipid compositions.

In Ref. [79], orientations of β-carotene and zeaxanthin in six mixed-lipid bilayers in different phases were compared. The bilayers were composed of POPC, DSPC and cholesterol in different proportions such that some bilayers were in the liquid-disordered and some in the liquid-ordered phase. In each phase, the distribution of the β-carotene orientations was broader than that of zeaxanthin. Moreover, in each bilayer, the carotenoids caused bilayer thinning. This result was at variance with the result in Ref. [80], which showed that both β-carotene and zeaxanthin increased the POPC bilayer thickness. The contradicting results might have stemmed from different lipid compositions of bilayers used in both studies, and possibly from too short an equilibration time of the bilayers in Ref. [79].

The MD simulations described in Ref. [81] were carried out to investigate the reorientational dynamics and preferred orientation of two xanthophylls in four POPC bilayers on the μs timescale. In the initial structures, six molecules of lutein or zeaxanthin were placed in the bilayer either vertically or horizontally. The molecules oriented horizontally were located in the region between POPC glycerol and phosphate groups (Figure 5, left-hand side); thus, no molecules aggregated in the water phase. Each system was MD simulated for 1.1 μs. The molecules could reorient freely, although none of the vertically placed xanthophyll molecules reoriented to the horizontal position. In contrast, all horizontally placed zeaxanthin molecules and five out of six horizontally placed lutein molecules reoriented to the vertical position during MD simulation times ranging from 10 to 500 ns. One lutein molecule remained in the horizontal position for the whole MD simulation time [81] (Figure 5, upper right).

To answer the question as to why none of the zeaxanthin molecules remained in the horizontal orientation, whereas one out of six lutein molecules did remain so, detailed conformational analyses of the C5-C6-C7-C8 (β ring) and C5′-C6′-C7′-C8′ (ε ring) torsion angles (called β-ring torsion and ε-ring torsion, respectively) were performed. The β-ring torsion of lutein and zeaxanthin in vacuum, water and the bilayer had two low-energy conformations that were close to each other, of 30° and −30°, and separated by a relatively low energy barrier; therefore, transitions between them were very frequent. The ε-ring torsion of lutein in vacuum had two low-energy conformations of 130° and −50°. Even though the angular distance between them was quite large and they were separated by a relatively high barrier, transitions between them were moderately frequent. However, for lutein in water and in the bilayer, the torsion occupied only the lower-energy conformation of 130°. H-bonds of lutein OH groups with polar groups of water and PC, together with a relatively high-energy barrier, hindered transitions between the conformational states. Moreover, analyses of the interactions between water molecules and the methyl groups of the polyene chain, as well as the OH groups of the ionone rings, of horizontally oriented lutein and zeaxanthin revealed which interactions with water and which orientations of the rings played a role in stabilising the horizontal orientation of the molecules in the bilayer. The result that the ε-ring torsion of lutein in the bilayer occupied only one conformation was contrary to a commonly made claim that "free" rotation of the ε ring about the single C6′–C7′ bond is the possible cause of the horizontal location of lutein in the bilayer. Conversely, the fixed conformation of the ε-ring torsion, the mutually perpendicular orientations of the lutein ε and β rings (the β ring is coplanar with the polyene chain plane, the ε ring is nearly perpendicular to it) and the orientation of the β ring OH group of the horizontal lutein were the key factors that made a horizontal lutein less likely to rotate to the vertical position than a horizontal zeaxanthin [81]. Of note was the observation that the plane of the polyene chain of any molecule that entered the bilayer from the horizontal orientation was perpendicular to the bilayer surface.

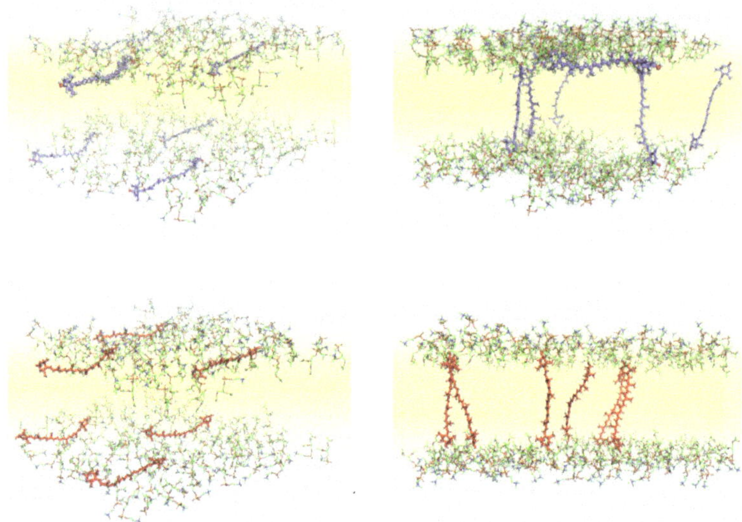

Figure 5. Initial (**left** column) and final (**right** column; after 1100 ns of MD simulations) structures of the POPC bilayer containing lutein (**upper** row) and zeaxanthin (**lower** row). In the initial structure, six xanthophyll molecules were placed parallel to the bilayer surface (horizontally). The lutein molecule, which remained in the horizontal position during the whole simulation time, is presented in the ball-and-stick model. In the figures, only the POPC head groups and xanthophyll molecules are shown to better illustrate the details. The atoms are represented in standard colours, except for the xanthophyll carbon atoms, which are *blue* for lutein and *red* for zeaxanthin. Water is shaded *light blue,* and the lipid nonpolar region is shaded *light yellow*. Figure. Form Makuch K, Hryc J, Markiewicz M and Pasenkiewicz-Gierula M (2021) Lutein and Zeaxanthin in the Lipid Bilayer–Similarities and Differences Revealed by Computational Studies. *Front. Mol. Biosci.* 8:768449. https://doi.org/10.3389/fmolb.2021.768449. [81] CC-BY.

The phospholipid bilayers used in the MD simulation studies described above were, broadly speaking, simple models of an unspecified animal cell membrane. In Ref. [82], the behaviour of two carotenoids, lycopene and zeaxanthin, in a computer model of the human *stratum corneum* (SC) cell membrane was investigated. In human skin, as in other organs, carotenoids play a protective role against photodamage [83]. The main lipid species of the lipid matrix of the SC cell membrane is ceramides with acyl chains of different lengths. In this study, the lipid bilayer was built of ceramide NS24 (a sphingolipid with a 24-carbon-atom fatty acid linked via an amide). The bilayer with one inserted molecule of lycopene or zeaxanthin was MD simulated for 100 ns. The preferred orientation of zeaxanthin in the ceramide bilayer was similar to that in the PC bilayer [75–77,81], i.e., vertical [82]. In contrast, the preferred orientation of lycopene was horizontal. The results of unbiased MD simulations of bilayers with a single carotenoid molecule were confirmed by umbrella sampling simulations [74]. The free energy profile for rotating zeaxanthin from 90 to 0° indicated that its preferred orientation in the bilayer was ~25°, and that due to rotation, the system gained over 3 kcal/mol; these results conform well to those in Refs. [75–77], even though the bilayers used in those studies had different lipid compositions. Rotating lycopene from 90 to 0° resulted in an almost monotonic increase in the system free energy, which indicated that the preferred orientation of lycopene in the bilayer was ~90°, i.e., horizontal [82].

MD simulation of β-carotene in a bilayer made of lipids typical for the brain, POPC and DMPS, was carried out in Ref. [84]. Details concerning the simulation system were nevertheless scarce; the study revealed that even though (an unspecified number of) β-carotene molecules located initially in the water hydrating the bilayer aggregated, they, over time, spontaneously intercalated into the bilayer as an aggregate. As can be seen in the video attached to Ref. [84], a β-carotene molecule that intercalated into the bilayer at the beginning of the simulation as a single species, during a (probably) 200 ns MD simulation, had considerable rotational freedom there but was in the vertical position quite often. Once the aggregate intercalated into the bilayer, the single molecule joined it.

A combined experimental and theoretical study of the aggregation of carotenoids in the bilayer was carried out in Ref. [85]. Experiments that employed several methods were performed on chiral carotenoids that differed in polarity, i.e., α-carotene (nonpolar), zeaxanthin and fucoxanthin (polar), in the DPPC bilayer. MD simulations and quantum-chemical calculations were made only on an α-carotene monomer and dimer inserted horizontally into the middle of the DPPC bilayer. Both in a monomeric and a dimeric state, the molecules moved about and rotated in the bilayer core rather freely; thus, the "dimer" they formed was not firm, and their tilts relative to the bilayer normal were broadly spread, between 30 and 140°, with an average of 100°. Even though the molecules in the "dimer" were, on average, ~21 Å apart from each other, they somehow affected each other, e.g., they became more bent than as monomers. The difference in the bending between monomeric and dimeric α-carotene expressed in Å was very small, although it was quite apparent in QM calculations. The stabilising effect of bending on the "dimer" structure shown in simulation and confirmed experimentally indicated the propensity of α-carotene to aggregate in the bilayer in contrast to more polar and vertically oriented zeaxanthin and fucoxanthin [85].

6. Carotenoids in the Membrane: A Summary of Computational Studies and Their Significance

What have we learnt from computer modelling studies about the behaviour of 40-carbon carotenoids near and in a lipid bilayer? Starting from the beginning: When carotenoid molecules are put into water hydrating the bilayer, some of them aggregate there, but some spontaneously intercalate into the bilayer. Both outcomes are due to their long nonpolar polyene chains. Unfortunately, simulations do not conclusively resolve whether the molecules intercalate the bilayer as monomers or as larger aggregates. While in some simulations, intercalation is monomeric, some others have indicated that multimers can also intercalate. In an MD simulation study [86], lutein intercalated into the POPC bilayer both as a monomer and a dimer, although in the subsequent MD simulations, the latter never happened [72,73,81]. A large β-carotene aggregate intercalating into the POPC-DMPS bilayer was shown in Ref. [84]. An experiment, in contrast to a simulation, has no means to tell what the multimeric form of carotenoids intercalating the bilayer is, but it is able to discriminate between carotenoid monomers and multimers in the bilayer as well as the type of aggregate formed, e.g., [85,87,88]. α-Carotene self-assembled readily in the gel-phase DPPC bilayer and formed "head-to-tail" J-aggregates of uncertain sizes that remained stable when temperature was increased [85]. Carotenoids with polar groups, namely fucoxanthin, zeaxanthin and lutein, formed "side-by-side" H-aggregates of rather small sizes were unstable at elevated temperatures, and in liquid-phase bilayers, they were in the monomeric form [85,87,88]. This experimental result conforms to the results of 1.1-μs MD simulations of zeaxanthin and lutein in the POPC bilayer at 310 K where the molecules did not show any sign of aggregation [81].

Spontaneity of xanthophyll intercalation into the lipid bilayer directly stems from the negative value of the change in free energy for the process. The calculated free energy gain in lutein (zeaxanthin) intercalating the bilayer is ~20 kcal/mol [73]. MD simulations and free energy calculations have also demonstrated that the intercalation of lutein from the β ring end is more probable than from the ε ring end; the barrier for the ε ring to pass through the PC glycerol groups region of the bilayer was ~2 kcal/mol higher than that for the β ring [73].

MD simulations have confirmed experimental results that the vertical position of xanthophyll molecules in the bilayer is far more probable than the horizontal one, whereas umbrella sampling simulations have given numerical values of the free energy gain due to the rotation of lutein and zeaxanthin from the horizontal to the vertical positions; the values were in the range <3, 5> kcal/mol [75–77,82]. The gain, which is altogether not very large, is smaller for lutein than zeaxanthin, which makes the horizontal orientation of lutein in the bilayer ~10 times more probable than that of zeaxanthin [76]. These energetical evaluations are in harmony with the results of MD simulations in Ref. [81], where out of 12 xanthophyll molecules placed horizontally in the bilayer, 11 spontaneously reoriented to vertical position, and the molecule that did not reorient was lutein. In contrast to polar, nonpolar carotenoids, β-carotene, α-carotene and lycopene are mainly in the horizontal position in the lipid bilayer; this is evident from experimental measurements, MD simulations [75,78,82,85] and free energy calculations [75,82]. Echinenone, with only one polar group, is in the PC-PE bilayer [78], orientated similarly to that of zeaxanthin in the PC bilayer [75].

The answer to the question as to why the vertical or horizontal orientation of carotenoids in the membrane is important can provide only experimental or quantum mechanical studies. Here, one can only speculate, bearing in mind that the main function of carotenoids is to protect the cell from the damaging effects of light and oxygen. Let us take macular pigments as an example. The maximum absorption of light by macular pigments is at 460 nm, the wavelength corresponding to blue light, which is damaging to retinal cells. The cornea and lens of the eye absorb 99.34% of incident UV light, so in practice, it does not reach the retina [89]. Macular pigments absorb ~60% of the incident blue light at 460 nm and 40–46% of the entire damaging wavelength range from 400 to 500 nm [90,91]. The highest concentration of macular pigments is in the photoreceptor axon layer, which precedes the photoreceptors; macular pigments thus act as an effective pre-receptoral light filter [91,92]. The transmembrane (vertical) position of lutein and zeaxanthin makes it possible to fulfil their task. However, the task would presumably be accomplished more effectively if the xanthophylls were in the horizontal position, which would allow them to absorb blue light from all directions [93,94]. MD simulations have indicated that this might be the case only for lutein and probably on a very short timescale [73,81]. However, recent results of the Gruszecki group [77] demonstrated that light-induced *cis*-isomerisation of xanthophylls in human retina samples and in model membranes triggered their reorientation from a vertical to a horizontal position. The consequences of the reorientation are briefly discussed above and in detail in Refs. [77,95].

Blue light that reaches the retina may initiate photosensitised production of singlet oxygen. Thus, xanthophylls, as pre-receptoral light filters, act as passive antioxidants [33,92]. They are also active antioxidants as they are able to quench singlet oxygen and other reactive oxygen species [33,92,94–96]. However, the ability of macular pigments to protect the retina stems mainly from their selective location in the domains of retinal cell membranes that are enriched in phospholipids with long polyunsaturated acyl chains, which are especially vulnerable to oxygen damage. In these domains, they act as a lipid-soluble antioxidant, and their vertical as well as horizontal position there plays a crucial role in phospholipid protection [95,97–100].

Computer simulations supplement experimental data not only by providing numerical values for the free energy gain on xanthophyll intercalation or rotation in the membrane, but also by predicting the timescales of the processes. Full intercalation of a lutein molecule into the POPC bilayer is very fast and ranges between ~7.5–15 and ~95 ns for either β or ε ring intercalation [73]. The same time range applies to zeaxanthin. The predicted time of lutein and zeaxanthin reorientation from the horizontal to the vertical position is in the range from 10 to 500 ns [81]. One lutein molecule did not reorient during the simulation time of 1.1 μs in Ref. [81]. In classical MD simulation, such an observation time is considered rather long, but it is very short compared to that of experimental measurements on macroscopic samples. As lutein in a horizontal position has not been observed experimentally [77], it is quite probable that in a longer MD simulation, the horizontal lutein reorients to the vertical position. However, one should then recall that photochemical reactions are very fast. Photoexcitation of lutein populates its second singlet excited state, S_2, that decays in a few hundred femtoseconds to the lowest singlet excited state, S_1, which then relaxes back to the ground state, S_0, on a picosecond timescale [101,102]. Thus, during the lifetime of its horizontal orientation (>1 μs), lutein may participate in many light-generated processes. Moreover, even though the horizontal position is much less probable than the vertical one, both orientations must be in some statistical equilibrium in the bilayer.

This review clearly shows that a combined experimental and computational approach in studying molecular-level biological processes provides us with the opportunity to know and understand them better. It also shows that our knowledge concerning how carotenoids are transported in the body and enter the cell membrane is rather fragmentary. However, hopefully, this review might help the reader in recognising topics that merit more in-depth study.

Author Contributions: M.P.-G. wrote the manuscript. J.H. wrote a draft of the paragraph on carotenoid absorption and availability in the human body; M.M. collected, summarised and critically evaluated the literature on the simulation studies on carotenoid–membrane interactions and prepared figures. All authors contributed to the final version of the manuscript. All authors have read and agreed to the published version of the manuscript.

Funding: This research received no external funding.

Acknowledgments: M.P.-G. thanks W. K. Subczynski for constructive discussions.

Conflicts of Interest: The authors declare no conflicts of interest.

References

1. Fuentes, P.; Pizarro, L.; Moreno, J.C.; Handford, M.; Rodriguez-Concepcion, M.; Stange, C. Light-dependent changes in plastid differentiation influence carotenoid gene expression and accumulation in carrot roots. *Plant Mol. Biol.* **2012**, *79*, 47–59. [CrossRef]
2. Sun, T.H.; Rao, S.M.; Zhou, X.S.; Li, L. Plant carotenoids: Recent advances and future perspectives. *Mol. Hortic.* **2022**, *2*, 3. [CrossRef] [PubMed]
3. Gómez-Sagasti, M.T.; López-Pozo, M.; Artetxe, U.; Becerril, J.M.; Hernández, A.; García-Plazaola, J.I.; Esteban, R. Carotenoids and their derivatives: A "Swiss Army knife-like" multifunctional tool for fine-tuning plant-environment interactions. *Environ. Exp. Bot.* **2023**, *207*, 105229. [CrossRef]
4. Demmig-Adams, B.; Stewart, J.J.; López-Pozo, M.; Polutchko, S.K.; Adams, W.W. Zeaxanthin, a molecule for photoprotection in many different environments. *Molecules* **2020**, *25*, 5825. [CrossRef] [PubMed]
5. Caferri, R.; Guardini, Z.; Bassi, R.; Dall'Osto, L. Assessing photoprotective functions of carotenoids in photosynthetic systems of plants and green algae. In *Methods Enzymol*; Wurtzel, T.E., Ed.; Academic Press: Cambridge, MA, USA, 2022; Volume 674, pp. 53–84.
6. Eggersdorfer, M.; Wyss, A. Carotenoids in human nutrition and health. *Arch. Biochem. Biophys.* **2018**, *652*, 18–26. [CrossRef] [PubMed]
7. Milani, A.; Basirnejad, M.; Shahbazi, S.; Bolhassani, A. Carotenoids: Biochemistry, pharmacology and treatment. *Br. J. Pharmacol.* **2017**, *174*, 1290–1324. [CrossRef] [PubMed]
8. Fiedor, J.; Burda, K. Potential role of carotenoids as antioxidants in human health and disease. *Nutrients* **2014**, *6*, 466–488. [CrossRef] [PubMed]
9. Zhao, Z.L.; Chen, J.; Ci, F.F.; Pang, H.; Cheng, N.; Xing, A.J. α-Carotene: A valuable carotenoid in biological and medical research. *J. Sci. Food Agric.* **2022**, *102*, 5606–5617. [CrossRef] [PubMed]
10. Stringham, J.M.; Johnson, E.J.; Hammond, B.R. Lutein across the Lifespan: From Childhood Cognitive Performance to the Aging Eye and Brain. *Curr. Dev. Nutr.* **2019**, *3*, nzz066. [CrossRef]
11. Lopresti, A.L.; Smith, S.J.; Drummond, P.D. The Effects of Lutein and Zeaxanthin Supplementation on Cognitive Function in Adults With Self-Reported Mild Cognitive Complaints: A Randomized, Double-Blind, Placebo-Controlled Study. *Front. Nutr.* **2022**, *9*, 843512. [CrossRef]

12. Parekh, R.; Hammond, B.R.; Chandradhara, D. Lutein and zeaxanthin supplementation improves dynamic visual and cognitive performance in children: A randomized, double-blind, parallel, placebo-controlled study. *Adv. Ther.* **2024**, *41*, 1496–1511. [CrossRef]
13. Yabuzaki, J. Carotenoids Database: Structures, chemical fingerprints and distribution among organisms. *Database-Oxford* **2017**, *2017*, bax004. [CrossRef] [PubMed]
14. Amengual, J. Bioactive Properties of Carotenoids in Human Health. *Nutrients* **2019**, *11*, 2388. [CrossRef]
15. Pitts, S.B.J.; Johnson, N.S.; Wu, Q.; Firnhaber, G.C.; Kaur, A.P.; Obasohan, J. A meta-analysis of studies examining associations between resonance Raman spectroscopy-assessed skin carotenoids and plasma carotenoids among adults and children. *Nutr. Rev.* **2022**, *80*, 230–241. [CrossRef] [PubMed]
16. Fiedor, L.; Dudkowiak, A.; Pilch, M. The origin of the dark S state in carotenoids: A comprehensive model. *J. R. Soc. Interface* **2019**, *16*, 20190191. [CrossRef] [PubMed]
17. Harrison, E.H. Mechanisms involved in the intestinal absorption of dietary vitamin A and provitamin A carotenoids. *Bba-Mol. Cell Biol. Lipids* **2012**, *1821*, 70–77. [CrossRef]
18. Lemmens, L.; Colle, I.; Van Buggenhout, S.; Palmero, P.; Van Loey, A.; Hendrickx, M. Carotenoid bioaccessibility in fruit- and vegetable-based food products as affected by product (micro) structural characteristics and the presence of lipids: A review. *Trends Food Sci. Tech.* **2014**, *38*, 125–135. [CrossRef]
19. Xavier, A.A.O.; Mercadante, A.Z. The bioaccessibility of carotenoids impacts the design of functional foods. *Curr. Opin. Food Sci.* **2019**, *26*, 1–8. [CrossRef]
20. Parker, R.S. Carotenoids.4. Absorption, metabolism, and transport of carotenoids. *FASEB J.* **1996**, *10*, 542–551. [CrossRef]
21. Chacón-Ordóñez, T.; Carle, R.; Schweiggert, R. Bioaccessibility of carotenoids from plant and animal foods. *J. Sci. Food Agric.* **2019**, *99*, 3220–3239. [CrossRef]
22. Reboul, E. Proteins involved in fat-soluble vitamin and carotenoid transport across the intestinal cells: New insights from the past decade. *Prog. Lipid Res.* **2023**, *89*, 101208. [CrossRef] [PubMed]
23. Furr, H.C.; Clark, R.M. Transport, uptake, and target. Tissue storage of carotenoids. In *Carotenoids in Health and Disease*; Krinsky, N.I., Mayne, S.T., Sies, H., Eds.; Marcel Dekker: New York, NY, USA, 2004; pp. 229–278.
24. Gugliucci, A. The chylomicron saga: Time to focus on postprandial metabolism. *Front. Endocrinol.* **2024**, *14*, 1322869. [CrossRef] [PubMed]
25. Shmarakov, I.O.; Yuen, J.J.; Blaner, W.S. Carotenoid Metabolism and Enzymology. In *Carotenoids and Human Health*; Tanumihardjo, S.A., Ed.; Humana Press (Springer): New York, NY, USA, 2013; pp. 29–56.
26. Canene-Adams, K.; Erdman, J.W., Jr. Absorption, transport, distribution in tissues and bioavailability. In *Carotenoids, Nutrition and Health*; Britton, G., Liaaen-Jensen, S., Pfander, H., Eds.; Birkhäuser: Basel, Switzerland, 2009; Volume 5, pp. 115–144.
27. Li, B.X.; Vachali, P.; Chang, F.Y.; Gorusupudi, A.; Arunkumar, R.; Shi, L.J.; Rognon, G.T.; Frederick, J.M.; Bernstein, P.S. HDL is the primary transporter for carotenoids from liver to retinal pigment epithelium in transgenic ApoA-I/Bco2 mice. *Arch. Biochem. Biophys.* **2022**, *716*, 109111. [CrossRef] [PubMed]
28. Arunkumar, R.; Gorusupudi, A.; Bernstein, P.S. The macular carotenoids: A biochemical overview. *Bba-Mol. Cell Biol. Lipids* **2020**, *1865*, 158617. [CrossRef] [PubMed]
29. Krinsky, N.I.; Cornwell, D.G.; Oncley, J.L. The transport of vitamin A and carotenoids in human plasma. *Arch. Biochem. Biophys.* **1958**, *73*, 233–246. [CrossRef]
30. Böhm, V.; Lietz, G.; Olmedilla-Alonso, B.; Phelan, D.; Reboul, E.; Bánati, D.; Borel, P.; Corte-Real, J.; de Lera, A.R.; Desmarchelier, C.; et al. From carotenoid intake to carotenoid blood and tissue concentrations—Implications for dietary intake recommendations. *Nutr. Rev.* **2021**, *79*, 544–573. [CrossRef] [PubMed]
31. Elvira-Torales, L.I.; García-Alonso, J.; Periago-Castón, M.J. Nutritional Importance of Carotenoids and Their Effect on Liver Health: A Review. *Antioxidants-Basel* **2019**, *8*, 229. [CrossRef]
32. Bohn, T.; Desmarchelier, C.; El, S.N.; Keijer, J.; van Schothorst, E.; Rühl, R.; Borel, P. β-Carotene in the human body: Metabolic bioactivation pathways—From digestion to tissue distribution and excretion. *P. Nutr. Soc.* **2019**, *78*, 68–87. [CrossRef] [PubMed]
33. Landrum, J.T. Reactive oxygen and nitrogen species in biological systems: Reactions and regulation by carotenoids. In *Carotenoids and Human Health*; Tanumihardjo, S.A., Ed.; Humana Press (Springer): New York, NY, USA, 2013; Volume 1, pp. 57–101.
34. Bandara, S.; von Lintig, J. Aster la vista: Unraveling the biochemical basis of carotenoid homeostasis in the human retina. *Bioessays* **2022**, *44*, e2200133. [CrossRef]
35. Thomas, S.E.; Harrison, E.H. Mechanisms of selective delivery of xanthophylls to retinal pigment epithelial cells by human lipoproteins. *J. Lipid Res.* **2016**, *57*, 1865–1878. [CrossRef]
36. Harrison, E.H. Mechanisms of transport and delivery of vitamin A and carotenoids to the retinal pigment epithelium. *Mol. Nutr. Food Res.* **2019**, *63*, e1801046. [CrossRef] [PubMed]
37. Shen, W.J.; Asthana, S.; Kraemer, F.B.; Azhar, S. Scavenger receptor B type 1: Expression, molecular regulation, and cholesterol transport function. *J. Lipid Res.* **2018**, *59*, 1114–1131. [CrossRef] [PubMed]
38. Yang, J.Y.; Anishchenko, I.; Park, H.; Peng, Z.L.; Ovchinnikov, S.; Baker, D. Improved protein structure prediction using predicted interresidue orientations. *Proc. Natl. Acad. Sci. USA* **2020**, *117*, 1496–1503. [CrossRef]
39. Powers, H.R.; Sahoo, D. SR-B1's next top model: Structural perspectives on the functions of the HDL receptor. *Curr. Atheroscler. Rep.* **2022**, *24*, 277–288. [CrossRef]

40. Quadro, L.; Giordano, E.; Costabile, B.K.; Nargis, T.; Iqbal, J.; Kim, Y.; Wassef, L.; Hussain, M.M. Interplay between β-carotene and lipoprotein metabolism at the maternal-fetal barrier. *Bba-Mol. Cell Biol. Lipids* **2020**, *1865*, 158591. [CrossRef] [PubMed]
41. Borel, P.; Moussa, M.; Reboul, E.; Lyan, B.; Defoort, C.; Vincent-Baudry, S.; Maillot, M.; Gastaldi, M.; Darmon, M.; Portugal, H.; et al. Human plasma levels of vitamin E and Carotenoids are associated with genetic polymorphisms in genes involved in lipid metabolism. *J. Nutr.* **2007**, *137*, 2653–2659. [CrossRef] [PubMed]
42. Reboul, E. Mechanisms of carotenoid intestinal absorption: Where do we stand? *Nutrients* **2019**, *11*, 838. [CrossRef]
43. Steinhoff, J.S.; Lass, A.; Schupp, M. Biological functions of RBP4 and its relevance for human diseases. *Front. Physiol.* **2021**, *12*, 659977. [CrossRef]
44. Ask, N.M.; Leung, M.; Radhakrishnan, R.; Lobo, G.P. Vitamin A transporters in visual function: A mini review on membrane receptors for dietary vitamin A uptake, storage, and transport to the eye. *Nutrients* **2021**, *13*, 3987. [CrossRef]
45. Chen, Y.; Clarke, O.B.; Kim, J.; Stowe, S.; Kim, Y.K.; Assur, Z.; Cavalier, M.; Godoy-Ruiz, R.; von Alpen, D.C.; Manzini, C.; et al. Structure of the STRA6 receptor for retinol uptake. *Science* **2016**, *353*, aad8266. [CrossRef]
46. Petkovich, M.; Chambon, P. Retinoic acid receptors at 35 years. *J. Mol. Endocrinol.* **2022**, *69*, T13–T24. [CrossRef] [PubMed]
47. Riabroy, N.; Dever, J.T.; Tanumihardjo, S.A. α-Retinol and 3,4-didehydroretinol support growth in rats when fed at equimolar amounts and α-retinol is not toxic after repeated administration of large doses. *Br. J. Nutr.* **2014**, *111*, 1373–1381. [CrossRef] [PubMed]
48. Likkei, K.; Moldenhauer, M.; Tavraz, N.N.; Maksimov, E.G.; Sluchanko, N.N.; Friedrich, T. Lipid composition and properties affect protein-mediated carotenoid uptake efficiency from membranes. *Biochim. Biophys. Acta-Biomem.* **2024**, *1866*, 184241. [CrossRef] [PubMed]
49. Shyam, R.; Vachali, P.; Gorusupudi, A.; Nelson, K.; Bernstein, P.S. All three human scavenger receptor class B proteins can bind and transport all three macular xanthophyll carotenoids. *Arch. Biochem. Biophys.* **2017**, *634*, 21–28. [CrossRef]
50. Bandara, S.; Ramkumar, S.; Imanishi, S.; Thomas, L.D.; Sawant, O.B.; Imanishi, Y.; von Lintig, J. Aster proteins mediate carotenoid transport in mammalian cells. *Proc. Natl. Acad. Sci. USA* **2022**, *119*, e2200068119. [CrossRef] [PubMed]
51. Slonimskiy, Y.B.; Egorkin, N.A.; Friedrich, T.; Maksimov, E.G.; Sluchanko, N.N. Microalgal protein AstaP is a potent carotenoid solubilizer and delivery module with a broad carotenoid binding repertoire. *Febs J.* **2022**, *289*, 999–1022. [CrossRef] [PubMed]
52. Bandara, S.; Moon, J.; Ramkumar, S.; von Lintig, J. ASTER-B regulates mitochondrial carotenoid transport and homeostasis. *J. Lipid Res.* **2023**, *64*, 100369. [CrossRef] [PubMed]
53. Powers, H.R.; Jenjak, S.E.; Volkman, B.F.; Sahoo, D. Development and validation of a purification system for functional full-length human SR-B1 and CD36. *J. Biol. Chem.* **2023**, *299*, 105187. [CrossRef]
54. Sluchanko, N.N.; Slonimskiy, Y.B.; Egorkin, N.A.; Varfolomeeva, L.A.; Kleymenov, S.Y.; Minyaev, M.E.; Faletrov, Y.V.; Moysenovich, A.M.; Parshina, E.Y.; Friedrich, T.; et al. Structural basis for the carotenoid binding and transport function of a START domain. *Structure* **2022**, *30*, 1647–1659. [CrossRef]
55. Kornilov, F.D.; Slonimskiy, Y.B.; Lunegova, D.A.; Egorkin, N.A.; Savitskaya, A.G.; Kleymenov, S.Y.; Maksimov, E.G.; Goncharuk, S.A.; Mineev, K.S.; Sluchanko, N.N. Structural basis for the ligand promiscuity of the neofunctionalized, carotenoid-binding fasciclin domain protein AstaP. *Commun. Biol.* **2023**, *6*, 471. [CrossRef]
56. Horvath, M.P.; George, E.W.; Tran, Q.T.; Baumgardner, K.; Zharov, G.; Lee, S.; Sharifzadeh, H.; Shihab, S.; Mattinson, T.; Li, B.X.; et al. Structure of the lutein-binding domain of human StARD3 at 1.74 angstrom resolution and model of a complex with lutein. *Acta Crystallogr. F* **2016**, *72*, 609–618. [CrossRef] [PubMed]
57. Shyam, R.; Gorusupudi, A.; Nelson, K.; Horvath, M.P.; Bernstein, P.S. RPE65 has an additional function as the lutein to meso-zeaxanthin isomerase in the vertebrate eye. *Proc. Natl. Acad. Sci. USA* **2017**, *114*, 10882–10887. [CrossRef] [PubMed]
58. Varfolomeeva, L.A.; Slonimskiy, Y.B.; Egorkin, N.A.; Minyaev, M.E.; Faletrov, Y.V.; Boyko, K.M.; Sluchanko, N.N. Preparation and structural studies of the silkworm carotenoid-binding protein complexed with a new pigment. *Crystallogr. Rep.* **2022**, *67*, 909–917. [CrossRef]
59. Hazai, E.; Bikádi, Z.; Zsila, F.; Lockwood, S.F. Molecular modeling of the non-covalent binding of the dietary tomato carotenoids lycopene and lycophyll, and selected oxidative metabolites with 5-lipoxygenase. *Bioorgan Med. Chem.* **2006**, *14*, 6859–6867. [CrossRef]
60. Neculai, D.; Schwake, M.; Ravichandran, M.; Zunke, F.; Collins, R.F.; Peters, J.; Neculai, M.; Plumb, J.; Loppnau, P.; Pizarro, J.C.; et al. Structure of LIMP-2 provides functional insights with implications for SR-BI and CD36. *Nature* **2013**, *504*, 172–176. [CrossRef] [PubMed]
61. Molteni, C.; La Motta, C.; Valoppi, F. Improving the bioaccessibility and bioavailability of carotenoids by means of nanostructured delivery systems: A comprehensive review. *Antioxidants-Basel* **2022**, *11*, 1931. [CrossRef]
62. Waterhouse, A.; Bertoni, M.; Bienert, S.; Studer, G.; Tauriello, G.; Gumienny, R.; Heer, F.T.; de Beer, T.A.P.; Rempfer, C.; Bordoli, L.; et al. SWISS-MODEL: Homology modelling of protein structures and complexes. *Nucleic Acids Res.* **2018**, *46*, W296–W303. [CrossRef]
63. Fiser, A.; Sali, A. Modeller: Generation and refinement of homology-based protein structure models. *Methods Enzymol.* **2003**, *374*, 461–491. [CrossRef]
64. Rizzuti, B. Molecular simulations of proteins: From simplified physical interactions to complex biological phenomena. *Biochim. Biophys. Acta Proteins Proteom.* **2022**, *1870*, 140757. [CrossRef]

65. Popova, A.V.; Andreeva, A.S. Chapter Eight—Carotenoid–lipid interactions. In *Advances in Planar Lipid Bilayers and Liposomes*; Iglič, A., Genova, J., Eds.; Academic Press: Cambridge, MA, USA, 2013; Volume 17, pp. 215–236.
66. Augustynska, D.; Jemiola-Rzeminska, M.; Burda, K.; Strzalka, K. Influence of polar and nonpolar carotenoids on structural and adhesive properties of model membranes. *Chem-Biol. Interact.* **2015**, *239*, 19–25. [CrossRef]
67. Cvetkovic, D.; Fiedor, L.; Wisniewska-Becker, A.; Markovic, D. Organization of carotenoids in models of biological membranes: Current status of knowledge and research. *Curr. Anal. Chem.* **2013**, *9*, 86–98. [CrossRef]
68. Perez-Lopez, M.I.; Mendez-Reina, R.; Trier, S.; Herrfurth, C.; Feussner, I.; Bernal, A.; Forero-Shelton, M.; Leidy, C. Variations in carotenoid content and acyl chain composition in exponential, stationary and biofilm states of Staphylococcus aureus, and their influence on membrane biophysical properties. *Bba-Biomembranes* **2019**, *1861*, 978–987. [CrossRef] [PubMed]
69. Bykowski, M.; Mazur, R.; Wójtowicz, J.; Suski, S.; Garstka, M.; Mostowska, A.; Kowalewska, L. Too rigid to fold: Carotenoid-dependent decrease in thylakoid fluidity hampers the formation of chloroplast grana. *Plant Physiol.* **2021**, *185*, 210–227. [CrossRef] [PubMed]
70. Manrique-Moreno, M.; Jemiota-Rzeminska, M.; Múnera-Jaramillo, J.; López, G.D.; Suesca, E.; Leidy, C.; Strzalka, K. Carotenoids modulate the thermotropic phase behavior of model systems that mimic its membrane composition. *Membranes* **2022**, *12*, 945. [CrossRef] [PubMed]
71. Jemiola-Rzeminska, M.; Pasenkiewicz-Gierula, M.; Strzalka, K. The behaviour of β-carotene in the phosphatidylcholine bilayer as revealed by a molecular simulation study. *Chem. Phys. Lipids* **2005**, *135*, 27–37. [CrossRef] [PubMed]
72. Pasenkiewicz-Gierula, M.; Baczynski, K.; Murzyn, K.; Markiewicz, M. Orientation of lutein in a lipid bilayer—Revisited. *Acta Biochim. Pol.* **2012**, *59*, 115–118. [CrossRef] [PubMed]
73. Makuch, K.; Markiewicz, M.; Pasenkiewicz-Gierula, M. Asymmetric spontaneous intercalation of lutein into a phospholipid bilayer, a computational study. *Comput. Struct. Biotec* **2019**, *17*, 516–526. [CrossRef] [PubMed]
74. Torrie, G.M.; Valleau, J.P. Nonphysical sampling distributions in Monte Carlo free-energy estimation: Umbrella sampling. *J. Comput. Phys.* **1977**, *23*, 187–199. [CrossRef]
75. Cerezo, J.; Zuniga, J.; Bastida, A.; Requena, A.; Ceron-Carrasco, J.P. Conformational changes of beta-carotene and zeaxanthin immersed in a model membrane through atomistic molecular dynamics simulations. *Phys. Chem. Chem. Phys.* **2013**, *15*, 6527–6538. [CrossRef]
76. Grudzinski, W.; Nierzwicki, L.; Welc, R.; Reszczynska, E.; Luchowski, R.; Czub, J.; Gruszecki, W.I. Localization and orientation of xanthophylls in a lipid bilayer. *Sci. Rep.* **2017**, *7*, 9619. [CrossRef]
77. Luchowski, R.; Grudzinski, W.; Welc, R.; Pinto, M.M.M.; Sek, A.; Ostrowski, J.; Nierzwicki, L.; Chodnicki, P.; Wieczor, M.; Sowinski, K.; et al. Light-modulated sunscreen mechanism in the retina of the human eye. *J. Phys. Chem. B* **2021**, *125*, 6090–6102. [CrossRef]
78. Semenov, A.N.; Gvozdev, D.A.; Zlenko, D.V.; Protasova, E.A.; Khashimova, A.R.; Parshina, E.Y.; Baizhumanov, A.A.; Lotosh, N.Y.; Kim, E.E.; Kononevich, Y.N.; et al. Modulation of Membrane Microviscosity by Protein-Mediated Carotenoid Delivery as Revealed by Time-Resolved Fluorescence Anisotropy. *Membranes* **2022**, *12*, 905. [CrossRef] [PubMed]
79. Johnson, Q.R.; Mostofian, B.; Gomez, G.F.; Smith, J.C.; Cheng, X.L. Effects of carotenoids on lipid bilayers. *Phys. Chem. Chem. Phys.* **2018**, *20*, 3795–3804. [CrossRef] [PubMed]
80. Mostofian, B.; Johnson, Q.R.; Smith, J.C.; Cheng, X.L. Carotenoids promote lateral packing and condensation of lipid membranes. *Phys. Chem. Chem. Phys.* **2020**, *22*, 12281–12293. [CrossRef] [PubMed]
81. Makuch, K.; Hryc, J.; Markiewicz, M.; Pasenkiewicz-Gierula, M. Lutein and zeaxanthin in the lipid bilayer-similarities and differences revealed by computational studies. *Front. Mol. Biosci.* **2021**, *8*, 768449. [CrossRef] [PubMed]
82. Ri, J.S.; Choe, C.S.; Choe, S.H.; Jong, K.H.; Hong, S.N.; Schleusener, J.; Lademann, J.; Darvin, M.E. Lycopene, but not zeaxanthin, serves as a skeleton for the formation of an orthorhombic organization of intercellular lipids within the lamellae in the stratum corneum: Molecular dynamics simulations of the hydrated ceramide NS bilayer model. *Biochim. Biophys. Acta-Biomem.* **2023**, *1865*, 184081. [CrossRef] [PubMed]
83. Flieger, J.; Raszewska-Famielec, M.; Radzikowska-Büchner, E.; Flieger, W. Skin protection by carotenoid pigments. *Int. J. Mol. Sci.* **2024**, *25*, 1431. [CrossRef]
84. Gastaldo, I.P.; Himbert, S.; Ram, U.; Rheinstädter, M.C. The effects of resveratrol, caffeine, β-carotene, and epigallocatechin gallate (EGCG) on amyloid-β aggregation in aynthetic brain membranes. *Mol. Nutr. Food Res.* **2020**, *64*, e2000632. [CrossRef]
85. Hachlica, N.; Stefanska, M.; Mach, M.; Kowalska, M.; Wydro, P.; Domagala, A.; Kessler, J.; Zajac, G.; Kaczor, A. Organization of carotenoid aggregates in membranes studied selectively using resonance Raman optical activity. *Small* **2024**, *20*, e2306707. [CrossRef]
86. Pasenkiewicz-Gierula, M.; Markiewicz, M. Orientation of lutein in a lipid bilayer—Revisited. In Proceedings of the 16th International Symposium on Carotenoids, Krakow, Poland, 17–22 July 2011; p. 105.
87. Sujak, A.; Okulski, W.; Gruszecki, W.I. Organisation of xanthophyll pigments lutein and zeaxanthin in lipid membranes formed with dipalmitoylphosphatidylcholine. *Biochim. Biophys. Acta-Biomem.* **2000**, *1509*, 255–263. [CrossRef]
88. Sujak, A.; Mazurek, P.; Gruszecki, W.I. Xanthophyll pigments lutein and zeaxanthin in lipid multibilayers formed with dimyristoylphosphatidylcholine. *J. Photochem. Photobiol. B* **2002**, *68*, 39–44. [CrossRef] [PubMed]
89. Fukuoka, H.; Gali, H.E.; Bu, J.J.; Sella, R.; Afshari, N.A. Ultraviolet light exposure and its penetrance through the eye in a porcine model. *Int. J. Ophthalmol-Chi* **2023**, *16*, 172–177. [CrossRef] [PubMed]

90. Snodderly, D.M.; Auran, J.D.; Delori, F.C. The macular pigment. 2. Spatial-distribution in primate retinas. *Invest. Ophth Vis. Sci.* **1984**, *25*, 674–685.
91. Schalch, W.; Landrum, J.T.; Bone, R.A. The Eye. In *Carotenoids, Nutrition and Health*; Britton, G., Liaaen-Jensen, S., Pfande, H., Eds.; Birkhäuser: Basel, Switzerland, 2009; Volume 5, pp. 301–330.
92. O'Connell, E.; Neelam, K.; Nolan, J.; Au Eong, K.G.; Beatty, S. Macular carotenoids and age-related maculopathy. *Ann. Acad. Med. Singap.* **2006**, *35*, 821–830. [CrossRef] [PubMed]
93. Ahmed, S.; Lott, M.; Marcus, D. The macular xantophyls. *Surv. Ophthalmol.* **2005**, *50*, 183–193. [CrossRef] [PubMed]
94. Vishwanathan, R.; Johnson, E.J. Lutein and Zeaxanthin and Eye Disease. In *Carotenoids and Human Health*; Tanumihardjo, S.A., Ed.; Humana Press (Springer): New York, NY, USA, 2013; Volume 1, pp. 215–235.
95. Welc-Stanowska, R.; Pietras, R.; Mielecki, B.; Sarewicz, M.; Luchowski, R.; Widomska, J.; Grudzinski, W.; Osyczka, A.; Gruszecki, W.I. How do xanthophylls protect lipid membranes from oxidative damage? *J. Phys. Chem. Lett.* **2023**, *14*, 7440–7444. [CrossRef]
96. Cantrell, A.; McGarvey, D.J.; Truscott, T.G.; Rancan, F.; Böhm, F. Singlet oxygen quenching by dietary carotenoids in a model membrane environment. *Arch. Biochem. Biophys.* **2003**, *412*, 47–54. [CrossRef]
97. Subczynski, W.K.; Wisniewska, A.; Widomska, J. Location of macular xanthophylls in the most vulnerable regions of photoreceptor outer-segment membranes. *Arch. Biochem. Biophys.* **2010**, *504*, 61–66. [CrossRef]
98. Widomska, J.; Gruszecki, W.I.; Subczynski, W.K. Factors differentiating the antioxidant activity of macular xanthophylls in the human eye retina. *Antioxidants* **2021**, *10*, 601. [CrossRef]
99. Widomska, J.; Zareba, M.; Subczynski, W.K. Can xanthophyll-membrane interactions explain their selective presence in the retina and brain? *Foods* **2016**, *5*, 7. [CrossRef]
100. Wisniewska, A.; Subczynski, W.K. Distribution of macular xanthophylls between domains in a model of photoreceptor outer segment membranes. *Free Radic. Biol. Med.* **2006**, *41*, 1257–1265. [CrossRef] [PubMed]
101. Polívka, T.; Sundström, V. Ultrafast dynamics of carotenoid excited states -: From solution to natural and artificial systems. *Chem. Rev.* **2004**, *104*, 2021–2071. [CrossRef] [PubMed]
102. Accomasso, D.; Arslancan, S.; Cupellini, L.; Granucci, G.; Mennucci, B. Ultrafast Excited-State Dynamics of Carotenoids and the Role of the S State. *J. Phys. Chem. Lett.* **2022**, *13*, 6762–6769. [CrossRef] [PubMed]

International Journal of
Molecular Sciences

MDPI

Review

Bioactivity and Bioavailability of Carotenoids Applied in Human Health: Technological Advances and Innovation

Tomas Gabriel Bas

Escuela de Ciencias Empresariales, Universidad Catolica del Norte, Coquimbo 1780000, Chile; tomas.bas@ucn.cl

Abstract: This article presents a groundbreaking perspective on carotenoids, focusing on their innovative applications and transformative potential in human health and medicine. Research jointly delves deeper into the bioactivity and bioavailability of carotenoids, revealing therapeutic uses and technological advances that have the potential to revolutionize medical treatments. We explore pioneering therapeutic applications in which carotenoids are used to treat chronic diseases such as cancer, cardiovascular disease, and age-related macular degeneration, offering novel protective mechanisms and innovative therapeutic benefits. Our study also shows cutting-edge technological innovations in carotenoid extraction and bioavailability, including the development of supramolecular carriers and advanced nanotechnology, which dramatically improve the absorption and efficacy of these compounds. These technological advances not only ensure consistent quality but also tailor carotenoid therapies to each patient's health needs, paving the way for personalized medicine. By integrating the latest scientific discoveries and innovative techniques, this research provides a prospective perspective on the clinical applications of carotenoids, establishing a new benchmark for future studies in this field. Our findings underscore the importance of optimizing carotenoid extraction, administration, bioactivity, and bioavailability methods to develop more effective, targeted, and personalized treatments, thus offering visionary insight into their potential in modern medical practices.

Keywords: carotenoids; advanced bioavailability; bioactivity; supramolecular carriers; personalized medicine; chronic disease prevention; extraction

1. Introduction

Carotenoids comprise a group of fat-soluble organic pigments synthesized by phototrophs (algae, cyanobacteria, and plants), non-photosynthetic bacteria, and fungi, with more than 1117 structures identified from 683 different sources within the hydrocarbon class [1]. These pigments exhibit highly attractive biological activities, such as antioxidant properties, anti-inflammatory effects, and the attributes of provitamin A, making them attractive for applications in medicine and the therapeutic treatment of specific diseases [2–6]. Carotenoids can be classified into two main categories according to their chemical and nutritional characteristics: carotenes and xanthophylls [7–10]. Carotenes are characterized by the presence of carbon and hydrogen atoms in their chemical structure and include β-carotene, α-carotene, γ-carotene, δ-carotene, lycopene, neurosporene, and torulene, known for their provitamin content. Meanwhile, the xanthophyll category mainly comprises astaxanthin, zeaxanthin, lutein, torularhodin, canthaxanthin, and violaxanthin [11]. Carotenoids, as vital secondary metabolites, play a crucial role in maintaining various functions of human organs, including eye, cardiovascular, and skin health, immune function, cognitive function and bone health. Its mechanisms of action involve neutralizing free radicals, modulating gene expression, and regulating inflammatory pathways [12–14]. The structure of these various phytochemicals manifests itself as tetraterpenes with eight five-carbon isoprenoid units (mainly C40), containing up to 15 conjugated double bonds and exhibiting a symmetric arrangement typified by β-carotene, resulting in red, yellow, and oranges [11].

Natural sources of carotenoids are diverse and significant amounts are found in brown algae and microalgae such as *Heterochlorella luteoviridis*, *Dunaliella salina* or *Chlorella vulgaris*, by-products of carrot juice and skin processing, tomatoes, corn gluten meal, citrus residue, pomegranate residue, red pepper, *Lycium*, *ferocissimum fruits*, guava pulp, the peel and pulp of thai gac fruit (*Momordica cochinchinensis Spreng*), apricot waste, orange and pumpkin peels, vegetables, shrimp and yeast such as *Rhodotorula glutinis*, *Xanthophyllomyces* or *Dendrorhous*, among others [15,16]. These abundant sources offer promising prospects for the extraction of valuable carotenoids that are applicable not only in traditional industries such as food dyes, cosmetics, and nutraceuticals, but also in the pharmaceutical industry for human health [17–20].

The objective of this research is to document advances related to cutting-edge techniques in the extraction of carotenoids, highlighting the impact of their use in medicine and the pharmaceutical industry. This article features the most innovative methodologies and technological advances related to carotenoid extraction, with special focus on improving their bioactivity and bioavailability for applications in human health. The transformative potential of these cutting-edge extraction techniques is highlighted by detailing how they improve the efficiency, sustainability, and quality of carotenoid extraction. By emphasizing the scientific rigor and novelty of these methods, this article sets a new benchmark for future research and practical applications in the fields of food, pharmaceuticals, and nutraceuticals. This article addresses the bioactivity of carotenoids in the biopharmaceutical sector, seeking to understand the therapeutic and innovative implications of their administration and bioavailability in medicine. Different scientific results are analyzed that seek to provide a more comprehensive understanding of the fundamental role that these molecules can play in the medical treatment and management of certain specific diseases such as cancer, heart and eye diseases (macula), or inflammatory ailments.

2. The Functionality of Carotenoids

Carotenoids participate in different biological functions in plant, aquatic, animal and human systems [21]. In plants, plastids regulate biosynthesis, improving light absorption. In nonphotosynthetic tissues, carotenoids act as dyes and precursors of isoprene [11]. In human cells, they contribute to skin health, improving defense against ultraviolet (UV) radiation and also providing protection against diabetes, cancer, inflammation, and quality of vision [22–24]. Recently, research has been conducted on sources of so-called unusual colorless carotenoids, such as phytoene and phytofluene from aquatic environments, such as the microalgae *D. salina* [25,26]. These compounds are part of human plasma, internal tissues, and skin and are considered a sustainable endogenous nutrient to maintain healthy skin and prevent the symptoms of photoaging, the main source of skin aging [27]. Similarly, studies recognize the importance of apocarotenoids and their possible involvement in biological actions against cancer [28]. However, despite promising data, it is crucial to have reliable information on the content and volume of intake of different types of carotenoids for each specific case, to establish the recommended prescription units. This is key, since excessive administration of these compounds could cause undesirable and even dangerous side effects to the patient's health [29,30].

Therefore, it is essential to know the molecular structure of carotenoids, because they determine their bioavailability, metabolism, and specific benefits for their use in medicine [31]. Table 1 shows the specificity of different molecular structures of seven carotenoids [32]. Recognizing these differences is essential to optimize their intake and take advantage of their potential in the prevention and health promotion.

Optimizing analytical methods at the molecular level involves several steps, including sample preparation, extraction, separation, detection, and quantification. This comprehensive approach to the distinction between different types of carotenoids and optimized analytical methods ensures accurate identification, quantification, and analysis of carotenoids at the molecular level. Table 2 shows a summary of the steps, samples, methods, and optimization.

Table 1. Specificity of the distinctive molecular structures of seven different carotenoids.

Carotenoid	Molecular Structure	Specific Features	Specificity
Beta-Carotene	$C_{40}H_{56}$	11 conjugated double bonds, cyclic end groups	Vitamin A precursor, antioxidant
Lycopene	$C_{40}H_{56}$	11 conjugated double bonds, linear structure, no cyclic end groups	Non-provitamin A, antioxidant
Lutein	$C_{40}H_{56}O_2$	10 conjugated double bonds, hydroxyl groups on terminal rings	Eye health, antioxidant
Zeaxanthin	$C_{40}H_{56}O_2$	10 conjugated double bonds, hydroxyl groups on terminal rings	Eye health, antioxidant
Alpha-Carotene	$C_{40}H_{56}$	Similar to beta-carotene, different double bond arrangement	Vitamin A precursor (less efficient), antioxidant
Beta-Cryptoxanthin	$C_{40}H_{56}O$	9 conjugated double bonds, one hydroxyl group	Vitamin A precursor, antioxidant
Astaxanthin	$C_{40}H_{52}O_4$	13 conjugated double bonds, keto groups on terminal rings	Non-provitamin A, strong antioxidant, benefits for skin health, inflammation, and immune function

Table 2. Summary of the steps, samples, methods, and optimization.

Sample	Method	Optimization
Sample Preparation	Homogenization, Cryogenic grinding	Use of low temperatures and proper homogenization techniques to preserve carotenoid integrity.
Extraction	Solvent extraction, Supercritical fluid extraction (SFE), Ultrasound-assisted extraction (UAE)	Optimization of solvent type, pressure, temperature, and ultrasonic power to enhance extraction.
Separation	high-performance liquid chromatography (HPLC), Ultra high-performance liquid chromatography (UHPLC)	Selection of appropriate columns, mobile phases, and temperature control for high resolution separation.
Detection	UV-Vis Spectrophotometry, Mmass spectrometry (MS), Diode array detection (DAD)	Use of advanced detection techniques to accurately identify and quantify carotenoids.
Quantification	External and Internal Standard Methods	Comparison of sample results with known standards for precise quantification.

2.1. *Carotenoid Production Methods*

Carotenoid production methods can be broadly classified into two types—synthetic and natural—although there are hybrid methods as well [33]. Synthetic carotenoids production is considerably faster and less expensive compared to natural carotenoids. However, the quality of synthetic carotenoids may not be the best option for developing molecules used in medicine and disease treatment [34,35]. In medicine, preference is shifting toward natural carotenoids over synthetic ones, as the latter can be harmful to human health in high doses and produce hazardous waste during manufacturing. However, carotenoids derived from natural compounds (plants, algae) are relatively more expensive to obtain, but offer complete traceability and safety of their components, crucial for human health [4].

Different Carotenoid Extraction Methods

The relentless pursuit of maximizing carotenoid extraction to increase recovery yields and reduce costs has led to numerous research efforts [36]. However, there is no single method that stands out above the rest. The method depends on the substrate used, the type of carotenoid sought, its final use (cosmetic, food or medicinal), and the level of purity required [12]. The choice of extraction method will depend on the different levels of polarity and susceptibility to oxidation of carotenoids, as well as the physical and chemical barriers of each carotenoid, which vary depending on the substrate and intended use [37]. Numerous studies address the diverse existing methods of extraction of carotenoid, from the most conventional to the most innovative [12]. In this range of innovative advances in carotenoid extraction methods, there has been a focus on enhancing yield, purity, and sustainability while minimizing environmental impact and ensuring the integrity of the bioactive compounds. We present a detailed description of cutting-edge techniques in carotenoid extraction, highlighting their scientific rigor and significance [36–41].

- SFE involves using carbon dioxide (CO_2) above its critical temperature and pressure, where it exhibits the properties of both a liquid and a gas. This state allows CO_2 to penetrate solid materials like a gas but dissolve substances like a liquid, making it highly effective for extracting carotenoids from plant matrices. Tunable parameters enable selective extraction, allowing for the targeting of specific carotenoids without the extraction of unwanted compounds. Recent innovations in SFE include the optimization of pressure and temperature parameters to improve carotenoid yield and selectivity. Integration with cosolvents such as ethanol has further enhanced the extraction efficiency for a broader range of carotenoids. This method is considered highly sustainable, as it eliminates the need for organic solvents, reduces the extraction time, and allows for easy separation and recycling of CO_2. It is particularly valued for its ability to preserve the bioactivity of carotenoids and maintain their functional properties for subsequent applications.
- Pressurized liquid extraction (PLE), also known as accelerated solvent extraction (ASE), utilizes solvents at high temperatures and pressures to dissolve and extract carotenoids more effectively than traditional solvent extraction methods. Increased solubility and diffusivity of carotenoids under these conditions result in higher extraction efficiencies and shorter processing times. Innovations in PLE involve the use of environmentally friendly solvents such as water and ethanol under subcritical conditions, which reduce the environmental footprint and improve safety. Additionally, multistep gradient extraction techniques have been developed to selectively extract different carotenoids based on their polarity and solubility. PLE offers rapid extraction with higher yields and reduced solvent consumption, making it an efficient and sustainable alternative to conventional methods. The ability to fine-tune extraction parameters allows selective isolation of specific carotenoids, enhancing the purity and quality of the extracts.
- Ionic liquid extraction (ILE) uses ionic liquids, which are composed of ions, to extract carotenoids. These solvents offer a unique combination of low volatility, high thermal stability, and the ability to dissolve both polar and nonpolar compounds. The versatility and tunability of ionic liquids make them ideal for selective and efficient carotenoid extraction. The application of ionic liquids, which are salts in a liquid state, as green solvents for carotenoid extraction. ILE offers an eco-friendly use with ionic liquids that are often recyclable and less volatile than traditional solvents. It is possible to dissolve a wide range of compounds, including carotenoids. It can be tailored to target specific carotenoids.
- UAE uses ultrasound waves to create cavitation, which disrupts cell walls and facilitates the release of carotenoids into the solvent. Recent advances in UAE include the development of dual-frequency ultrasonic systems and the use of pulsed ultrasound, which improve extraction efficiency and reduce processing time. The combination of UAE with green solvents such as aqueous ethanol has further minimized the environmental impact. The UAE is recognized for its ability to enhance mass transfer, leading to higher carotenoid yields with lower energy consumption and shorter extraction times. The nonthermal nature of ultrasound preserves the chemical structure and bioactivity of carotenoids, making this method ideal for sensitive compounds.
- Microwave-assisted extraction (MAE) uses microwave energy to heat solvents and samples, causing rapid cell-wall disruption and the efficient release of carotenoids. Innovations in MAE include the use of ionic liquids and deep eutectic solvents, which offer superior solubilizing capabilities and enhance the extraction of carotenoids from

complex matrices. The development of controlled microwave systems has enabled precise tuning of the energy input to optimize extraction conditions. MAE provides a rapid, efficient, and energy-saving approach to carotenoid extraction. The technique's ability to target specific molecular interactions results in high yields and selectivity, preserving the integrity of carotenoids, and reducing the use of hazardous solvents.
- Enzyme-assisted extraction (EAE) uses specific enzymes to hydrolyze cell walls, facilitating the release of carotenoids into the extraction medium. Recent progress in EAE includes the identification and application of novel enzymes that specifically target cell wall components in different carotenoid sources. The combination of enzymes with mild extraction conditions has significantly improved the yield and purity of carotenoid extracts. EAE is particularly advantageous for its mild operating conditions, which preserve the bioactivity of carotenoids. The use of specific enzymes allows selective extraction, the minimization of the degradation of sensitive compounds, and improvements in the overall quality of the extract.
- Pulsed electric field (PEF) extraction involves the application of short pulses of high voltage to plant materials, creating temporary pores in cell membranes. This process, known as electroporation, enhances the permeability of cells, allowing for a more efficient release of carotenoids into the extraction solvent. PEF is particularly advantageous for extracting carotenoids from heat-sensitive materials. The development of advanced PEF systems with precise control over pulse parameters has enhanced the efficiency of extracting carotenoid from various matrices. The integration of PEF with subsequent extraction steps, such as solvent extraction, has further improved the carotenoid yield and purity. PEF is a nonthermal technique that preserves the structural integrity and bioactivity of carotenoids. It is considered to be environmentally friendly due to its low energy consumption and minimal use of solvents, making it an attractive option for sustainable carotenoid extraction.
- Hydrothermal extraction utilizes water at high temperatures and pressures to extract carotenoids, utilizing the solubilizing power of subcritical and supercritical water. Innovations in hydrothermal extraction include the use of continuous flow systems and the combination with supercritical CO_2 to enhance extraction efficiency. The development of novel reactor designs has improved the scalability and application range of this technique. Hydrothermal extraction is valued for its use of water, a non-toxic and readily available solvent, making it an eco-friendly alternative. Its ability to operate under mild conditions ensures the preservation of carotenoid bioactivity and provides high extraction efficiency for both polar and non-polar carotenoids.

These advancements represent a significant leap in the efficiency, sustainability, and efficacy of carotenoid extraction processes. The integration of these cutting-edge techniques into industrial applications promises to enhance the availability and quality of carotenoid-rich products, supporting their use in health-promoting foods, supplements, and therapeutic agents. The scientific rigor behind these methodologies ensures that the integrity and bioactivity of carotenoids are maintained, maximizing their potential health benefits, and aligning with the growing demand for environmentally sustainable extraction practices.

Regarding chemical methods, organic solvents such as hexane and ethanol have been conventionally used, which, although reducing costs, attract criticism for their lack of environmental sustainability and potential health risks in medical use [16]. This has led to a shift towards more environmentally sustainable methods, with recent developments moving toward the use of environmentally friendly solvents, called 'green solvents' or supercritical fluids [39]. As seen, there are numerous pigment extraction techniques, but enzyme-assisted extraction stands out as an interesting strategy that allows much more efficient, sustainable, and pure pigment recovery [42]. Each method uses different mechanisms for cell disruption and applies different temperature and pressure conditions [41,43]. In the relentless search for improving the quality of the final product for use in medicine, biotechnological advances have been achieved through genetic engineering techniques, which involve the modification of organisms such as bacteria, yeast, or plants to produce carotenoids of higher purity and with better concentrations of greater and better quality [12]. This method typically involves the introduction of genes that encode carotenoid biosynthetic enzymes into the genome of the host organism [38]. By manipulating genetic pathways, it is possible to improve the production of specific carotenoids. The efficient extraction of carotenoids in medicine presents a significant challenge due to the complex nature of human health and the various diseases that need to be treated [12].

3. Carotenoids in Medicine and Therapeutic Treatment of Diseases

Numerous studies around the world have established the fundamental role of carotenoids in medicine and their prevalence in the therapeutic treatment of diseases, not only as a source of antioxidants, but also as a source of provitamin 'A' and as components of macular pigment, which underlines its potential in the prevention and treatment of various specific diseases [7,22,44–49]. Beyond their crucial role in ocular health, carotenoids have been associated with improving cognitive function, promoting cardiovascular wellness, and even potential anticancer properties, thus emphasizing their broader impact on human physiology [50,51]. This highlights the intricate and multifaceted nature of the functionality of carotenoid within the human body, showing its involvement in essential cellular processes and their potential as therapeutic agents. Ongoing research in this area continues to shed light on the therapeutic potential of carotenoids and their potential applications in the field of medicine.

3.1. Antioxidant Activity

Carotenoids play a critical role in the prevention and treatment of specific diseases, specifically in combating the harmful effects of reactive oxygen species (ROS) and reactive nitrogen species (RNS) in many pathophysiological conditions [52]. Their exceptional ability to act as potent free-radical inhibitors underscores their importance in preventing cellular damage and protecting against a spectrum of chronic diseases, including cardiovascular diseases, cancer, and neurodegenerative disorders.

The different molecular configurations of carotenoids, characterized by conjugated double bonds, give them the ability to effectively counteract the detrimental impact of oxidative stress, thus preserving cell integrity and mitigating the risk of chronic inflammatory diseases [53,54]. Carotenoids function as potent free-radical scavengers, preventing the spread of oxidative chain reactions and preventing the resulting cell damage [55]. Their multifaceted actions as singlet oxygen inhibitors and lipid peroxidation inhibitors highlight their critical role in maintaining redox homeostasis and preserving cell function, thus promoting general well-being and longevity [56,57]. In addition to their antioxidant capacity, they exert protective effects on other crucial biomolecules, including proteins and deoxyribonucleic acid (DNA), thereby reducing the progression of degenerative disorders induced by oxidative stress [58–60].

3.2. Anti-Inflammatory Effects

Carotenoids also demonstrate significant anti-inflammatory effects by targeting key signaling pathways and downregulating the expression of critical pro-inflammatory mediators implicated in various inflammatory conditions [61,62]. These potent phytochemicals exert their anti-inflammatory actions by modulating the activity of critical transcription factors, such as the "nuclear factor kappa light chain enhancer of activated B cells" (NF-κB), which serve as central regulators of the inflammatory cascade [63–66]. By attenuating nuclear translocation and DNA binding activity of NF-κB, carotenoids effectively inhibit the expression of a spectrum of pro-inflammatory cytokines, chemokines, and adhesion molecules, thus reducing the inflammatory response and alleviating the progression of inflammatory disorders [67–69].

The intricate interaction between carotenoids and various signaling cascades underscores their potential to mitigate chronic inflammatory diseases, such as rheumatoid arthritis and inflammatory bowel disease, characterized by aberrant immune responses and dysregulated inflammatory pathways [70,71]. Its ability to target multiple points along the inflammatory signaling axis highlights its versatility in damping the production of pro-inflammatory mediators, thus restoring immune homeostasis and improving clinical manifestations of inflammatory pathologies [72]. Furthermore, the immunomodulatory activities of carotenoids extend to the regulation of immune cell function, including the modulation of macrophage polarization, lymphocyte proliferation, and T cell differentiation, thus orchestrating a balanced immune response and promoting immune tolerance [73]. By influencing the interaction between immune cells and their microenvironment, carotenoids contribute to maintaining immune homeostasis and preventing excessive immune activation, ultimately reducing the pathogenesis of chronic inflammatory diseases [74,75].

3.3. Immunomodulatory Effects of Carotenoids

Carotenoids, including β-carotene and astaxanthin, exhibit multifaceted interactions with immune cells, exerting profound immunomodulatory effects that significantly impact immune cell proliferation, differentiation, and production of cytokines, chemokines, and growth factors. This fine-tunes the immune response and maintains immune homeostasis [76–79]. These bioactive compounds demonstrate a remarkable ability to modulate the functional activities of various subsets of immune cells, including lymphocytes, macrophages, and dendritic cells, orchestrating a balanced immune response and promoting immune homeostasis [80]. Due to their immunomodulatory properties, carotenoids show great promise in the management and treatment of various immune-related disorders, including autoimmune diseases and immunodeficiencies [81–83]. Its diverse actions include, as observed, T cell differentiation, B cell activation, and natural killer cell function, resulting in a balanced immune environment that effectively protects against aberrant and immune-mediated pathologies [84,85].

3.4. Carotenoids and Cardiovascular Health

Phytochemicals such as lycopene and β-carotene have attracted much attention for their crucial role in promoting heart health and mitigating the risk of cardiovascular disease (CVD) through their multifaceted biological activities [86,87]. Lycopene, found in tomatoes and their products, has emerged as a prominent cardioprotective agent, attributed to its notable antiatherogenic properties, particularly its ability to inhibit the oxidation of low-density lipoproteins (LDL), a key step in the progression of the onset and onset of the disease of atherosclerosis [88–90]. By preventing LDL oxidation, lycopene effectively reduces the formation of atherosclerotic plaques, thus reducing the risk of coronary artery disease and improving cardiovascular health [91–93]. Furthermore, β-carotene, a potent precursor of vitamin A, exerts important cardioprotective effects through its anti-inflammatory properties, thus contributing to improvements in various cardiovascular diseases [76,94–96]. Its conversion to vitamin 'A' within the body allows the modulation of immune responses and negative regulation of inflammatory mediators, thus attenuating the progression of chronic inflammatory conditions and reducing the risk of cardiovascular pathologies, including atherosclerosis and myocardial infarction [97,98].

3.5. Carotenoids and Implications for Visual Health

Lutein and zeaxanthin play an indispensable role in preserving visual health and preventing age-related macular degeneration (AMD), the leading cause of visual impairment and blindness in the elderly population [99,100]. Its antioxidant properties are essential to combat the harmful effects of reactive oxygen species (ROS) and mitigate cumulative damage caused by light-induced oxidative stress, thus safeguarding the integrity of the macular region and preserving visual acuity [101].

3.6. Carotenoids and Cancer Prevention

Several studies indicate that carotenoids also play an important role in cancer prevention, mostly through their potent chemopreventive properties, which involve the suppression of mutagenesis, reduction in oxidative stress, induction of apoptosis in cancer cells, and inhibition of angiogenesis, a critical process in tumor development and progression [102,103]. Their strong antioxidant capabilities allow them to neutralize reactive oxygen species (ROS) and curb oxidative damage, thus mitigating the risk of DNA mutations and cell abnormalities that can culminate in carcinogenesis [103,104].

3.6.1. Chemopreventive Properties

Carotenoids' chemopreventive potential is observed by inducing apoptosis in cancer cells, triggering programmed cell death, and preventing the uncontrolled proliferation of malignant cells, thereby stopping tumor progression and metastasis spread [105–107]. Its ability to modulate signaling pathways and regulatory mechanisms involved in cell growth and survival underlines its fundamental role in preventing the onset and progression of various malignancies [108,109]. Furthermore, the ability of carotenoids to suppress angiogenesis, a critical step in tumor progression, underscores their potential to prevent tumor growth and metastasis [110–112]. Their regulatory influence on the formation of new blood vessels and their ability to alter the microenvironment leading to tumor expansion exemplifies their multifaceted role in preventing the angiogenic cascade, which is critical for limiting tumor development and reducing the spread of cancer cells [110,113].

3.6.2. Epidemiological and Clinical Evidence

A large body of epidemiological and clinical evidence underscores the important role of carotenoids in cancer prevention, particularly in reducing the risk of lung, breast, and prostate cancer. Different studies have established a strong correlation between a high dietary intake of carotenoids and a reduced susceptibility to these prevalent forms of cancer [114,115]. Specifically, diets rich in β-carotene and lycopene have shown strong protective effects against the development and progression of these malignancies, thus highlighting the vital role of these bioactive compounds in preventing cancer incidence [116,117]. The documented association between higher carotenoid consumption and lower risk of lung cancer underscores the important potential of these compounds to mitigate the detrimental impact of various carcinogens on lung tissue [118–120]. Furthermore, the protective influence of carotenoids against breast cancer has been elucidated through extensive clinical research, shedding light on their critical role in modulating hormonal imbalances and reducing cellular aberrations that can precipitate breast carcinogenesis [121,122]. On the other hand, there is compelling clinical evidence for the preventive properties of carotenoids against prostate cancer, highlighting their potential as critical agents to stop the onset and progression of this prevalent malignant disease [123–125]. Its ability to modulate cell signaling pathways and prevent uncontrolled proliferation of prostate cancer cells underscores its important role in reducing the incidence and burden of this widespread form of cancer [126–128].

3.7. Emerging Applications in Medicine

As science advances, carotenoids are increasingly recognized for their emerging applications in therapeutic medicine, particularly in the realms of wound healing and metabolic health [129]. Through their multifaceted properties, specific carotenoids such as β-carotene and astaxanthin have shown promising wound healing attributes by orchestrating modulation of inflammatory responses and oxidative stress [130]. Their role in accelerating tissue repair and regeneration has shown the potential to improve wound closure rates and minimize scar formation, underscoring their importance as key therapeutic agents in the treatment of various wound-related complications [131,132]. Furthermore, the distinctive contributions of lutein and zeaxanthin to metabolic health have attracted significant attention due to their potential to improve insulin sensitivity and regulate glucose metabolism [29,133]. Their documented influence in reducing markers associated with metabolic syndrome, a complex interplay of metabolic abnormalities related to increased cardiovascular risk, underscores their critical role in promoting metabolic homeostasis and protecting against the progression of metabolic disorders [134–136].

3.8. Carotenoids and Nanotechnology

The use of nanotechnology in the formulation of carotenoids represents a significant advance in the fields of nutritional and pharmaceutical sciences and emerges as a promising avenue to improve the bioavailability and non-degradation of active carotenoid compounds, which are essential nutritional components essential with important health benefits [137,138]. By reducing carotenoid particles to nanometer sizes, their absorption and availability within the body can be significantly improved [139,140]. This approach offers a potential solution to the long-standing challenge of poor bioactivity and bioavailability associated with conventional carotenoid formulations [141]. Studies have shown that nanosized carotenoid particles, typically in the range of a few nanometers, exhibit greater solubility and stability, allowing easier absorption in the gastrointestinal tract [142]. The smaller particle size allows for better dispensability and interaction with body cells, facilitating more efficient utilization [143]. Furthermore, the higher surface-to-volume ratio of these nanosized particles promotes a better interaction with biological membranes, leading to greater bioavailability and greater tissue accumulation [137,144,145]. Thanks to continued advances in R&D, nanosized carotenoid particles have the potential to revolutionize the delivery and efficacy of these essential dietary compounds, contributing to better health outcomes and more precise and timely disease prevention [142].

Various techniques, such as supergravity rotating packed-bed crystallization, have been developed to produce nanodisperse microcapsules with a high content of E-carotenoid, ensuring a high content of transisomers and optimal bioactivity and bioavailability [146,147]. This innovative approach not only improves carotenoids but also contributes to their sustained release over time, resulting in longer and more effective biological activity. There is already a documented and patented process for producing microcapsules with a high E-carotenoid content [148]. In parallel, work is underway on new encapsulation techniques, with gelatin-based protective colloids that allow greater stability and prevent capsule breakage. This is mainly due to the instability of carotenoids, especially with respect to heat treatments and oxygen sensitivity [149–152].

Supramolecular Carriers

Supramolecular transporters, also known as supramolecular carriers, are advanced delivery systems designed to improve the absorption and bioavailability of various compounds, including carotenoids [153,154]. These transporters use principles of supramolecular chemistry, which involves the assembly of molecules through noncovalent interactions, to improve the solubility, stability, bioavailability, and cellular uptake of carotenoids in humans. These advanced delivery systems protect carotenoids from degradation, facilitate their absorption by intestinal cells, and provide controlled release, ensuring that the body effectively absorbed and utilizes carotenoids [155–157]. The fundamental property of supramolecular transporters lies in their ability to alter their surface by introducing additional functional groups, facilitating targeted delivery to specific organs or cellular receptors. The formulation of functional nanostructured carriers allows the release of drugs as needed in specific microscopic environments associated with various diseases [158,159]. The same authors point out that nanostructured carriers include lipid nanoparticles, polymers, micelles, inorganic nanoparticles, hybrid nanoparticles, and other types of particles. Applications of supramolecular transporters have been used in various domains, including biomedicine and materials science [160,161]. Their unique optical and electronic properties make them promising candidates for use in sensors, biomaterials, and optoelectronic devices, offering interesting perspectives for the development of advanced technological solutions [162,163]. Table 3 shows a more detailed explanation of the mechanisms, how supramolecular transporters improve carotenoids absorption in humans, their advantages, some examples and potential applications.

Table 3. How do supramolecular transporters improve the absorption of carotenoids in humans?

Characteristic	Mechanism	Advantages	Examples of Carotenoids	Applications
Enhanced Solubility and Dispersibility	Micelles: The hydrophobic core and hydrophilic exterior solubilize carotenoids.	Improved solubility in aqueous environments.	Beta-carotene, lutein, astaxanthin	Nutraceuticals, functional foods
	Cyclodextrins: The hydrophobic cavity and the hydrophilic outer surface form host-guest complexes.	Increased solubility of water.	Beta-carotene, lycopene	Dietary supplements, food fortification
	Liposomes: The lipid bilayer encapsulates carotenoids.	Enhanced dispersibility in gastrointestinal fluids.	Astaxanthin, lutein	Pharmaceuticals, cosmetic formulations
Increased Stability and Protection	Encapsulation: Protect carotenoids from oxidative degradation, light, and heat.	Maintain bioactivity during processing, storage, and digestion.	Beta-carotene, lycopene, astaxanthin	Nutraceuticals, pharmaceuticals
	Antioxidant Coencapsulation: Additional protection against oxidative damage.	Further, it improves the stability of carotenoids.	Various carotenoids	Dietary supplements, functional foods
Enhanced Cellular Uptake	Size and Surface Modification: Optimal size for endocytosis, surface ligands for targeting.	Increased efficiency of intestinal cell uptake.	Beta-carotene, lutein, astaxanthin	Pharmaceuticals, nutraceuticals
	Targeting Ligands: Enhanced specificity for intestinal cells.	Improved targeted delivery and absorption.	Lutein, zeaxanthin	Eye health supplements, pharmaceuticals
Controlled Release Profiles	pH-Responsive Carriers: Carotenoids release in response to gastrointestinal pH.	Sustained and controlled carotenoid release.	Beta-carotene, lutein	Pharmaceuticals, nutraceuticals
	Enzyme-Responsive Carriers: Release upon exposure to specific gastrointestinal enzymes.	Ensures gradual release and absorption.	Various carotenoids	Dietary supplements, functional foods
Overcoming Biological Barriers	Mucus Penetration: Nanocarriers penetrate the mucus layer in the intestinal tract.	Enhanced interaction with intestinal epithelial cells.	Various carotenoids	Pharmaceuticals, nutraceuticals
	Paracellular transport: Temporary opening of tight junctions between epithelial cells.	Facilitates the passage of carotenoids through the paracellular route.	Various carotenoids	Pharmaceuticals, functional foods
Interaction with Lipid Absorption Pathways	Chylomicron Formation: Lipid-based carriers incorporated into chylomicrons.	Efficient transport of carotenoids through lipid transport mechanisms.	Beta-carotene, astaxanthin	Nutraceuticals, functional foods

Table 3. *Cont.*

Characteristic	Mechanism	Advantages	Examples of Carotenoids	Applications
	Lipid Transport Proteins: Interaction with transport proteins such as SR-BI.	Facilitates carotenoid uptake and transport.	Various carotenoids	Pharmaceuticals, nutraceuticals

4. Innovation and Technological Advances in Carotenoids

Innovation in carotenoid biosynthesis, metabolic pathways, regulatory mechanisms, and bioavailability of biomolecules are essential to develop efficient production processes that obtain high-quality carotenoid-based compounds specific for specific diseases [164]. In the medical sector, the continuous generation of knowledge related to carotenoids drives innovation in extraction methodologies, purification techniques, and biotechnological approaches, leading to higher product yields, greater bioavailability, traceability, specificity, and fundamentally better preventive and curative actions [165].

The dynamic nature of innovation generation in the carotenoid production sector encourages the development of new drug formulations and delivery systems that improve the bioactivity, stability, solubility, and targeted delivery of carotenoid-based pharmaceuticals [166]. In the same vein, the integration of interdisciplinary research efforts promotes a deeper understanding of the interactions between carotenoids and cellular processes, thus providing information on their mechanisms of action and their possible applications in the prevention and treatment of diseases [167]. Collaborative knowledge exchange between researchers, biochemists, pharmacologists, physicians, nanotechnologists, bioinformatics specialists, artificial intelligence specialists and biotechnologists plays a critical role in accelerating the development of innovative strategies for the production, purification, and formulation of high-impact carotenoids in medicine [168,169]. A comprehensive understanding of carotenoids factors that influence the bioactivity, bioavailability, stability, and pharmacokinetics allows the design of optimized drug delivery systems that improve their therapeutic efficacy and minimize adverse effects, as seen in the case of supramolecular carriers [170,171].

Generating innovative products is crucial in the context of the production of carotenoids for medicinal purposes. It requires a deep understanding of its biosynthetic pathways, cutting-edge extraction techniques related to biotechnological approaches, and genetic engineering techniques aimed at improving the production efficiency, purity, and bioavailability of the compounds [164,165]. Knowledge about the intricate biochemical processes involved in carotenoid biosynthesis is essential in developing cutting-edge production strategies that ensure high yields and superior product quality [157]. Advances in bioprocessing technologies and fermentation techniques have facilitated the scalable and cost-effective production of carotenoids, enabling their widespread application in pharmaceutical formulations and nutraceuticals [172,173].

Furthermore, the integration of analytical techniques and quality-control measures is vital to ensure the safety, efficacy and stability of carotenoid-based medications [174]. The use of innovative delivery systems, such as nanoformulations and liposomal carriers, has revolutionized the field of carotenoid therapy, allowing targeted delivery and increased bioavailability of these bioactive compounds [166]. A comprehensive understanding of the biological mechanisms underlying the therapeutic effects of carotenoids, together with innovative research efforts and cooperation, may pave the way for the development of new pharmacotherapies and preventive interventions, thus addressing the growing global burden of chronic diseases [175,176].

The Dissemination of New Carotenoid Development Techniques

Different methods of carotenoid extraction are essential for the subsequent use of this type of molecule. However, there are numerous facets related to the quality of the carotenoid obtained and that have to do with its purity and which are important to address. Along these lines, we can say that there are different techniques, already used in other nearby industries, such as biopharmaceuticals, that could be of great help in the quality control of carotenoids obtained for medical purposes related to the purity of the carotenoid obtained.

To improve the analytical rigor and guarantee constant quality of the carotenoid, there are, among others, two important methods to consider, the one referring to advanced quality by design (QbD) and multiple attribute methods (MAM) [177,178]. The application of QbD in analytical sciences, particularly in method development, involves systematic approaches such as 'Design of experiments' (DoE) and risk assessments to optimize analytical performance and ensure the robustness of the results [179–181]. These principles guide the establishment of the 'Method Operable Design Region' (MODR), essential to achieve the desired quality in the measurement of specific properties of carotenoids. Furthermore, the integration of QbD into the development phases supports the regulatory flexibility and improves methodological consistency, reducing the variability in the results [182]. Furthermore, MAMs offer significant advantages in carotenoid analysis, providing a comprehensive and integrative approach to study various attributes simultaneously [183]. These methods would enable the accurate and synchronous quantification of multiple carotenoids in complex matrices, improving both efficiency and analytical performance. This is particularly beneficial in strict regulatory environments such as those of the Food and Drug Administration, where accurate evaluation of carotenoid concentrations in health products is the key to approval [184]. In addition, MAM methods facilitate the integration of multiomics data, providing information on the interactions between carotenoids and other biomolecules, thus improving the understanding of their biological functions and effects on lipid profiles [185,186]. MAMs also contribute to the improvement of carotenoid analysis by using different methods such as HPLC combined with MS and DAD, UV and visible spectrophotometry, thin-layer chromatography (TLC), and supercritical fluid chromatography, offering a robust approach to accurate identification and quantification of carotenoids even in the presence of other similar compounds [187,188]. Advanced MS technologies, such as ultra high-performance liquid chromatography–MS (UHPLC-MS) and data-independent MS (MSE) workflows, improve the detection and characterization of carotenoids in complex biological matrices [189]. These methods allow the precise quantification of hormones derived from carotenoids and apocarotenoids (APOs) from plant tissues, helping to understand their metabolism and regulatory functions [18,190].

The use of high-resolution hybrid mass spectrometers coupled with HPLC enables the precise determination of carotenoid profiles in plant tissues and food samples, generating valuable data to unravel complex results [191,192]. The time required to perform HPLC analysis of carotenoids has been drastically reduced [193–195]. This is due to a number of factors, such as smaller-particle-size columns operating at lower pressures, the development of isocratic methods, and more recently, the use of a C30 column [196]. The C30 column is capable of isocratically separating a wider range of carotenoids, eliminating the need to run a gradient and, therefore, reducing the analysis time [197]. Another development is the production of superficially porous particles (SPP) that have an efficiency similar to that of small particles but use lower pressures [198]. Although SPP columns have not yet been tested for carotenoids, they are likely to become the column of choice due to their efficiency and cost-effectiveness. Because of these advances, it appears that the choice of column and the time it takes to run the analysis would no longer be a limiting factor for carotenoid HPLC.

However, the implementation of MAMs for the analysis of carotenoids is not always optimal, as it also presents some technological challenges that can disturb the results in some specific cases. A very important one is the possible interference of complex biological matrices, which can hinder the detection of carotenoids, requiring advanced purification and isolation techniques with additional costs and time [199,200]. Furthermore, the structural characterization of carotenoid metabolites requires robust analytical protocols and equipment parameters, as even minor deviations can lead to significant errors in their quantification [201]. However, there are different technologies for overcoming biological interferences in carotenoid analysis.

Traditional methods to assess carotenoid bioavailability have provided valuable information, but recent technological advances have paved the way for innovative approaches that offer more detailed and accurate evaluations. These cutting-edge methodologies range from in vitro models and advanced analytical techniques to in vivo studies and novel in silico tools. They also include emerging technologies such as single-cell transcriptomics, CRISPR-based gene editing, and microfluidics-based gut-on-a-chip systems, which together are transforming our understanding of how carotenoids are processed in the body [202]. These innovations not only improve our knowledge of carotenoid bioavailability, but also have the potential to inform the development of more effective nutritional interventions and supplements. Table 4 highlights the key aspects of different innovative methods, including their contributions to the characterization of carotenoid bioavailability, their specific advantages and the techniques involved.

Table 4. The innovative methods for characterizing carotenoid bioavailability.

Method	Innovation	Advantages	Techniques
Single-Cell Transcriptomics [203]	Study of metabolism at individual cell level	High-resolution data on cell-specific responses	Drop-seq, 10× Genomics
CRISPR/Cas9 Gene Editing [202]	Precise genetic modifications	Identifies specific genes affecting bioavailability	CRISPR Knockout Models
Organoid Cultures [204]	3D miniaturized organ models	Mimics human intestine closely	Gut Organoids
Microfluidics-Based Gut-on-a-Chip [205]	Simulates human gut environment	Real-time observation in controlled setting	Organ-on-Chip Devices
Super-Resolution Imaging [206]	Visualization at nanometer scale	Detailed carotenoid distribution within cells	STED Microscopy
Metabolic Flux Analysis [207]	Measures metabolic reaction rates	Dynamic view of carotenoid metabolism	Stable Isotope Labeling
Multi-Omics Integration [208]	Combines genomics, proteomics, and metabolomics	Comprehensive biological insights	Integrative Analysis Platforms
Spatial Transcriptomics [209]	Maps gene expression in tissue context	Spatially resolved insights in gut	Spatially Resolved RNA Sequencing
Synthetic Biology Approaches [210]	Engineering of microbes or cells	Precise control over metabolic processes	Engineered Probiotics
Bioprinting of Tissue Models [211]	3D bioprinting for complex tissue structures	Accurate tissue models for absorption studies	3D Bioprinting
AI and Machine Learning [212]	Predicts bioavailability from complex data	Enhances data analysis and pattern recognition	Predictive Modeling

5. Discussion

This research presents an analysis related to innovative formulations for the development of new drugs and combination therapies that incorporate different carotenoids in medicine and in the treatment of diseases. Recent research findings on carotenoids in medicine have focused on their metabolism, the use of biotechnology, and their nutritional and health benefits. Their consumption has been associated with a reduced risk of vitamin A deficiency, age-related macular degeneration, cancer, and cardiovascular and skin diseases [99,213–215]. Significant progress has been made in understanding the bioavailability of carotenoids and their potential use in the cosmetic industry and skin-related diseases [216,217]. The close integration of knowledge and innovation in the production of carotenoids for medicinal applications has the potential to drive advances in personalized medicine and precision healthcare [218]. In other words, advances in the knowledge of carotenoids in the prevention and treatment of diseases underscore their potential as indispensable therapeutic agents, capable of mitigating the pathological consequences of oxidative stress and inflammation, thus offering promising perspectives for the development of new strategies in the therapy and improvement of chronic diseases [219–222].

Comprehensive research on how carotenoids modulate the immune system highlights their critical role in shaping immune cell function and orchestrating a balanced immune response, indicating their therapeutic potential in the management and treatment of various blood-related disorders of the immune system [95]. A comprehensive understanding of the anti-inflammatory effects produced by carotenoids underlines their crucial role in modulating inflammatory signaling pathways and regulating immune responses, thus offering promising perspectives for the development of new therapeutic interventions in the management and treatment of a wide range of inflammatory disorders [71]. The influence of carotenoids on the proliferation and differentiation of immune cells further enhances their potential to promote immune tolerance and improve severity cases related to autoimmune diseases characterized by dysregulated immune responses and altered immune tolerance [223,224]. Their multifaceted contributions to redox signaling pathways, immune modulation, and inflammatory responses highlight their critical role in orchestrating cellular homeostasis and strengthening defense mechanisms against a variety of diseases [225,226]. The multifaceted chemopreventive attributes of carotenoids enhance their potential as valuable therapeutic agents in cancer prevention, emphasizing the importance of their incorporation into diet regimens and their integration into new therapeutic strategies that aim to curb the incidence and progression of various malignancies [227,228]. The accumulation of epidemiological and clinical evidence supporting the cancer prevention attributes of carotenoids serves as a compelling impetus to integrate carotenoid rich diets and supplementation strategies into the prevention and management regimens of certain cancers, highlighting their potential as therapeutic agents in the fight against cancer [229]. The accumulation of evidence also highlights the benefits of carotenoids in some emerging applications in medicine, such as wound healing and metabolic health, where it emphasizes their enormous potential as valuable therapeutic adjuncts to address a variety of complex medical conditions [230,231].

This study also emphasizes the importance of research on the use of carotenoids, particularly lutein. This compound is of particular interest due to its demonstrated effectiveness in minimizing the progression of age-related macular degeneration and cataracts, highlighting the importance of promoting its consumption to maintain adequate ocular health [49,232]. Its role as an antioxidant and blue-light filter, which amplifies its concentration in the macula lutea of the retina, is key. This presence significantly reduces chromatic aberration and serves as a defense against light-induced oxidative stress [233,234]. Second, specific carotenoids, such as β-carotene, serve as precursors of vital retinoids and participate in critical processes such as retinoid cycling within the eye [235–237]. These retinoids are essential in several processes, including retinoid cycling within the eye, where visual pigments facilitate phototransduction. These pigments, closely aligned with G protein-coupled receptors, play a critical role in the translation of light into nerve signals [238,239]. Furthermore, retinoid oxidation leads to the synthesis of all-trans retinoic acid (RA), a vital substance that binds to retinoic acid receptors (RAR) to influence gene expression in a broad spectrum of physiological processes [240–242]. This underscores the imperative need to incorporate these potent carotenoids into dietary regimens and therapeutic interventions aimed at promoting ocular well-being and mitigating the risk of vision-related disorders [243]. By acting as a protective barrier against blue light and oxidative stress, these carotenoids maintain the structural and functional integrity of photoreceptor cells, thus preserving optimal visual function and improving overall ocular health [244,245].

Scientific research has made significant progress in the prevention of cardiovascular diseases through concerted actions of lycopene and β-carotene, underscoring their fundamental role in mitigating the risk of cardiovascular events (CVD) and reducing the progression of cardiovascular disease atherosclerosis through its synergistic antiatherogenic and anti-inflammatory properties [246,247]. These profoundly bioactive compounds offer promising prospects for developing innovative therapeutic strategies aimed at preventing and treating many cardiovascular disorders, highlighting their importance in promoting general cardiovascular well-being [248].

A crucial point concerns the extraction of carotenoids. This is key in estimating the quality and traceability of the origin of these compounds, depending on the needs sought in the different fields of medicine. As we have seen, this could be due to unsustainable chemicals or environmentally friendly solvents. However, genetic engineering offers the best advantage of producing carotenoids with high purity and structural characteristics similar to those found in nature, with better impact and optimal results in combating specific diseases and at a reasonable cost compared to synthetic origins [249]. The extraction of carotenoids is essential not only because of the way they are incorporated into nutritional and medical products and into the body itself. In this sense, it is imperative to improve the interpretation of the data reported on the net content of carotenoids after the different technological treatments to which they are subjected, in addition to establishing appropriate calculation formulas for each specific case [95]. Therefore, the contributions of other studies that have deepened the structural knowledge and properties of supramolecular carriers are important, highlighting their various self-assembly mechanisms and molecular architectures [250]. In this regard, innovative strategies have been employed to design and synthesize increasingly complex supramolecular carriers, leading to a better understanding of their unique properties and functionalities [74,157,251]. Collaborative efforts between supramolecular chemists and carotenoid researchers will be crucial to unlocking the full potential of supramolecular carotenoids to address complex technological challenges related to therapeutics, medicine, and the prevention and treatment of complex diseases [252].

The key to reflecting the use of carotenoids in medicine and their application in the therapeutic treatment of diseases in a highly efficient and innovative product depends on our understanding of their properties, reactions, and benefits, as well as their drawbacks at organic, molecular, technological and strategic levels, as well as their behavior in various commercial markets [29]. Continuous advancement in scientific research in the field of carotenoids and the design of new molecular architectures drives the development of new drug formulations and delivery systems that improve the stability, solubility, and targeted delivery of carotenoid-based pharmaceuticals, thus improving its effectiveness and clinical applicability [253]. This continuous progress in scientific research and innovation (R&D) in the field of carotenoids drives the emergence and refinement of new carotenoid formulations in medicine, increasing product stability and solubility, and targeting the delivery of specific carotenoid-based therapeutic drug treatments, thus improving their clinical effect, efficacy, and applicability in the treatment of specific diseases [166].

This research highlights a growing emphasis on the development of carotenoid-rich formulations for the treatment of age-related macular degeneration, cognitive disorders, oncology treatments, and other chronic diseases, highlighting their importance in contemporary medical research.

A major challenge for the future of carotenoids is related to the dynamic interaction between improving their bioavailability and adopting personalized approaches to carotenoid consumption, exemplifying the evolving landscape of medical applications. In this sense, the integration of MS into multiattribute methods would significantly improve the identification, quantification, and characterization of carotenoids, essential for studying their functions and metabolic pathways. This integration allows carotenoids from complex backgrounds without interference from other substances, ensuring more accurate measurements. Furthermore, the application of techniques such as matrix solid-phase dispersion (MSPD) further exemplifies how MAM could optimize carotenoid preparation and extraction processes, leading to more efficient analysis workflows [254]. These methods have been crucial in determining the content of carotenoids in different food products, ensuring precision and reliability in the evaluation of the presence of carotenoids such as lutein and zeaxanthin in supplements and food products. These methodologies not only reduce the sample preparation time, but also increase throughput, which ultimately makes it feasible to handle multiple samples simultaneously and more efficiently [255]. As can be seen, MAMs in the analysis of carotenoids use several technologies to overcome biological interference. Techniques such as linked scanning mass spectrometry help identify carotenoid compounds accurately, even in the presence of contamination [256]. For isolation and identification, methods that involve protection from light, heat, and oxygen are crucial to prevent degradation and maintain integrity [257]. The use of online supercritical fluid extraction in conjunction with supercritical fluid chromatography and mass spectrometry enables an efficient and accurate analysis of carotenoids and their cleavage products, reducing extraction time and improving analytical precision [258,259].

Using strategies aimed at maximizing carotenoid absorption and tailoring therapeutic regimens based on individual genetic and health profiles, the scope of carotenoid-based interventions continues to expand, fostering new avenues for precise and effective healthcare delivery along with the increasing popularity of artificial intelligence. The diverse applications of carotenoids in medicine extend to their role in improving bioavailability and the potential for personalized approaches, thus contributing to the burgeoning field of personalized medicine. Given the variability in carotenoid bioavailability influenced by bioactivity, individual genetic predispositions, and dietary patterns, strategies aimed at increasing their bioavailability have gained importance. Incorporating dietary fats along with carotenoid-rich foods and using innovative delivery systems, such as nanoparticles or emulsions, are recognized as effective strategies to improve the absorption and bioavailability of carotenoids, thereby maximizing their therapeutic potential.

Table 5 summarizes the topics analyzed in this investigation and some of the highlighted references for each of the identified grouped keywords, while Figures 1 and 2 show a complementary and more detailed view of carotenoids based on their bioactivity, bioavailability, different extraction methods, and the most important benefits for human health.

Table 5. Comprehensive analysis of grouped keywords; topic summary; outstanding references in the use of carotenoids in therapeutics and medicine for diseases.

Gruped Keywords	Topic Summary	Featured References
1. Carotenoids. Anti-inflammatory; Cancer and Chemopreventive Functions	Carotenoids, by modulating inflammatory pathways and regulating immune responses, show promise in treating inflammatory disorders. Additionally, their chemopreventive attributes position them as valuable agents in cancer prevention and treatment, urging their incorporation into diets.	[2–6,22–24,28,50,51,61–66,71,76,94–96,99,102–107,113–128,201,213–215,227,228,246,247]
2. Eye Health and Cardiovascular Well-being	Specific carotenoids like lutein and zeaxanthin are proven to slow age-related macular degeneration and cataract progression, emphasizing their consumption for visual health preservation. Moreover, lycopene and beta-carotene's combined actions contribute to cardiovascular health by reducing cardiovascular disease risk and atherosclerosis progression.	[12–14,50,51,76,86,87,91–99,134–136,213,215,235–239,246–248]
3. Supramolecular Carriers and Applications in Medicine	Recent studies on supramolecular carriers shed light on their self-assembly mechanisms, suggesting potential collaboration between supramolecular chemists and carotenoid researchers. Carotenoids' multifaceted applications in wound healing and metabolic health underline their significance in addressing various complex medical conditions, hinting at their potential to revolutionize medical interventions.	[29,71,95,129–136,153–163,168–176,218–222,230,231,250–252]

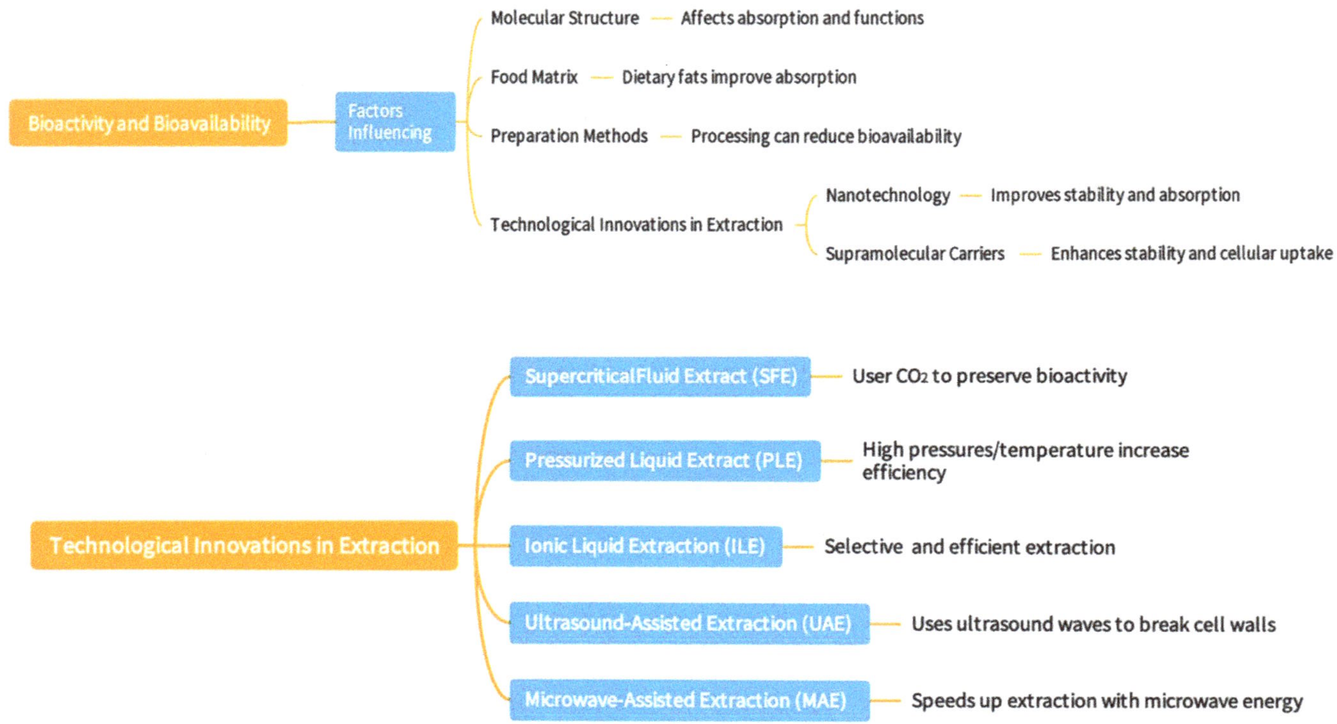

Figure 1. Overview of carotenoids related to their bioactivity, bioavailability, and technological extractions.

Figure 2. Overview of carotenoids related to their benefits for human health.

6. Conclusions

This research presents a detailed and methodical assessment of the role and applications of carotenoids in therapeutic medicine, highlighting their fundamental role in the treatment of a variety of diseases. This analysis synthesizes the findings of an extensive review and offers a comprehensive perspective on the multifunctional nature of carotenoids and their emerging applications in medical science. Schemes 1 and 2 summarize the most notable elements related to the findings of this research. Scheme 1 shows some technologies related to the extraction methods and bioavailability of carotenoids used in human health. On the other hand, the innovative applications of carotenoids in human health can be seen in Scheme 2. A detailed explanation of the Schemes 1 and 2 is given in the following.

Biological bioactivity and therapeutic potential: Carotenoids, mainly synthesized by phototrophs, bacteria, and fungi, are identified by their antioxidant, anti-inflammatory properties, and provitamin A activity. These pigments, divided into carotenes and xanthophylls, are essential for the maintenance of human health and influence eye health, cardiovascular health, skin health, immune and cognitive functions, and bone health.

Mechanisms of action: Research looks at the mechanisms through which carotenoids exert their effects. They neutralize free radicals, modulate gene expression, and regulate inflammatory pathways. This multifaceted action makes them effective against chronic diseases such as cancer, cardiovascular disease, and age-related macular degeneration.

Advances in carotenoid research: This study highlights significant advances in the understanding of carotenoid metabolism, absorption, and bioavailability, particularly through the advancement of supramolecular transporters. These advances have opened new avenues to improve the therapeutic efficacy of carotenoids.

Application in medicine: We emphasize the growing recognition of carotenoids in medicine, particularly in wound healing and metabolic health. Specific carotenoids, such as β-carotene and astaxanthin, have shown promising attributes in wound healing by modulating inflammatory responses and oxidative stress. The distinctive contributions of lutein and zeaxanthin to metabolic health have attracted significant attention, illustrating their potential to improve insulin sensitivity and regulate glucose metabolism.

Nanotechnology and carotenoids: The use of nanotechnology in the formulation of carotenoids is a significant advance in the nutritional and pharmaceutical sciences. It addresses the long-standing challenge of poor carotenoid bioavailability by demonstrating that carotenoid nanoparticles exhibit increased solubility, stability, and absorption, enhancing their therapeutic potential.

Finally, this study meticulously illustrates the intricate interplay between the bioactive properties of carotenoids and their applications in medicine. It reveals their growing potential in disease prevention and treatment, guided by ongoing scientific research and innovation. The interconnection of carotenoids with various cellular processes and their potential as therapeutic agents is evident. Continued innovation with the use of cutting-edge technologies in this field is essential to advance the medical applications of carotenoids, which contribute significantly to the field of precision healthcare and personalized medicine

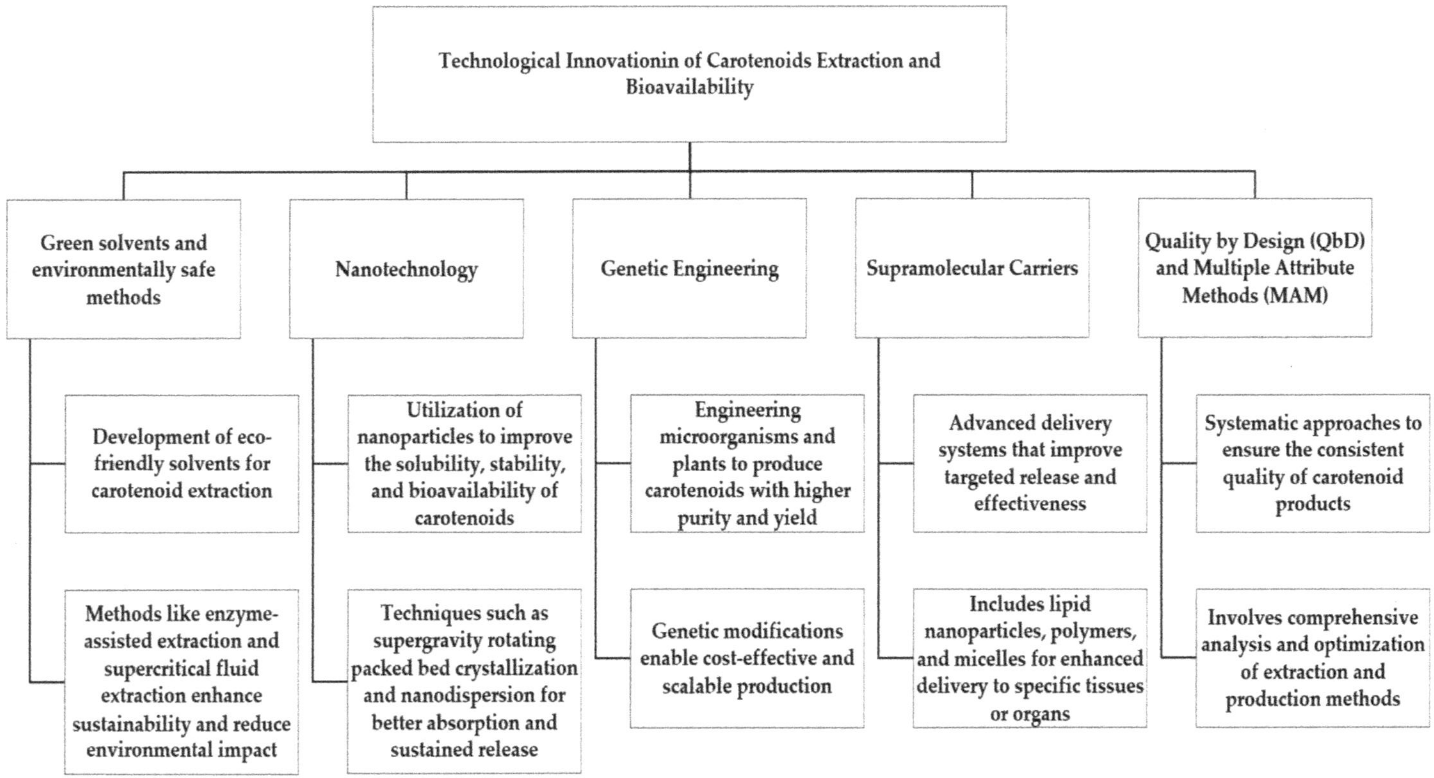

Scheme 1. Different modes of technological innovation in carotenoid extraction and bioavailability.

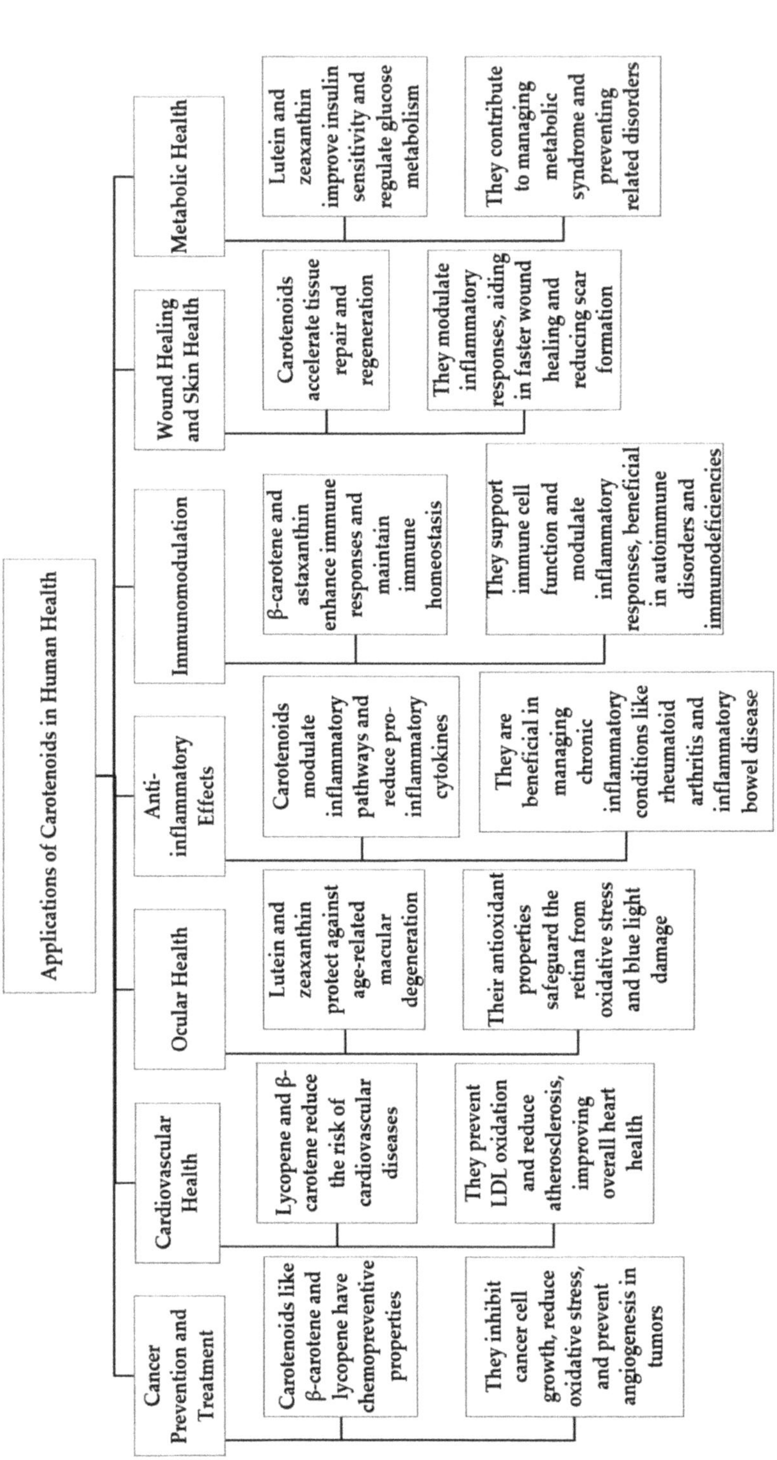

Scheme 2. Applications of carotenoids in human health (cancer prevention and treatment; cardiovascular health; ocular health; anti-inflammatory effects; immunomodulation; skin health; and metabolic health).

7. Future Directions

Carotenoids, colorful pigments abundant in fruits and vegetables, are poised to play an even greater role in the promotion of human health in the coming years. Here is a closer look at some of the exciting research directions:

- Targeting Chronic Diseases: Researchers are excited about the potential of carotenoids to combat a variety of chronic health issues. Studies are exploring their effectiveness in preventing and managing conditions such as metabolic disorders (think diabetes), cardiovascular diseases (heart disease and stroke), and even neurodegenerative disorders (Alzheimer's and Parkinson's).
- Unlocking Gut Microbiome Connections: A fascinating new area of research is investigating the link between gut bacteria and how our bodies process carotenoids. Understanding how these tiny intestinal residents influence carotenoid metabolism and health effects can pave the way for personalized approaches. Imagine creating dietary or supplement plans tailored to an individual's gut microbiome to maximize the benefits they receive from carotenoids!
- Engineering Carotenoids for Enhanced Health: The future might involve not just consuming naturally occurring carotenoids, but also potentially creating new ones with even greater health benefits. Genetic engineering holds promise for modifying plants to produce higher levels of existing carotenoids or even creating entirely new varieties with specific health-promoting properties.
- Beyond Disease Prevention: Carotenoids are not just disease fighters. Research is also investigating their broader roles in plant health and function. This deeper understanding could lead to advancements in agriculture, allowing us to cultivate crops with enhanced nutritional value and resilience.
- Advanced Detection Methods: Imagine a non-invasive way to measure your body's carotenoid levels! Techniques such as Raman spectroscopy are being explored for their potential to provide real-time information on carotenoid status. This could be a game-changer for personalized nutrition plans, allowing adjustments based on individual needs.

In essence, the future of carotenoid research is bursting with possibilities. By exploring their potential to fight disease, personalizing their benefits through connections with the gut microbiome, and developing new production and detection methods, carotenoids hold the key to unlocking a new era of proactive health management.

Funding: This research received no external funding.

Conflicts of Interest: The author declares no conflicts of interest.

References

1. Anwar, H.; Hussain, G.; Mustafa, I. Antioxidants from Natural Sources. In *Antioxidants in Foods and Their Applications*; Intechopen: London, UK, 2018; Volume 3.
2. Ashokkumar, V.; Flora, G.; Sevanan, M.; Sripriya, R.; Chen, W.-H.; Park, J.-H.; Kumar, G. Technological Advances in the Production of Carotenoids and Their Applications–A Critical Review. *Bioresour. Technol.* **2018**, *367*, 128215. [CrossRef] [PubMed]
3. Foong, L.C.; Loh, C.W.L.; Ng, H.S.; Lan, J.C.W. Recent Development in the Production Strategies of Microbial Carotenoids. *World J. Microbiol. Biotechnol.* **2021**, *37*, 12. [CrossRef]
4. Tamaki, S.; Mochida, K.; Suzuki, K. Diverse Biosynthetic Pathways and Protective Functions against Environmental Stress of Antioxidants in Microalgae. *Plants* **2021**, *10*, 1250. [CrossRef] [PubMed]
5. Muhammad, G.; Butler, T.O.; Chen, B.; Lv, Y.; Xiong, W.; Zhao, X.; Solovchenko, A.E.; Zhao, A.; Mofijur, M.; Xu, J.; et al. Sustainable Production of Lutein—An Underexplored Commercially Relevant Pigment from Microalgae. *Biomass Conv. Bioref.* **2022**, *14*, 7255–7276. [CrossRef]
6. Milani, A.; Basirnejad, M.; Shahbazi, S.; Bolhassani, A. Carotenoids: Biochemistry, Pharmacology and Treatment. *Br. J. Pharmacol.* **2017**, *174*, 1290–1324. [CrossRef] [PubMed]
7. Barreiro, C.; Barredo, J.L. Carotenoids Production: A Healthy and Profitable Industry. *Methods Mol. Biol.* **2018**, *1852*, 45–55. [CrossRef] [PubMed]
8. Johra, F.T.; Bepari, A.K.; Bristy, A.T.; Reza, H.M. A Mechanistic Review of β-Carotene, Lutein, and Zeaxanthin in Eye Health and Disease. *Antioxidants* **2020**, *9*, 1046. [CrossRef]

9. Maoka, T. Carotenoids as Natural Functional Pigments. *J. Nat. Med.* **2020**, *74*, 1–16. [CrossRef]
10. Zia-Ul-Haq, M. Past, Present and Future of Carotenoids Research. In *Carotenoids: Structure and Function in the Human Body*; Springer Nature: London, UK, 2021; pp. 827–854. [CrossRef]
11. Mussagy, C.U.; Winterburn, J.; Santos-Ebinuma, V.C.; Pereira, J.F.B. Production and Extraction of Carotenoids Produced by Microorganisms. *Appl. Microbiol. Biotechnol.* **2019**, *103*, 1095–1114. [CrossRef]
12. Papapostolou, H.; Kachrimanidou, V.; Alexandri, M.; Plessas, S.; Papadaki, A.; Kopsahelis, N. Natural Carotenoids: Recent Advances on Separation from Microbial Biomass and Methods of Analysis. *Antioxidants* **2023**, *12*, 1030. [CrossRef]
13. Liaqat, F.; Khazi, M.I.; Bahadar, A.; Li, H.; Aslam, A.; Liaquat, R.; Agathos, S.N.; Li, J. Mixotrophic Cultivation of Microalgae for Carotenoid Production. *Rev. Aquacult.* **2023**, *15*, 35–61. [CrossRef]
14. Prajapati, R.A.; Jadeja, G.C. Natural Food Colorants: Extraction and Stability Study. *Mater. Today* **2022**, *57*, 2381–2395. [CrossRef]
15. Bas, T.G.; Fariña, R.; Gallardo, F.; Vilches, M. Economic–Financial Assessment of Seawater Desalination Plants in Northern Chile to Reduce Hydric Scarcity and a Proposal for the Environmental and Sustainable Use of Brine Waste by Cultivating the Microalga Dunaliella salina to Produce β-Carotene. *Processes* **2023**, *11*, 1668. [CrossRef]
16. Olatunde, A.; Tijjani, H.; Ishola, A.A.; Egbuna, C.; Hassan, S.; Akram, M. Carotenoids as Functional Bioactive Compounds. In *Functional Foods and Nutraceuticals. Bioactive Components, Formulations and Innovations*; Springer: Berlin/Heidelberg, Germany, 2020; pp. 415–444. [CrossRef]
17. Galasso, C.; Corinaldesi, C.; Sansone, C. Carotenoids from Marine Organisms: Biological Functions and Industrial Applications. *Antioxidants* **2017**, *6*, 96. [CrossRef]
18. Kolašinac, S.M.; Stevanović, Z.P.D.; Kilibarda, S.N.; Kostić, A.Ž. Carotenoids: New Applications of "Old" Pigments. *Phyton* **2021**, *90*, 1041–1062. [CrossRef]
19. Gallini, N.; Hollis, A. To Sell or Scale Up: Canada's Patent Strategy in a Knowledge Economy. 2019. Available online: https://irpp.org/research-studies/to-sell-or-scale-up-canadas-patent-strategy-in-a-knowledge-economy/ (accessed on 18 September 2023).
20. Swapnil, P.; Meena, M.; Singh, S.K.; Dhuldhaj, U.P.; Marwal, A. Vital Roles of Carotenoids in Plants and Humans to Deteriorate Stress with Its Structure, Biosynthesis, Metabolic Engineering and Functional Aspects. *Curr. Plant Biol.* **2021**, *26*, 100203. [CrossRef]
21. Eggersdorfer, M.; Wyss, A. Carotenoids in Human Nutrition and Health. *Arch. Biochem. Biophys.* **2018**, *652*, 18–26. [CrossRef]
22. Darvin, M.E.; Lademann, J.; von Hagen, J.; Lohan, S.B.; Kolmar, H.; Meinke, M.C.; Jung, S. Carotenoids in Human Skin In Vivo: Antioxidant and Photo-Protectant Role against External and Internal Stressors. *Antioxidants* **2022**, *11*, 1451. [CrossRef] [PubMed]
23. Almagro, L.; Correa-Sabater, J.M.; Sabater-Jara, A.B.; Pedreño, M.A. Biotechnological Production of β-Carotene Using Plant In Vitro Cultures. *Planta* **2022**, *256*, 41. [CrossRef] [PubMed]
24. Havas, F.; Krispin, S.; Meléndez-Martínez, A.J.; von Oppen-Bezalel, L. Preliminary Data on the Safety of Phytoene and Phytofluene-Rich Products for Human Use Including Topical Application. *J. Toxicol.* **2018**, *2018*, 5475784. [CrossRef]
25. Harvey, P.J.; Ben-Amotz, A. Towards a Sustainable Dunaliella salina Microalgal Biorefinery for 9-cis β-Carotene Production. *Algal Res.* **2020**, *50*, 102002. [CrossRef]
26. So, B.; Kwon, K.H. Nutritional Approaches of the Changing Consumer after the Pandemic: Sustainable Potential of Phytoene and Phytofluene for Photoprotection and Skin Health. *Sustainability* **2023**, *15*, 4416. [CrossRef]
27. Khalid, M.; Bilal, M.; Iqbal, H.M.; Huang, D. Biosynthesis and Biomedical Perspectives of Carotenoids with Special Reference to Human Health-Related Applications. *Biocatal. Agric. Biotechnol.* **2019**, *17*, 399–407. [CrossRef]
28. Rodriguez-Concepcion, M.; Avalos, J.; Bonet, M.L.; Boronat, A.; Gomez-Gomez, L.; Hornero-Mendez, D.; Zhu, C. A Global Perspective on Carotenoids: Metabolism, Biotechnology, and Benefits for Nutrition and Health. *Prog. Lipid Res.* **2018**, *70*, 62–93. [CrossRef] [PubMed]
29. Marhuenda-Muñoz, M.; Hurtado-Barroso, S.; Tresserra-Rimbau, A.; Lamuela-Raventós, R.M. A Review of Factors That Affect Carotenoid Concentrations in Human Plasma: Differences between Mediterranean and Northern Diets. *Eur. J. Clin. Nutr.* **2019**, *72* (Suppl. S1), 18–25. [CrossRef] [PubMed]
30. Torregrosa-Crespo, J.; Montero, Z.; Fuentes, J.L.; Reig García-Galbis, M.; Garbayo, I.; Vílchez, C.; Martínez-Espinosa, R.M. Exploring the Valuable Carotenoids for the Large-Scale Production by Marine Microorganisms. *Mar. Drugs* **2018**, *16*, 203. [CrossRef]
31. Widjaja-Adhi, M.A.K.; Golczak, M. The Molecular Aspects of Absorption and Metabolism of Carotenoids and Retinoids in Vertebrates. *Biochim. Biophys. Acta, Mol. Cell Biol. Lipids* **2020**, *1865*, 158571. [CrossRef] [PubMed]
32. Meléndez-Martínez, A.J.; Mapelli-Brahm, P.; Hornero-Méndez, D.; Vicario, I.M. Chapter 1. Structures, Nomenclature and General Chemistry of Carotenoids and Their Esters. In *Food Chemistry, Function and Analysis*; Royal Society of Chemistry (RSC): London, UK, 2019; Volume 1, pp. 1–50. [CrossRef]
33. Goswami, R.K.; Agrawal, K.; Verma, P. An Overview of Microalgal Carotenoids: Advances in the Production and Its Impact on Sustainable Development. In *Bioenergy Research: Evaluating Strategies for Commercialization and Sustainability*; Wiley: Hoboken, NJ, USA, 2021; pp. 105–128. [CrossRef]
34. Bas, T.; Rengel, J.; Contreras, A.; Oliu, C.; Abarca, A. Determinants of Astaxanthin Industrial-Scale Production under Stress Caused by Light Photoperiod Management of Haematococcus pluvialis Cultivation. *Lat. Am. J. Aquat. Res.* **2021**, *49*, 725–738.
35. Martins, N.; Ferreira, I.C. Wastes and By-Products: Upcoming Sources of Carotenoids for Biotechnological Purposes and Health-Related Applications. *Trends Food Sci. Technol.* **2017**, *62*, 33–48. [CrossRef]

36. Saini, R.K.; Keum, Y.S. Carotenoid Extraction Methods: A Review of Recent Developments. *Food Chem.* **2018**, *240*, 90–103. [CrossRef] [PubMed]
37. Eun, H.; Lee, S.Y. Metabolic Engineering and Fermentation of Microorganisms for Carotenoids Production. *Curr. Opin. Biotechnol.* **2024**, *87*, 103104. [CrossRef] [PubMed]
38. Adadi, P.; Barakova, N.V.; Krivoshapkina, E.F. Selected Methods of Extracting Carotenoids, Characterization, and Health Concerns: A Review. *J. Agric. Food Chem.* **2018**, *66*, 5925–5947. [CrossRef] [PubMed]
39. Chemat, F.; Abert Vian, M.; Ravi, H.K.; Khadhraoui, B.; Hilali, S.; Perino, S.; Fabiano Tixier, A.-S. Review of Alternative Solvents for Green Extraction of Food and Natural Products: Panorama, Principles, Applications and Prospects. *Molecules* **2019**, *24*, 3007. [CrossRef] [PubMed]
40. Nutrizio, M.; Chemat, F.; Muangrat, R.; Seesuriyachan, P.; Phimolsiripol, Y.; Donsi, F.; Jambrak, A.R. Toward Green Extraction Processes. In *Sustainable Separation Engineering. Materials, Techniques and Process Development*; Wiley: Hoboken, NJ, USA, 2022; pp. 519–561. [CrossRef]
41. Benucci, I.; Lombardelli, C.; Mazzocchi, C.; Esti, M. Natural Colorants from Vegetable Food Waste: Recovery, Regulatory Aspects, and Stability—A Review. *Compr. Rev. Food Sci. Food Saf.* **2022**, *21*, 2715–2737. [CrossRef] [PubMed]
42. Liu, C.; Hu, B.; Cheng, Y.; Guo, Y.; Yao, W.; Qian, H. Carotenoids from Fungi and Microalgae: A Review on Their Recent Production, Extraction, and Developments. *Bioresour. Technol.* **2021**, *337*, 125398. [CrossRef] [PubMed]
43. Giordano, E.; Quadro, L. Lutein, Zeaxanthin and Mammalian Development: Metabolism, Functions and Implications for Health. *Arch. Biochem. Biophys.* **2018**, *647*, 33–40. [CrossRef] [PubMed]
44. Miller, A.P.; Coronel, J.; Amengual, J. The Role of β-Carotene and Vitamin A in Atherogenesis: Evidences from Preclinical and Clinical Studies. *Biochim. Biophys. Acta, Mol. Cell Biol. Lipids* **2020**, *1865*, 158635. [CrossRef] [PubMed]
45. Dewanjee, S.; Bhattacharjee, N.; Chakraborty, P.; Bhattacharjee, S. Carotenoids as Antioxidants. In *Carotenoids: Structure and Function in the Human Body*; Springer Nature: London, UK, 2021; pp. 447–473. [CrossRef]
46. Hammond, B.R.; Renzi-Hammond, L. The Influence of the Macular Carotenoids on Women's Eye and Brain Health. *Nutr. Neurosci.* **2023**, *26*, 720–726. [CrossRef] [PubMed]
47. Terao, J. Revisiting Carotenoids as Dietary Antioxidants for Human Health and Disease Prevention. *Food Funct.* **2023**, *14*, 7799–7824. [CrossRef]
48. Li, X.; Holt, R.R.; Keen, C.L.; Morse, L.S.; Zivkovic, A.M.; Yiu, G.; Hackman, R.M. Potential Roles of Dietary Zeaxanthin and Lutein in Macular Health and Function. *Nutr. Rev.* **2023**, *81*, 670–683. [CrossRef] [PubMed]
49. Abuajah, C.I. Functional Components and Medicinal Properties of Food. In *Bioactive Molecules in Food. Reference Series in Phytochemistry*; Mérillon, J.M., Ramawat, K., Eds.; Springer: Berlin/Heidelberg, Germany, 2017. [CrossRef]
50. Caponio, G.R.; Lippolis, T.; Tutino, V.; Gigante, I.; De Nunzio, V.; Milella, R.A.; Gasparro, M.; Notarnicola, M. Nutraceuticals: Focus on Anti-Inflammatory, Anti-Cancer, Antioxidant Properties in Gastrointestinal Tract. *Antioxidants* **2022**, *11*, 1274. [CrossRef] [PubMed]
51. Bardaweel, S.K.; Gul, M.; Alzweiri, M.; Ishaqat, A.; ALSalamat, H.A.; Bashatwah, R.M. Reactive Oxygen Species: The Dual Role in Physiological and Pathological Conditions of the Human Body. *Eurasian J. Med.* **2018**, *50*, 193–201. [CrossRef] [PubMed]
52. Anbualakan, K.; Tajul Urus, N.Q.; Makpol, S.; Jamil, A.; Mohd Ramli, E.S.; Md Pauzi, S.H.; Muhammad, N. A Scoping Review on the Effects of Carotenoids and Flavonoids on Skin Damage Due to Ultraviolet Radiation. *Nutrients* **2023**, *15*, 92. [CrossRef] [PubMed]
53. Muzumdar, S.; Ferenczi, K. Nutrition and Youthful Skin. *Clin. Dermatol.* **2021**, *39*, 796–808. [CrossRef] [PubMed]
54. Stra, A.; Almarwaey, L.O.; Alagoz, Y.; Moreno, J.C.; Al-Babili, S. Carotenoid Metabolism: New Insights and Synthetic Approaches. *Front. Plant Sci.* **2023**, *13*, 1072061. [CrossRef] [PubMed]
55. Gómez-Sagasti, M.T.; López-Pozo, M.; Artetxe, U.; Becerril, J.M.; Hernández, A.; García-Plazaola, J.I.; Esteban, R. Carotenoids and Their Derivatives: A "Swiss Army Knife-like" Multifunctional Tool for Fine-Tuning Plant-Environment Interactions. *Environ. Exp. Bot.* **2023**, *207*, 105229. [CrossRef]
56. Åhlberg, M.K. A Profound Explanation of Why Eating Green (Wild) Edible Plants Promote Health and Longevity. *Food Front.* **2021**, *2*, 240–267. [CrossRef]
57. Young, A.J.; Lowe, G.L. Carotenoids—Antioxidant Properties. *Antioxidants* **2018**, *7*, 28. [CrossRef] [PubMed]
58. Nass, P.P.; do Nascimento, T.C.; Fernandes, A.S.; Caetano, P.A.; de Rosso, V.V.; Jacob-Lopes, E.; Zepka, L.Q. Guidance for Formulating Ingredients/Products from Chlorella vulgaris and Arthrospira platensis Considering Carotenoid and Chlorophyll Bioaccessibility and Cellular Uptake. *Food Res. Int.* **2022**, *157*, 111469. [CrossRef]
59. Zafar, J.; Aqeel, A.; Shah, F.I.; Ehsan, N.; Gohar, U.F.; Moga, M.A.; Festila, D.; Ciurea, C.; Irimie, M.; Chicea, R. Biochemical and Immunological Implications of Lutein and Zeaxanthin. *Int. J. Mol. Sci.* **2021**, *22*, 10910. [CrossRef]
60. Xu, J.; Liu, J.; Yue, G.; Sun, M.; Li, J.; Xiu, X.; Gao, Z. Therapeutic Effect of the Natural Compounds Baicalein and Baicalin on Autoimmune Diseases. *Mol. Med. Rep.* **2018**, *18*, 1149–1154. [CrossRef] [PubMed]
61. Imran, M.; Ghorat, F.; Ul-Haq, I.; Ur-Rehman, H.; Aslam, F.; Heydari, M.; Rebezov, M. Lycopene as a Natural Antioxidant Used to Prevent Human Health Disorders. *Antioxidants* **2020**, *9*, 706. [CrossRef] [PubMed]
62. Chauhan, A.; Islam, A.U.; Prakash, H.; Singh, S. Phytochemicals Targeting NF-κB Signaling. *J. Pharm. Anal.* **2022**, *12*, 394–405. [CrossRef] [PubMed]

63. Seo, E.J.; Fischer, N.; Efferth, T. Phytochemicals as Inhibitors of NF-κB for Treatment of Alzheimer's Disease. *Pharmacol. Res.* **2018**, *129*, 262–273. [CrossRef] [PubMed]
64. Vo, T.T.T.; Chu, P.-M.; Tuan, V.P.; Te, J.S.-L.; Lee, I.-T. The Promising Role of Antioxidant Phytochemicals in the Prevention and Treatment of Periodontal Disease via the Inhibition of Oxidative Stress Pathways: Updated Insights. *Antioxidants* **2020**, *9*, 1211. [CrossRef] [PubMed]
65. Desai, S.J.; Prickril, B.; Rasooly, A. Mechanisms of Phytonutrient Modulation of Cyclooxygenase-2 (COX-2) and Inflammation Related to Cancer. *Nutr. Cancer* **2018**, *70*, 350–375. [CrossRef] [PubMed]
66. Pisoschi, A.M.; Iordache, F.; Stanca, L.; Gajaila, I.; Ghimpeteanu, O.M.; Geicu, O.I.; Serban, A.I. Antioxidant, Anti-Inflammatory, and Immunomodulatory Roles of Nonvitamin Antioxidants in Anti-SARS-CoV-2 Therapy. *J. Med. Chem.* **2022**, *65*, 12562–12593. [CrossRef] [PubMed]
67. Fakhri, S.; Yarmohammadi, A.; Yarmohammadi, M.; Farzaei, M.H.; Echeverria, J. Marine Natural Products: Promising Candidates in the Modulation of Gut-Brain Axis Towards Neuroprotection. *Mar. Drugs* **2021**, *19*, 165. [CrossRef] [PubMed]
68. Atlas, D. Emerging Therapeutic Opportunities of Novel Thiol-Amides, NAC-Amide (AD4/NACA) and Thioredoxin Mimetics (TXM-Peptides) for Neurodegenerative-Related Disorders. *Free Radic. Biol. Med.* **2021**, *176*, 120–141. [CrossRef]
69. Direito, R.; Rocha, J.; Sepodes, B.; Eduardo-Figueira, M. Phenolic Compounds Impact on Rheumatoid Arthritis, Inflammatory Bowel Disease and Microbiota Modulation. *Pharmaceutics* **2021**, *13*, 145. [CrossRef]
70. Ávila-Román, J.; García-Gil, S.; Rodríguez-Luna, A.; Motilva, V.; Talero, E. Anti-Inflammatory and Anticancer Effects of Microalgal Carotenoids. *Mar. Drugs* **2021**, *19*, 531. [CrossRef]
71. Benameur, T.; Porro, C.; Twfieg, M.-E.; Benameur, N.; Panaro, M.A.; Filannino, F.M.; Hasan, A. Emerging Paradigms in Inflammatory Disease Management: Exploring Bioactive Compounds and the Gut Microbiota. *Brain Sci.* **2023**, *13*, 1226. [CrossRef] [PubMed]
72. Huang, Z.; Liu, Y.; Qi, G.; Brand, D.; Zheng, S.G. Role of Vitamin A in the Immune System. *J. Clin. Med.* **2018**, *7*, 258. [CrossRef]
73. Gao, Y.; Focsan, A.L.; Kispert, L.D. Antioxidant Activity in Supramolecular Carotenoid Complexes Favored by Nonpolar Environment and Disfavored by Hydrogen Bonding. *Antioxidants* **2020**, *9*, 625. [CrossRef] [PubMed]
74. Munteanu, C.; Schwartz, B. The Relationship Between Nutrition and the Immune System. *Front. Nutr.* **2022**, *9*, 1082500. [CrossRef]
75. Mitra, S.; Rauf, A.; Tareq, A.M.; Jahan, S.; Emran, T.B.; Shahriar, T.G.; Dhama, K.; Alhumaydhi, F.A.; Aljohani, A.S.; Rebezov, M.; et al. Potential Health Benefits of Carotenoid Lutein: An Updated Review. *Food Chem. Toxicol.* **2021**, *154*, 112328. [CrossRef]
76. Toti, E.; Chen, C.Y.O.; Palmery, M.; Villaño Valencia, D.; Peluso, I. Non-Provitamin A and Provitamin A Carotenoids as Immunomodulators: Recommended Dietary Allowance, Therapeutic Index, or Personalized Nutrition? *Oxid. Med. Cell. Longev.* **2018**, *2018*, 4637861. [CrossRef] [PubMed]
77. Pan, P.; Huang, Y.W.; Oshima, K.; Yearsley, M.; Zhang, J.; Arnold, M.; Yu, J.; Wang, L.S. The Immunomodulatory Potential of Natural Compounds in Tumor-Bearing Mice and Humans. *Crit. Rev. Food Sci. Nutr.* **2019**, *59*, 992–1007. [CrossRef]
78. Anjani, G.; Ayustaningwarno, F.; Eviana, R. Critical Review on the Immunomodulatory Activities of Carrot's β-Carotene and Other Bioactive Compounds. *J. Funct. Foods* **2022**, *99*, 105303. [CrossRef]
79. Tieu, S.; Charchoglyan, A.; Wagter-Lesperance, L.; Karimi, K.; Bridle, B.W.; Karrow, N.A.; Mallard, B.A. Immunoceuticals: Harnessing Their Immunomodulatory Potential to Promote Health and Wellness. *Nutrients* **2022**, *14*, 4075. [CrossRef] [PubMed]
80. Costagliola, G.; Nuzzi, G.; Spada, E.; Comberiati, P.; Verduci, E.; Peroni, D.G. Nutraceuticals in Viral Infections: An Overview of the Immunomodulating Properties. *Nutrients* **2021**, *13*, 2410. [CrossRef] [PubMed]
81. Ansari, P.; Samia, J.F.; Khan, J.T.; Rafi, M.R.; Rahman, M.S.; Rahman, A.B.; Abdel-Wahab, Y.H.A.; Seidel, V. Protective Effects of Medicinal Plant-Based Foods against Diabetes: A Review on Pharmacology, Phytochemistry, and Molecular Mechanisms. *Nutrients* **2023**, *15*, 3266. [CrossRef] [PubMed]
82. Jain, P.; Darji, P.; Thakur, B.S.; Jain, A.; Jain, P.K.; Khare, B. Immunostimulants: Concepts, Types and Functions. *Asian J. Dent. Health Sci.* **2022**, *2*, 26–34. [CrossRef]
83. Sukri, A.; Hanafiah, A.; Kosai, N.R. The Roles of Immune Cells in Gastric Cancer: Anti-Cancer or Pro-Cancer? *Cancers* **2022**, *14*, 3922. [CrossRef] [PubMed]
84. Thirunavukarasu, A.J.; Ross, A.C.; Gilbert, R.M. Vitamin A, Systemic T-Cells, and the Eye: Focus on Degenerative Retinal Disease. *Front. Nutr.* **2022**, *9*, 914457. [CrossRef] [PubMed]
85. Mozos, I.; Stoian, D.; Caraba, A.; Malainer, C.; Horbańczuk, J.O.; Atanasov, A.G. Lycopene and Vascular Health. *Front. Pharmacol.* **2018**, *9*, 521. [CrossRef] [PubMed]
86. Wang, M.; Tang, R.; Zhou, R.; Qian, Y.; Di, D. The Protective Effect of Serum Carotenoids on Cardiovascular Disease: A Cross-Sectional Study from the General US Adult Population. *Front. Nutr.* **2023**, *10*, 1154239. [CrossRef] [PubMed]
87. Kulawik, A.; Cielecka-Piontek, J.; Zalewski, P. The Importance of Antioxidant Activity for the Health-Promoting Effect of Lycopene. *Nutrients* **2023**, *15*, 3821. [CrossRef] [PubMed]
88. Kwatra, B. A Review on Potential Properties and Therapeutic Applications of Lycopene. *Int. J. Med. Biomed. Stud.* **2020**, *4*, 33–44. [CrossRef]
89. Przybylska, S.; Tokarczyk, G. Lycopene in the Prevention of Cardiovascular Diseases. *Int. J. Mol. Sci.* **2022**, *23*, 1957. [CrossRef] [PubMed]
90. Zhang, S.; Li, L.; Chen, W.; Xu, S.; Feng, X.; Zhang, L. Natural Products: The Role and Mechanism in Low-Density Lipoprotein Oxidation and Atherosclerosis. *Phytother. Res.* **2021**, *35*, 2945–2967. [CrossRef] [PubMed]

91. Bin-Jumah, M.N.; Nadeem, M.S.; Gilani, S.J.; Mubeen, B.; Ullah, I.; Alzarea, S.I.; Ghoneim, M.M.; Alshehri, S.; Al-Abbasi, F.A.; Kazmi, I. Lycopene: A Natural Arsenal in the War Against Oxidative Stress and Cardiovascular Diseases. *Antioxidants* **2022**, *11*, 232. [CrossRef]
92. Rejali, L.; Ozumerzifon, S.; Nayeri, H.; Asgary, S. Risk Reduction and Prevention of Cardiovascular Diseases: Biological Mechanisms of Lycopene. *Bioact. Compd. Health Dis.* **2022**, *5*, 202–221. [CrossRef]
93. Coelho, M.C.; Rodrigues, A.S.; Teixeira, J.A.; Pintado, M.E. Integral Valorisation of Tomato By-Products Towards Bioactive Compounds Recovery: Human Health Benefits. *Food Chem.* **2023**, *410*, 135319. [CrossRef] [PubMed]
94. Kabir, M.T.; Rahman, M.H.; Shah, M.; Jamiruddin, M.R.; Al-Harrasi, A.; Bhatia, S.; Ashraf, G.M.; Najda, A.; El-kott, A.F.; Mohamed, H.R.H.; et al. Therapeutic Promise of Carotenoids as Antioxidants and Anti-Inflammatory Agents in Neurodegenerative Disorders. *Biomed. Pharmacother.* **2022**, *146*, 112610. [CrossRef]
95. Mohd Hassan, N.; Yusof, N.A.; Yahaya, A.F.; Mohd Rozali, N.N.; Othman, R. Carotenoids of Capsicum Fruits: Pigment Profile and Health-Promoting Functional Attributes. *Antioxidants* **2019**, *8*, 469. [CrossRef]
96. Malekmohammad, K.; Sewell, R.D.; Rafieian-Kopaei, M. Antioxidants and Atherosclerosis: Mechanistic Aspects. *Biomolecules* **2019**, *9*, 301. [CrossRef]
97. Bohn, T.; Bonet, M.L.; Borel, P.; Keijer, J.; Landrier, J.; Milisav, I.; Ribot, J.; Riso, P.; Winklhofer-Roob, B.M.; Sharoni, Y.; et al. Mechanistic Aspects of Carotenoid Health Benefits–Where Are We Now? *Nutr. Res. Rev.* **2021**, *34*, 276–302. [CrossRef] [PubMed]
98. Tang, Q.; Wei, S.; He, X.; Zheng, X.; Tao, F.; Tu, P.; Gao, B. Lutein-Rich Beverage Alleviates Visual Fatigue in the Hyperglycemia Model of Sprague-Dawley Rats. *Metabolites* **2023**, *13*, 1110. [CrossRef]
99. Dziedziak, J.; Kasarełło, K.; Cudnoch-Jędrzejewska, A. Dietary Antioxidants in Age-Related Macular Degeneration and Glaucoma. *Antioxidants* **2021**, *10*, 1743. [CrossRef] [PubMed]
100. George, S.; Abrahamse, H. Redox Potential of Antioxidants in Cancer Progression and Prevention. *Antioxidants* **2020**, *9*, 1156. [CrossRef] [PubMed]
101. Gandla, K.; Babu, A.K.; Unnisa, A.; Sharma, I.; Singh, L.P.; Haque, M.A.; Dashputre, N.L.; Baig, S.; Siddiqui, F.A.; Khandaker, M.U.; et al. Carotenoids: Role in Neurodegenerative Diseases Remediation. *Brain Sci.* **2023**, *13*, 457. [CrossRef] [PubMed]
102. Briyal, S.; Ranjan, A.K.; Gulati, A. Oxidative Stress: A Target to Treat Alzheimer's Disease and Stroke. *Neurochem. Int.* **2023**, *165*, 105509. [CrossRef] [PubMed]
103. Lopresti, A.L.; Smith, S.J.; Drummond, P.D. The Effects of Lutein and Zeaxanthin Supplementation on Cognitive Function in Adults with Self-Reported Mild Cognitive Complaints: A Randomized, Double-Blind, Placebo-Controlled Study. *Front. Nutr.* **2022**, *9*, 843512. [CrossRef] [PubMed]
104. Faraone, I.; Sinisgalli, C.; Ostuni, A.; Armentano, M.F.; Carmosino, M.; Milella, L.; Russo, D.; Labanca, F.; Khan, H. Astaxanthin Anticancer Effects Are Mediated Through Multiple Molecular Mechanisms: A Systematic Review. *Pharmacol. Res.* **2020**, *155*, 104689. [CrossRef] [PubMed]
105. Ahmed, K.; Zaidi, S.F.; Cui, Z.G.; Zhou, D.; Saeed, S.A.; Inadera, H. Potential Proapoptotic Phytochemical Agents for the Treatment and Prevention of Colorectal Cancer. *Oncol. Lett.* **2019**, *18*, 487–498. [CrossRef] [PubMed]
106. Starska-Kowarska, K. Dietary Carotenoids in Head and Neck Cancer—Molecular and Clinical Implications. *Nutrients* **2022**, *14*, 531. [CrossRef]
107. Saini, R.K.; Keum, Y.S.; Daglia, M.; Rengasamy, K.R. Dietary Carotenoids in Cancer Chemoprevention and Chemotherapy: A Review of Emerging Evidence. *Pharmacol. Res.* **2020**, *157*, 104830. [CrossRef]
108. Choudhary, N.; Bawari, S.; Burcher, J.T.; Sinha, D.; Tewari, D.; Bishayee, A. Targeting Cell Signaling Pathways in Lung Cancer by Bioactive Phytocompounds. *Cancers* **2023**, *15*, 3980. [CrossRef]
109. Zhai, K.; Siddiqui, M.; Abdellatif, B.; Liskova, A.; Kubatka, P.; Büsselberg, D. Natural Compounds in Glioblastoma Therapy: Preclinical Insights, Mechanistic Pathways, and Outlook. *Cancers* **2021**, *13*, 2317. [CrossRef]
110. Koch, W.; Wawruszak, A.; Kukula-Koch, W.; Zdziebło, M.; Helon, P.; Almarhoon, Z.M.; Al-Omari, B.; Calina, D.; Sharifi-Rad, J. Exploring the Therapeutic Efficacy of Crocetin in Oncology: An Evidence-Based Review. *Naunyn-Schmiedeberg's Arch. Pharmacol.* **2023**, *397*, 1455–1476. [CrossRef] [PubMed]
111. Koklesová, L.; Líšková, A.; Samec, M.; Zhai, K.; Abotaleb, M.; Ashrafizadeh, M.; Brockmueller, A.; Shakibaei, M.; Biringer, K.; Bugos, O.; et al. Carotenoids in Cancer Metastasis—Status Quo and Outlook. *Biomolecules* **2020**, *10*, 1653. [CrossRef] [PubMed]
112. Crupi, P.; Faienza, M.F.; Naeem, M.Y.; Corbo, F.; Clodoveo, M.L.; Muraglia, M. Overview of the Potential Beneficial Effects of Carotenoids on Consumer Health and Well-Being. *Antioxidants* **2023**, *12*, 1069. [CrossRef] [PubMed]
113. Alsharairi, N.A. Quercetin Derivatives as Potential Therapeutic Agents: An Updated Perspective on the Treatment of Nicotine-Induced Non-Small Cell Lung Cancer. *Int. J. Mol. Sci.* **2023**, *24*, 15208. [CrossRef] [PubMed]
114. Amintas, S.; Dupin, C.; Boutin, J.; Beaumont, P.; Moreau-Gaudry, F.; Bedel, A.; Krisa, S.; Vendrely, V.; Dabernat, S. Bioactive Food Components for Colorectal Cancer Prevention and Treatment: A Good Match. *Crit. Rev. Food Sci. Nutr.* **2023**, *63*, 6615–6629. [CrossRef] [PubMed]
115. Konecki, T.; Juszczak, A.; Cichocki, M. Can Diet Prevent Urological Cancers? An Update on Carotenoids as Chemopreventive Agents. *Nutrients* **2022**, *14*, 1367. [CrossRef] [PubMed]
116. Lin, M.H.; Chen, H.C.; Liu, K.S. A Study of the Effects of Digital Learning on Learning Motivation and Learning Outcome. *Eurasia J. Math. Sci. Technol. Educ.* **2017**, *13*, 3553–3564. [CrossRef]

117. Erzurumlu, Y.; Catakli, D.; Dogan, H.K. Potent Carotenoid Astaxanthin Expands the Anti-Cancer Activity of Cisplatin in Human Prostate Cancer Cells. *J. Nat. Med.* **2023**, *77*, 572–583. [CrossRef]
118. Rudzińska, A.; Juchaniuk, P.; Oberda, J.; Wiśniewska, J.; Wojdan, W.; Szklener, K.; Mańdziuk, S. Phytochemicals in Cancer Treatment and Cancer Prevention—Review on Epidemiological Data and Clinical Trials. *Nutrients* **2023**, *15*, 1896. [CrossRef] [PubMed]
119. Vijay, K.; Sowmya, P.R.; Arathi, B.P.; Shilpa, S.; Shwetha, H.J.; Raju, M.; Baskaran, V.; Lakshminarayana, R. Low-Dose Doxorubicin with Carotenoids Selectively Alters Redox Status and Upregulates Oxidative Stress-Mediated Apoptosis in Breast Cancer Cells. *Food Chem. Toxicol.* **2018**, *118*, 675–690. [CrossRef]
120. Giani, M.; Montoyo-Pujol, Y.G.; Peiró, G.; Martínez-Espinosa, R.M. Halophilic Carotenoids and Breast Cancer: From Salt Marshes to Biomedicine. *Mar. Drugs* **2021**, *19*, 594. [CrossRef]
121. Gong, X.; Smith, J.R.; Swanson, H.M.; Rubin, L.P. Carotenoid Lutein Selectively Inhibits Breast Cancer Cell Growth and Potentiates the Effect of Chemotherapeutic Agents Through ROS-Mediated Mechanisms. *Molecules* **2018**, *23*, 905. [CrossRef] [PubMed]
122. Van Hoang, D.; Pham, N.M.; Lee, A.H.; Tran, D.N.; Binns, C.W. Dietary Carotenoid Intakes and Prostate Cancer Risk: A Case-Control Study from Vietnam. *Nutrients* **2018**, *10*, 70. [CrossRef] [PubMed]
123. Sekhoacha, M.; Riet, K.; Motloung, P.; Gumenku, L.; Adegoke, A.; Mashele, S. Prostate Cancer Review: Genetics, Diagnosis, Treatment Options, and Alternative Approaches. *Molecules* **2022**, *27*, 5730. [CrossRef] [PubMed]
124. Russo, G.L.; Moccia, S.; Russo, M.; Spagnuolo, C. Redox Regulation by Carotenoids: Evidence and Conflicts for Their Application in Cancer. *Biochem. Pharmacol.* **2021**, *194*, 114838. [CrossRef] [PubMed]
125. Koklesová, L.; Líšková, A.; Samec, M.; Buhrmann, C.; Samuel, S.M.; Varghese, E.; Ashrafizadeh, M.; Najafi, M.; Shakibaei, M.; Büsselberg, D.; et al. Carotenoids in Cancer Apoptosis—The Road from Bench to Bedside and Back. *Cancers* **2020**, *12*, 2425. [CrossRef] [PubMed]
126. El-Masry, T.A.; El-Nagar, M.M.F.; El Mahdy, N.A.; Alherz, F.A.; Taher, R.; Osman, E.Y. Potential Antitumor Activity of Combined Lycopene and Sorafenib Against Solid Ehrlich Carcinoma via Targeting Autophagy and Apoptosis and Suppressing Proliferation. *Pharmaceuticals* **2024**, *17*, 527. [CrossRef] [PubMed]
127. Abreu, A.P.; Martins, R.; Nunes, J. Emerging Applications of *Chlorella* sp. and *Spirulina* (*Arthrospira*) sp. *Bioengineering* **2023**, *10*, 955. [CrossRef] [PubMed]
128. Leyane, T.S.; Jere, S.W.; Houreld, N.N. Oxidative Stress in Ageing and Chronic Degenerative Pathologies: Molecular Mechanisms Involved in Counteracting Oxidative Stress and Chronic Inflammation. *Int. J. Mol. Sci.* **2022**, *23*, 7273. [CrossRef] [PubMed]
129. Miguel, S.P.; Ribeiro, M.P.; Otero, A.; Coutinho, P. Application of Microalgae and Microalgal Bioactive Compounds in Skin Regeneration. *Algal Res.* **2021**, *58*, 102395. [CrossRef]
130. Comino-Sanz, I.M.; López-Franco, M.D.; Castro, B.; Pancorbo-Hidalgo, P.L. The Role of Antioxidants on Wound Healing: A Review of the Current Evidence. *J. Clin. Med.* **2021**, *10*, 3558. [CrossRef]
131. Hosokawa, M. Health-Promoting Functions of the Marine Carotenoid Fucoxanthin. In *Carotenoids: Advances in Experimental Medicine and Biology*; Springer: Berlin/Heidelberg, Germany, 2021; Volume 1261, pp. 273–284. [CrossRef]
132. Kern, H.J.; Mitmesser, S.H. Role of Nutrients in Metabolic Syndrome: A 2017 Update. *Nutr. Diet. Suppl.* **2018**, *10*, 13–26. [CrossRef]
133. Hajizadeh-Sharafabad, F.; Ghoreishi, Z.; Maleki, V.; Tarighat-Esfanjani, A. Mechanistic Insights into the Effect of Lutein on Atherosclerosis, Vascular Dysfunction, and Related Risk Factors: A Systematic Review of In Vivo, Ex Vivo and In Vitro Studies. *Pharmacol. Res.* **2019**, *149*, 104477. [CrossRef] [PubMed]
134. Aguilera-Méndez, A.; Boone-Villa, D.; Nieto-Aguilar, R.; Villafaña-Rauda, S.; Molina, A.S.; Sobrevilla, J.V. Role of Vitamins in the Metabolic Syndrome and Cardiovascular Disease. *Pflugers Arch.* **2022**, *474*, 117–140. [CrossRef]
135. Teng, H.; Zheng, Y.; Cao, H.; Huang, Q.; Xiao, J.; Chen, L. Enhancement of Bioavailability and Bioactivity of Diet-Derived Flavonoids by Application of Nanotechnology: A Review. *Crit. Rev. Food Sci. Nutr.* **2023**, *63*, 378–393. [CrossRef] [PubMed]
136. Polat, D.Ç.; Karadağ, A.E.; Köprülü, R.E.P.; Karantas, I.D.; Mutlu, G.; Çağlar, E.Ş.; Okur, M.; Okur, N.Ü.; Sıafaka, P.I. Phytochemical Compounds Loaded to Nanocarriers as Potential Therapeutic Substances for Alzheimer's Disease—Could They Be Effective? *Curr. Pharm. Des.* **2022**, *28*, 2437–2460. [CrossRef] [PubMed]
137. Jafari, S.M.; McClements, D.J. Nanotechnology Approaches for Increasing Nutrient Bioavailability. *Adv. Food Nutr. Res.* **2017**, *81*, 1–30. [CrossRef] [PubMed]
138. Focsan, A.L.; Polyakov, N.E.; Kispert, L.D. Supramolecular Carotenoid Complexes of Enhanced Solubility and Stability—The Way of Bioavailability Improvement. *Molecules* **2019**, *24*, 3947. [CrossRef] [PubMed]
139. Tarhan, O.; Spotti, M.J. Nutraceutical Delivery Through Nano-Emulsions: General Aspects, Recent Applications and Patented Inventions. *Colloids Surf. B* **2021**, *200*, 111526. [CrossRef] [PubMed]
140. Pateiro, M.; Gómez, B.; Munekata, P.E.; Barba, F.J.; Putnik, P.; Kovačević, D.B.; Lorenzo, J.M. Nanoencapsulation of promising bioactive compounds to improve their absorption, stability, functionality and the appearance of the final food products. *Molecules* **2021**, *26*, 1547. [CrossRef]
141. Molteni, C.; La Motta, C.; Valoppi, F. Improving the Bioaccessibility and Bioavailability of Carotenoids by Means of Nanostructured Delivery Systems: A Comprehensive Review. *Antioxidants* **2022**, *11*, 1931. [CrossRef]
142. Yang, C.; Zhang, L.; Tsao, R. Chemistry and Biochemistry of Dietary Carotenoids: Bioaccessibility, Bioavailability and Bioactivities. *J. Food Bioact.* **2020**, *10*, 32–46. [CrossRef]

143. Khalid, K.; Tan, X.; Zaid, H.F.M.; Tao, Y.; Chew, C.L.; Chu, D.T.; Lam, M.K.; Ho, Y.; Lai, C.W. Advanced in Developmental Organic and Inorganic Nanomaterial: A Review. *Bioengineered* **2020**, *11*, 328–355. [CrossRef] [PubMed]
144. Soukoulis, C.; Bohn, T. A Comprehensive Overview on the Micro- and Nano-Technological Encapsulation Advances for Enhancing the Chemical Stability and Bioavailability of Carotenoids. *Crit. Rev. Food Sci. Nutr.* **2018**, *58*, 1–36. [CrossRef] [PubMed]
145. Mantovani, R.A.; Xavier, A.A.O.; Tavares, G.M.; Mercadante, A.Z. Lutein Bioaccessibility in Casein-Stabilized Emulsions is Influenced by the Free to Acylated Carotenoid Ratio, But Not by the Casein Aggregation State. *Food Res. Int.* **2022**, *161*, 111778. [CrossRef] [PubMed]
146. Livingstone, T.L.; Beasy, G.; Mills, R.D.; Plumb, J.; Needs, P.W.; Mithen, R.; Traka, M.H. Plant Bioactives and the Prevention of Prostate Cancer: Evidence from Human Studies. *Nutrients* **2019**, *11*, 2245. [CrossRef] [PubMed]
147. Dille, M.J.; Hattrem, M.N.; Draget, K.I. Bioactively Filled Gelatin Gels; Challenges and Opportunities. *Food Hydrocoll.* **2018**, *76*, 17–29. [CrossRef]
148. de Alcantara Lemos, J.; Oliveira, A.E.M.F.M.; Araujo, R.S.; Townsend, D.M.; Ferreira, L.A.M.; de Barros, A.L.B. Recent Progress in Micro and Nano-Encapsulation of Bioactive Derivatives of the Brazilian Genus Pterodon. *Biomed. Pharmacother.* **2021**, *143*, 112137. [CrossRef] [PubMed]
149. Elvira-Torales, L.I.; García-Alonso, J.; Periago-Castón, M.J. Nutritional Importance of Carotenoids and Their Effect on Liver Health: A Review. *Antioxidants* **2019**, *8*, 229. [CrossRef] [PubMed]
150. Bera, S.; Mitra, R.; Singh, J. Recent Advancement in Protected Delivery Methods for Carotenoid: A Smart Choice in Modern Nutraceutical Formulation Concept. *Biotechnol. Genet. Eng. Rev.* **2023**, *17*, 1–57. [CrossRef]
151. Schmidt, H.W.; Würthner, F. A Periodic System of Supramolecular Elements. *Angew. Chem. Int. Ed.* **2020**, *59*, 8766–8775. [CrossRef]
152. Hou, Y.; Zou, L.; Li, Q.; Chen, M.; Ruan, H.; Sun, Z.; Xu, X.; Yang, J.; Ma, G. Supramolecular Assemblies Based on Natural Small Molecules: Union Would Be Effective. *Mater. Today Bio* **2022**, *15*, 100327. [CrossRef] [PubMed]
153. Chen, H.; Liu, Z.; Li, L.; Cai, X.; Xiang, L.; Wang, S. Peptide Supramolecular Self-Assembly: Regulatory Mechanism, Functional Properties, and Its Application in Foods. *J. Agric. Food Chem.* **2024**, *72*, 5526–5541. [CrossRef] [PubMed]
154. Zhang, Z.; Hao, J. Bioinspired Organohydrogels with Heterostructures: Fabrications, Performances, and Applications. *Adv. Colloid Interface Sci.* **2021**, *292*, 102408. [CrossRef] [PubMed]
155. Polyakov, N.E.; Focsan, A.L.; Gao, Y.; Kispert, L.D. The Endless World of Carotenoids—Structural, Chemical and Biological Aspects of Some Rare Carotenoids. *Int. J. Mol. Sci.* **2023**, *24*, 9885. [CrossRef] [PubMed]
156. Majumder, J.; Taratula, O.; Minko, T. Nanocarrier-Based Systems for Targeted and Site-Specific Therapeutic Delivery. *Adv. Drug Deliv. Rev.* **2019**, *144*, 57–77. [CrossRef] [PubMed]
157. Cai, S.S.; Li, T.; Akinade, T.; Zhu, Y.; Leong, K.W. Drug Delivery Carriers with Therapeutic Functions. *Adv. Drug Deliv. Rev.* **2021**, *176*, 113884. [CrossRef] [PubMed]
158. Manivasagan, P.; Bharathiraja, S.; Santha Moorthy, M.; Mondal, S.; Seo, H.; Dae Lee, K.; Oh, J. Marine Natural Pigments as Potential Sources for Therapeutic Applications. *Crit. Rev. Biotechnol.* **2018**, *38*, 745–761. [CrossRef] [PubMed]
159. Lombardo, D.; Calandra, P.; Pasqua, L.; Magazù, S. Self-Assembly of Organic Nanomaterials and Biomaterials: The Bottom-Up Approach for Functional Nanostructures Formation and Advanced Applications. *Materials* **2020**, *13*, 1048. [CrossRef] [PubMed]
160. Fan, S.; Zhang, Y.; Huang, X.; Geng, L.; Shao, H.; Hu, X.; Zhang, Y. Silk Materials for Medical, Electronic and Optical Applications. *Sci. China Technol. Sci.* **2019**, *62*, 903–918. [CrossRef]
161. Huang, J.; Liu, J.; Wang, J. Optical Properties of Biomass-Derived Nanomaterials for Sensing, Catalytic, Biomedical and Environmental Applications. *TrAC Trends Anal. Chem.* **2020**, *124*, 115800. [CrossRef]
162. López, G.D.; Álvarez-Rivera, G.; Carazzone, C.; Ibáñez, E.; Leidy, C.; Cifuentes, A. Bacterial Carotenoids: Extraction, Characterization, and Applications. *Crit. Rev. Anal. Chem.* **2023**, *53*, 1239–1262. [CrossRef] [PubMed]
163. Wu, Y.; Yan, P.; Li, Y.; Liu, X.; Wang, Z.; Chen, T.; Zhao, X. Enhancing β-Carotene Production in *Escherichia coli* by Perturbing Central Carbon Metabolism and Improving the NADPH Supply. *Front. Bioeng. Biotechnol.* **2020**, *8*, 585. [CrossRef] [PubMed]
164. Snell, J.A.; Jandova, J.; Wondrak, G.T. Hypochlorous Acid: From Innate Immune Factor and Environmental Toxicant to Chemopreventive Agent Targeting Solar UV-Induced Skin Cancer. *Front. Oncol.* **2022**, *12*, 887220. [CrossRef] [PubMed]
165. Nephali, L.; Piater, L.A.; Dubery, I.A.; Patterson, V.; Huyser, J.; Burgess, K.; Tugizimana, F. Biostimulants for Plant Growth and Mitigation of Abiotic Stresses: A Metabolomics Perspective. *Metabolites* **2020**, *10*, 505. [CrossRef] [PubMed]
166. Esmonde-White, K.A.; Cuellar, M.; Lewis, I.R. The Role of Raman Spectroscopy in Biopharmaceuticals from Development to Manufacturing. *Anal. Bioanal. Chem.* **2022**, *414*, 969–991. [CrossRef] [PubMed]
167. Chong, J.W.R.; Tang, D.Y.Y.; Leong, H.Y.; Khoo, K.S.; Show, P.L.; Chew, K.W. Bridging Artificial Intelligence and Fucoxanthin for the Recovery and Quantification from Microalgae. *Bioengineered* **2023**, *14*, 2244232. [CrossRef] [PubMed]
168. Sridhar, K.; Inbaraj, B.S.; Chen, B.H. Recent Advances on Nanoparticle Based Strategies for Improving Carotenoid Stability and Biological Activity. *Antioxidants* **2021**, *10*, 713. [CrossRef] [PubMed]
169. Singh, A.K.; Pal, P.; Pandey, B.; Goksen, G.; Sahoo, U.K.; Lorenzo, J.M.; Sarangi, P.K. Development of "Smart Foods" for Health by Nanoencapsulation: Novel Technologies and Challenges. *Food Chem. X* **2023**, *20*, 100910. [CrossRef] [PubMed]
170. Rodríguez-Sifuentes, L.; Marszalek, J.E.; Hernández-Carbajal, G.; Chuck-Hernández, C. Importance of Downstream Processing of Natural Astaxanthin for Pharmaceutical Application. *Front. Chem. Eng.* **2021**, *2*, 601483. [CrossRef]

171. Brahma, D.; Dutta, D.; Mukherjee, S. Mass Production of Valuable Pro-Vitamin a Pigment from a Microbe, Cost Analysis and Targeting It for Health Benefiting Purpose. In *Industrial Microbiology Based Entrepreneurship: Making Money from Microbes*; Springer: Berlin/Heidelberg, Germany, 2022; pp. 147–178. [CrossRef]
172. Bellavita, R.; Braccia, S.; Falanga, A.; Galdiero, S. An Overview of Supramolecular Platforms Boosting Drug Delivery. *Bioinorg. Chem. Appl.* **2023**, *2023*, 8608428. [CrossRef]
173. Parikh, R.B.; Obermeyer, Z.; Navathe, A.S. Regulation of Predictive Analytics in Medicine. *Science* **2019**, *363*, 810–812. [CrossRef] [PubMed]
174. Liu, Y.; Liu, Y. Construction of Lipid-Biomacromolecular Compounds for Loading and Delivery of Carotenoids: Preparation Methods, Structural Properties, and Absorption-Enhancing Mechanisms. *Crit. Rev. Food Sci. Nutr.* **2024**, *64*, 1653–1676. [CrossRef] [PubMed]
175. Metibemu, D.S.; Ogungbe, I.V. Carotenoids in Drug Discovery and Medicine: Pathways and Molecular Targets Implicated in Human Diseases. *Molecules* **2022**, *27*, 6005. [CrossRef] [PubMed]
176. Auclair, J.R.; Rathore, A.S. The Multi-Attribute Method (MAM) for the Characterization of Biopharmaceuticals. *LCGC North Am.* **2021**, *39*, 28–32. [CrossRef]
177. Millán-Martín, S.; Jakes, C.; Carillo, S.; Bones, J. Multi-Attribute Method (MAM) Analytical Workflow for Biotherapeutic Protein Characterization from Process Development to QC. *Curr. Protoc.* **2023**, *3*, e927. [CrossRef] [PubMed]
178. Chiarentin, L.; Gonçalves, C.; Augusto, C.; Miranda, M.; Cardoso, C.; Vitorino, C. Drilling into "Quality by Design" Approach for Analytical Methods. *Crit. Rev. Anal. Chem.* **2023**, *4*, 1–42. [CrossRef] [PubMed]
179. Grangeia, H.B.; Silva, C.; Simões, S.; Reis, M.S. Quality by Design in Pharmaceutical Manufacturing: A Systematic Review of Current Status, Challenges and Future Perspectives. *Eur. J. Pharm. Biopharm.* **2020**, *147*, 19–37. [CrossRef]
180. Park, G.; Kim, M.K.; Go, S.H.; Choi, M.; Jang, Y.P. Analytical Quality by Design (AQbD) Approach to the Development of Analytical Procedures for Medicinal Plants. *Plants* **2022**, *11*, 2960. [CrossRef]
181. Katekar, V.; Sangule, D.; Bhurbhure, O.; Ingle, P.; Dhage, S.; Jadhav, K. A Review on Quality by Design Approach in Analytical Methods. *J. Drug Deliv. Ther.* **2022**, *12*, 255–261. [CrossRef]
182. Djande, C.Y.H.; Pretorius, C.J.; Tugizimana, F.; Piater, L.A.; Dubery, I.A. Metabolomics: A Tool for Cultivar Phenotyping and Investigation of Grain Crops. *Agronomy* **2020**, *10*, 831. [CrossRef]
183. García-Gómez, B.E.; Salazar, J.; Nicolás-Almansa, M.; Razi, M.; Rubio, M.; Ruiz, D.; Martínez-Gómez, P. Molecular Bases of Fruit Quality in *Prunus* Species: An Integrated Genomic, Transcriptomic, and Metabolic Review with a Breeding Perspective. *Int. J. Mol. Sci.* **2020**, *22*, 333. [CrossRef] [PubMed]
184. Roychowdhury, R.; Das, S.K.; Gupta, A.; Parihar, P.; Chandrasekhar, K.; Sarker, U.; Sudhakar, C. Multi-Omics Pipeline and Omics-Integration Approach to Decipher Plant's Abiotic Stress Tolerance Responses. *Genes* **2023**, *14*, 1281. [CrossRef] [PubMed]
185. Pandita, D.; Pandita, A.; Wani, S.H.; Abdelmohsen, S.A.M.; Alyousef, H.A.; Abdelbacki, A.M.M.; Elansary, H.O. Crosstalk of Multi-Omics Platforms with Plants of Therapeutic Importance. *Cells* **2021**, *10*, 1296. [CrossRef] [PubMed]
186. Donno, D.; Mellano, M.G.; Gamba, G.; Riondato, I.; Beccaro, G.L. Analytical Strategies for Fingerprinting of Antioxidants, Nutritional Substances, and Bioactive Compounds in Foodstuffs Based on High Performance Liquid Chromatography–Mass Spectrometry: An Overview. *Foods* **2020**, *9*, 1734. [CrossRef] [PubMed]
187. Mahato, N.; Sinha, M.; Sharma, K.; Koteswararao, R.; Cho, M.H. Modern Extraction and Purification Techniques for Obtaining High Purity Food-Grade Bioactive Compounds and Value-Added Co-Products from Citrus Wastes. *Foods* **2019**, *8*, 523. [CrossRef] [PubMed]
188. Gerhardtova, I.; Jankech, T.; Majerova, P.; Piestansky, J.; Olesova, D.; Kovac, A.; Jampilek, J. Recent Analytical Methodologies in Lipid Analysis. *Int. J. Mol. Sci.* **2024**, *25*, 2249. [CrossRef] [PubMed]
189. Zheng, X.; Zhu, K.; Sun, Q.; Zhang, W.; Wang, X.; Cao, H.; Deng, X. Natural Variation in CCD4 Promoter Underpins Species-Specific Evolution of Red Coloration in Citrus Peel. *Mol. Plant* **2019**, *12*, 1294–1307. [CrossRef] [PubMed]
190. Gebregziabher, B.S.; Zhang, S.; Qi, J.; Azam, M.; Ghosh, S.; Feng, Y.; Sun, J. Simultaneous Determination of Carotenoids and Chlorophylls by the HPLC-UV-Vis Method in Soybean Seeds. *Agronomy* **2021**, *11*, 758. [CrossRef]
191. Zhang, R.; Chen, G.; Yang, B.; Du, M.; Kan, J. Insights into the Stability of Carotenoids and Capsaicinoids in Water-Based or Oil-Based Chili Systems at Different Processing Treatments. *Food Chem.* **2021**, *342*, 128308. [CrossRef]
192. Britton, G. Getting to Know Carotenoids. In *Carotenoids: Carotenoid and Apocarotenoid Analysis*; Elsevier: Amsterdam, The Netherlands, 2022; pp. 1–56. [CrossRef]
193. Rodríguez-Rodríguez, E.; Sánchez-Prieto, M.; Olmedilla-Alonso, B. Assessment of Carotenoid Concentrations in Red Peppers (*Capsicum annuum*) under Domestic Refrigeration for Three Weeks as Determined by HPLC-DAD. *Food Chem. X*, 2000; 6, 100092. [CrossRef]
194. Salomón, M.V.; Píccoli, P.; Fontana, A. Simultaneous Determination of Carotenoids with Different Polarities in Tomato Products Using a C30 Core-Shell Column Based Approach. *Microchem. J.* **2020**, *159*, 105390. [CrossRef]
195. Saha, S.; Walia, S.; Sharma, K.; Banerjee, K. Suitability of Stationary Phase for LC Analysis of Biomolecules. *Crit. Rev. Food Sci. Nutr.* **2019**, *60*, 2856–2873. [CrossRef] [PubMed]
196. Priyadarsani, S.; Patel, A.S.; Sharma, Y.; Kar, A. Process Development for the Instant Quantification of Lycopene from Agricultural Produces Using Supercritical Fluid Chromatography-Diode Array Detector (SFC-DAD). *Foods* **2022**, *11*, 522. [CrossRef] [PubMed]

197. Luo, C.; DeStefano, J.; Langlois, T.J.; Boyes, B.E.; Schuster, S.A.; Godinho, J.M. Fundamental to Achieving Fast Separations with High Efficiency: A Review of Chromatography with Superficially Porous Particles. *Biomed. Chromatogr.* **2021**, *35*, e5087. [CrossRef] [PubMed]
198. Schmitt, K.; Woiwode, U.; Kohout, M.; Zhang, T.; Lindner, W.; Lämmerhofer, M. Comparison of small size fully porous particles and superficially porous particles of chiral anion-exchange type stationary phases in ultra-high performance liquid chromatography: Effect of particle and pore size on chromatographic efficiency and kinetic performance. *J. Chromatogr. A.* **2018**, *1569*, 149–159. [CrossRef] [PubMed]
199. Molina, A.K.; Corrêa, R.C.G.; Prieto, M.A.; Pereira, C.; Barros, L. Bioactive Natural Pigments' Extraction, Isolation, and Stability in Food Applications. *Molecules* **2023**, *28*, 1200. [CrossRef] [PubMed]
200. Aguilar-Espinosa, M.; Ek-Ku, J.E.; Rivera-Madrid, R.; Vera-Ku, M. Advancing Carotenoid Quantification: A New Method for Semi-Quantitative Assessment of β-Carotene and Lycopene Content in Food Extracts. *J. Chromatogr. B* **2023**, *1231*, 123929. [CrossRef] [PubMed]
201. Lim, K.C.; Yusoff, F.M.; Karim, M.; Natrah, F.M. Carotenoids Modulate Stress Tolerance and Immune Responses in Aquatic Animals. *Rev. Aquacult.* **2023**, *15*, 872–894. [CrossRef]
202. Santos-Beneit, F. What is the Role of Microbial Biotechnology and Genetic Engineering in Medicine? *MicrobiolOpen* **2024**, *13*, e1406. [CrossRef]
203. Bohn, T.; Balbuena, E.; Ulus, H.; Iddir, M.; Wang, G.; Crook, N.; Eroglu, A. Carotenoids in Health as Studied by Omics-Related Endpoints. *Adv. Nutr.* **2023**, *14*, 1538–1578. [CrossRef]
204. Liang, D.; Su, W.; Tan, M. Advances of Microfluidic Intestine-on-a-Chip for Analyzing Anti-Inflammation of Food. *Crit. Rev. Food Sci. Nutr.* **2021**, *62*, 4418–4434. [CrossRef]
205. Wang, H.; Xu, C.; Tan, M.; Su, W. Advanced Gut-on-Chips for Assessing Carotenoid Absorption, Metabolism, and Transport. *Crit. Rev. Food Sci. Nutr.* **2023**, 1–19. [CrossRef] [PubMed]
206. Miranda, A.; Gómez-Varela, A.I.; Stylianou, A.; Hirvonen, L.; Sánchez, H.; Beule, P.D. How Did Correlative Atomic Force Microscopy and Super-Resolution Microscopy Evolve in the Quest for Unraveling Enigmas in Biology? *Nanoscale* **2021**, *13*, 2082–2099. [CrossRef] [PubMed]
207. Babele, P.K.; Young, J.D. Applications of Stable Isotope-Based Metabolomics and Fluxomics Toward Synthetic Biology of Cyanobacteria. *WIREs Syst. Biol. Med.* **2019**, *12*, e1472. [CrossRef] [PubMed]
208. Wörheide, M.A.; Krumsiek, J.; Kastenmüller, G.; Arnold, M. Multi-Omics Integration in Biomedical Research—A Metabolomics-Centric Review. *Anal. Chim. Acta* **2021**, *1141*, 144–162. [CrossRef] [PubMed]
209. Jain, S.; Eadon, M.T. Spatial Transcriptomics in Health and Disease. *Nat. Rev. Nephrol.* **2024**, 1–13. [CrossRef] [PubMed]
210. Yuan, S.; Alper, H.S. Metabolic Engineering of Microbial Cell Factories for Production of Nutraceuticals. *Microb. Cell Fact.* **2019**, *18*, 46. [CrossRef]
211. Hwang, D.S.; Choi, Y.; Jang, J. 3D Bioprinting-Based Vascularized Tissue Models Mimicking Tissue-Specific Architecture and Pathophysiology for In Vitro Studies. *Front. Bioeng. Biotechnol.* **2021**, *9*, 685507. [CrossRef] [PubMed]
212. Fatokun, A.A. Artificial Intelligence (AI) to the Rescue: Deploying Machine Learning to Bridge the Biorelevance Gap in Antioxidant Assays. *SLAS Technol.* **2021**, *26*, 16–25. [CrossRef]
213. Böhm, V.; Lietz, G.; Olmedilla-Alonso, B.; Phelan, D.; Reboul, E.; Bánáti, D.; Borel, P.; Corte-Real, J.; Lera, Á.R.; Desmarchelier, C.; et al. From Carotenoid Intake to Carotenoid Blood and Tissue Concentrations—Implications for Dietary Intake Recommendations. *Nutr. Rev.* **2021**, *79*, 544–573. [CrossRef]
214. Tan, B.L.; Norhaizan, M.E. Carotenoids: How Effective Are They to Prevent Age-Related Diseases? *Molecules* **2019**, *24*, 1801. [CrossRef]
215. Black, H.S.; Boehm, F.; Edge, R.; Truscott, T.G. The Benefits and Risks of Certain Dietary Carotenoids That Exhibit Both Anti- and Pro-Oxidative Mechanisms—A Comprehensive Review. *Antioxidants* **2020**, *9*, 264. [CrossRef] [PubMed]
216. Geng, T.; Bao, S.; Sun, X.; Ma, D.; Zhang, H.; Ge, Q.; Liu, X.; Ma, T. A Clarification of Concepts Related to the Digestion and Absorption of Carotenoids and a New Standardized Carotenoids Bioavailability Evaluation System. *Food Chem.* **2023**, *400*, 134060. [CrossRef]
217. Matsumoto, M.; Suganuma, H.; Shimizu, S.; Hayashi, H.; Sawada, K.; Tokuda, I.; Ihara, K.; Nakaji, S. Skin Carotenoid Level as an Alternative Marker of Serum Total Carotenoid Concentration and Vegetable Intake Correlates with Biomarkers of Circulatory Diseases and Metabolic Syndrome. *Nutrients* **2020**, *12*, 1825. [CrossRef] [PubMed]
218. Zerres, S.; Stahl, W. Carotenoids in Human Skin. *Biochim. Biophys. Acta Mol. Cell Biol. Lipids* **2020**, *1865*, 158588. [CrossRef]
219. Hotea, I.; Sirbu, C.; Plotuna, A.M.; Tîrziu, E.; Badea, C.; Berbecea, A.; Dragomirescu, M.; Radulov, I. Integrating (Nutri-) Metabolomics into the One Health Tendency—The Key for Personalized Medicine Advancement. *Metabolites* **2023**, *13*, 800. [CrossRef]
220. Vollmer, D.L.; West, V.A.; Lephart, E.D. Enhancing Skin Health: By Oral Administration of Natural Compounds and Minerals with Implications to the Dermal Microbiome. *Int. J. Mol. Sci.* **2018**, *19*, 3059. [CrossRef] [PubMed]
221. Bhatt, T.; Patel, K. Carotenoids: Potent to Prevent Diseases—Review. *Nat. Prod. Bioprospect.* **2020**, *10*, 109–117. [CrossRef]
222. Pereira, L.; Valado, A. Algae-Derived Natural Products in Diabetes and Its Complications—Current Advances and Future Prospects. *Life* **2023**, *13*, 1831. [CrossRef] [PubMed]

223. Marino, P.; Pepe, G.; Basilicata, M.G.; Vestuto, V.; Marzocco, S.; Autore, G.; Procino, A.; Gomez-Monterrey, I.M.; Manfra, M.; Campiglia, P. Potential Role of Natural Antioxidant Products in Oncological Diseases. *Antioxidants* **2023**, *12*, 704. [CrossRef]
224. Yao, Y.; Goh, H.M.; Kim, J.E. The Roles of Carotenoid Consumption and Bioavailability in Cardiovascular Health. *Antioxidants* **2021**, *10*, 1978. [CrossRef]
225. Riley, W.W.; Nickerson, J.G.; Mogg, T.J.; Burton, G.W. Oxidized β-Carotene Is a Novel Phytochemical Immune Modulator That Supports Animal Health and Performance for Antibiotic-Free Production. *Animals* **2023**, *13*, 289. [CrossRef] [PubMed]
226. Spahis, S.; Borys, J.M.; Levy, E. Metabolic Syndrome as a Multifaceted Risk Factor for Oxidative Stress. *Antioxid. Redox Signal.* **2017**, *26*, 445–461. [CrossRef] [PubMed]
227. Varghese, R.; Efferth, T.; Ramamoorthy, S. Carotenoids for Lung Cancer Chemoprevention and Chemotherapy: Promises and Controversies. *Phytomedicine* **2023**, *116*, 154850. [CrossRef] [PubMed]
228. Rocha, H.R.; Coelho, M.C.; Gomes, A.M.; Pintado, M.E. Carotenoids Diet: Digestion, Gut Microbiota Modulation, and Inflammatory Diseases. *Nutrients* **2023**, *15*, 2265. [CrossRef]
229. Eroglu, A.; Al'Abri, I.S.; Kopec, R.E.; Crook, N.; Bohn, T. Carotenoids and Their Health Benefits as Derived via Their Interactions with Gut Microbiota. *Adv. Nutr.* **2023**, *14*, 238–255. [CrossRef] [PubMed]
230. Ghoshal, G.; Jain, A.; Katare, O.P. Harnessing Nanotechnology Using Nutraceuticals for Cancer Therapeutics and Intervention. In *NanoNutraceuticals*; CRC Press: Boca Raton, FL, USA, 2018; pp. 57–70, ISBN 9781351138949.
231. Agaj, A.; Peršurić, Ž.; Pavelić, S.K. Mediterranean Food Industry By-Products as a Novel Source of Phytochemicals with a Promising Role in Cancer Prevention. *Molecules* **2022**, *27*, 8655. [CrossRef] [PubMed]
232. Maghsoudi, S.; Shahraki, B.T.; Rabiee, N.; Fatahi, Y.; Bagherzadeh, M.; Dinarvand, R.; Ahmadi, S.; Rabiee, M.; Tahriri, M.; Hamblin, M.R.; et al. The Colorful World of Carotenoids: A Profound Insight on Therapeutics and Recent Trends in Nano Delivery Systems. *Crit. Rev. Food Sci. Nutr.* **2022**, *62*, 3658–3697. [CrossRef] [PubMed]
233. Michalak, M.; Pierzak, M.; Kręcisz, B.; Suliga, E. Bioactive Compounds for Skin Health: A Review. *Nutrients* **2021**, *13*, 203. [CrossRef] [PubMed]
234. Roshanravan, N.; Ghaffari, S. The Therapeutic Potential of *Crocus sativus* Linn.: A Comprehensive Narrative Review of Clinical Trials. *Phytother. Res.* **2022**, *36*, 98–111. [CrossRef]
235. Lem, D.W.; Davey, P.G.; Gierhart, D.L.; Rosen, R.B. A Systematic Review of Carotenoids in the Management of Age-Related Macular Degeneration. *Antioxidants* **2021**, *10*, 1255. [CrossRef]
236. Jia, Y.P.; Sun, L.; Yu, H.S.; Liang, L.P.; Li, W.; Ding, H.; Song, X.; Zhang, L.J. The Pharmacological Effects of Lutein and Zeaxanthin on Visual Disorders and Cognition Diseases. *Molecules* **2017**, *22*, 610. [CrossRef]
237. Mrowicka, M.; Mrowicki, J.; Kucharska, E.; Majsterek, I. Lutein and Zeaxanthin and Their Roles in Age-Related Macular Degeneration—Neurodegenerative Disease. *Nutrients* **2022**, *14*, 827. [CrossRef] [PubMed]
238. Han, X.; Zhao, R.; Zhang, G.; Jiao, Y.; Wang, Y.; Wang, D.; Cai, H. Association of Retinol and Carotenoids Content in Diet and Serum with Risk for Colorectal Cancer: A Meta-Analysis. *Front. Nutr.* **2022**, *9*, 918777. [CrossRef]
239. Bohn, T.; Hellman-Regen, J.; de Lera, A.R.; Böhm, V.; Rühl, R. Human Nutritional Relevance and Suggested Nutritional Guidelines for Vitamin A5/X and Provitamin A5/X. *Nutr. Metab.* **2023**, *20*, 34. [CrossRef] [PubMed]
240. Takahashi, N.; Saito, D.; Hasegawa, S.; Yamasaki, M.; Imai, M. Vitamin A in Health Care: Suppression of Growth and Induction of Differentiation in Cancer Cells by Vitamin A and Its Derivatives and Their Mechanisms of Action. *Pharmacol. Ther.* **2022**, *230*, 107942. [CrossRef] [PubMed]
241. Moon, J.; Ramkumar, S.; von Lintig, J. Genetic Tuning of β-Carotene Oxygenase-1 Activity Rescues Cone Photoreceptor Function in STRA6-Deficient Mice. *Hum. Mol. Genet.* **2023**, *32*, 798–809. [CrossRef] [PubMed]
242. Choi, E.H.; Daruwalla, A.; Suh, S.; Leinonen, H.; Palczewski, K. Retinoids in the Visual Cycle: Role of the Retinal G Protein-Coupled Receptor. *J. Lipid Res.* **2021**, *62*, 100040. [CrossRef] [PubMed]
243. Zenkel, M.; Hoja, U.; Gießl, A.; Berner, D.; Hohberger, B.; Weller, J.M.; König, L.; Hübner, L.; Ostermann, T.A.; Gusek-Schneider, G.; et al. Dysregulated Retinoic Acid Signaling in the Pathogenesis of Pseudoexfoliation Syndrome. *Int. J. Mol. Sci.* **2022**, *23*, 5977. [CrossRef] [PubMed]
244. Abdelaal, M.R.; Ibrahim, E.; Elnagar, M.R.; Soror, S.H.; Haffez, H. Augmented Therapeutic Potential of EC-Synthetic Retinoids in Caco-2 Cancer Cells Using an In Vitro Approach. *Int. J. Mol. Sci.* **2022**, *23*, 9442. [CrossRef]
245. Jiang, Y.; Chen, L.; Taylor, R.N.; Li, C.; Zhou, X. Physiological and Pathological Implications of Retinoid Action in the Endometrium. *J. Endocrinol.* **2018**, *236*, R169–R188. [CrossRef]
246. Sahin, E.; Orhan, C.; Sahin, N.; Padigaru, M.; Morde, A.; Lal, M.; Dhavan, N.; Erten, F.; Bilgic, A.A.; Ozercan, I.H.; et al. Lutein/Zeaxanthin Isomers and Quercetagetin Combination Safeguards the Retina from Photo-Oxidative Damage by Modulating Neuroplasticity Markers and the Nrf2 Pathway. *Pharmaceuticals* **2023**, *16*, 1543. [CrossRef]
247. Jiang, H.; Wang, L.; Li, J.; Fan, Y.; Li, Z.; Ma, M.; Liu, S.; Li, B.; Shi, J.; Li, C.; et al. Dietary Vitamins, Carotenoids and Their Sources in Relation to Age-Related Macular Degeneration Risk in China: A Population-Based Case-Control Study. *Br. J. Nutr.* **2023**, *129*, 1804–1811. [CrossRef] [PubMed]
248. Lem, D.W.; Gierhart, D.L.; Davey, P.G. Carotenoids in the Management of Glaucoma: A Systematic Review of the Evidence. *Nutrients* **2021**, *13*, 1949. [CrossRef] [PubMed]

249. Xie, Z.; Yang, F. The Effects of Lycopene Supplementation on Serum Insulin-Like Growth Factor 1 (IGF-1) Levels and Cardiovascular Disease: A Dose-Response Meta-Analysis of Clinical Trials. *Complement. Ther. Med.* **2021**, *56*, 102632. [CrossRef] [PubMed]
250. Collins, E.J.; Bowyer, C.; Tsouza, A.; Chopra, M. Tomatoes: An Extensive Review of the Associated Health Impacts of Tomatoes and Factors That Can Affect Their Cultivation. *Biology* **2022**, *11*, 239. [CrossRef]
251. Saini, R.K.; Prasad, P.; Lokesh, V.; Shang, X.; Shin, J.; Keum, Y.S.; Lee, J.H. Carotenoids: Dietary Sources, Extraction, Encapsulation, Bioavailability, and Health Benefits—A Review of Recent Advancements. *Antioxidants* **2022**, *11*, 795. [CrossRef] [PubMed]
252. Avinash, M.B.; Govindaraju, T. Architectonics: Design of Molecular Architecture for Functional Applications. *Acc. Chem. Res.* **2018**, *51*, 414–426. [CrossRef] [PubMed]
253. Duong, V.; Nguyen, T.; Maeng, H. Preparation of Solid Lipid Nanoparticles and Nanostructured Lipid Carriers for Drug Delivery and the Effects of Preparation Parameters of Solvent Injection Method. *Molecules* **2020**, *25*, 4781. [CrossRef] [PubMed]
254. Wianowska, D.; Bryshten, I. New Insights into Vitamin K—From Its Natural Sources Through Biological Properties and Chemical Methods of Quantitative Determination. *Crit. Rev. Anal. Chem.* **2022**, 1–23. [CrossRef] [PubMed]
255. Liu, Q.; Martínez-Jarquín, S.; Zenobi, R. Recent Advances in Single-Cell Metabolomics Based on Mass Spectrometry. *CCS Chem.* **2023**, *5*, 310–324. [CrossRef]
256. Wijesekara, T.; Xu, B. A Critical Review on the Stability of Natural Food Pigments and Stabilization Techniques. *Food Res. Int.* **2024**, *179*, 114011. [CrossRef]
257. Bohn, T. Determinants and Determination of Carotenoid Bioavailability from Infant Food Formulas and Adult Nutritionals Including Liquid Dairy Products. *J. AOAC Int.* **2019**, *102*, 1044–1058. [CrossRef] [PubMed]
258. Cerrato, A.; Aita, S.E.; Cannazza, G.; Capriotti, A.L.; Cavaliere, C.; Citti, C.; Bosco, C.D.; Gentili, A.; Montone, C.M.; Paris, R.; et al. Evaluation of the Carotenoid and Fat-Soluble Vitamin Profile of Industrial Hemp Inflorescence by Liquid Chromatography Coupled to Mass Spectrometry and Photodiode-Array Detection. *J. Chromatogr. A* **2023**, *1692*, 463838. [CrossRef] [PubMed]
259. Gaur, V.; Bera, S. Microbial Canthaxanthin: An Orange-Red Keto Carotenoid with Potential Pharmaceutical Applications. *BioTechnologia* **2023**, *104*, 315–328. [CrossRef] [PubMed]

International Journal of
Molecular Sciences

MDPI

Article

Regulation of Carotenoid Biosynthesis and Degradation in Lettuce (*Lactuca sativa* L.) from Seedlings to Harvest

Galina Brychkova [1,*], Cleiton Lourenço de Oliveira [1,2], Luiz Antonio Augusto Gomes [3], Matheus de Souza Gomes [4], Antoine Fort [1,5], Alberto Abrantes Esteves-Ferreira [6], Ronan Sulpice [6], Peter C. McKeown [1] and Charles Spillane [1,*]

1 Genetics & Biotechnology Laboratory, Agriculture, Food Systems & Bioeconomy Research Centre, Ryan Institute, School of Biological & Chemical Sciences, University of Galway, University Road, H91 REW4 Galway, Ireland; cleiton.oliveira@ufla.br (C.L.d.O.)
2 Department of Agriculture, Federal University of Lavras (DAG/ESAL), Aquenta Sol, Lavras 37200-000, MG, Brazil
3 Departamento de Agricultura, Universidade Federal de Lavras (UFLA), Lavras 37203-202, MG, Brazil
4 Laboratory of Bioinformatics and Molecular Analysis, Institute of Genetics and Biochemistry, Campus Patos de Minas, Federal University of Uberlandia, Av. Getúlio Vargas, 230, Patos de Minas 38700-103, MG, Brazil
5 Department of Life & Physical Science, Technological University of the Shannon: Midlands Midwest, N37 HD68 Athlone, Ireland
6 Plant Systems Biology Laboratory, Agriculture, Food Systems & Bioeconomy Research Centre, Ryan Institute, School of Biological & Chemical Sciences, University of Galway, University Road, H91 REW4 Galway, Ireland
* Correspondence: galina.brychkova@universityofgalway.ie (G.B.); charles.spillane@universityofgalway.ie (C.S.)

Abstract: Lettuce (*Lactuca sativa* L.) is one of the commercially important leafy vegetables worldwide. However, lettuce cultivars vary widely in their carotenoid concentrations at the time of harvest. While the carotenoid content of lettuce can depend on transcript levels of key biosynthetic enzymes, genes that can act as biomarkers for carotenoid accumulation at early stages of plant growth have not been identified. Transcriptomic and metabolomic analysis was performed on the inner and outer leaves of the six cultivars at different developmental stages to identify gene-to-metabolite networks affecting the accumulation of two key carotenoids, β-carotene and lutein. Statistical analysis, including principal component analysis, was used to better understand variations in carotenoid concentration between leaf age and cultivars. Our results demonstrate that key enzymes of carotenoid biosynthesis pathway can alter lutein and β-carotene biosynthesis across commercial cultivars. To ensure high carotenoids content in leaves, the metabolites sink from β-carotene and lutein to zeaxanthin, and subsequently, abscisic acid needs to be regulated. Based on 2–3-fold carotenoids increase at 40 days after sowing (DAS) as compared to the seedling stage, and 1.5–2-fold decline at commercial stage (60 DAS) compared to the 40 DAS stage, we conclude that the value of lettuce for human nutrition would be improved by use of less mature plants, as the widely-used commercial stage is already at plant senescence stage where carotenoids and other essential metabolites are undergoing degradation.

Keywords: carotenoids; carotenoid synthesis; gene expression; *Lactuca sativa*; lettuce development; biomarkers; biofortification

1. Introduction

An important objective of sustainable crop production is to increase the production of crops that offer enhanced nutritional values. Consumption of bioactive compounds such as carotenoids, especially β-carotene and lutein, is linked to improvements in human health and a reduced incidence of non-communicable diseases [1,2]. In animals, carotenoids act as antioxidants and prevent oxidative damage to cells [3,4]. Increased consumption of pro-vitamin β-carotene enriched food reduces Vitamin A deficiency [5,6], which remains one of the major health non-communicative problems in developing countries: those affected suffer increased disease susceptibility, ocular degeneration and even permanent blindness [7,8]. Lutein-enriched nutrition can also prevent eye diseases, improve photoprotection and reduce the risks of age-related macular degeneration by 80% [9,10].

The shift toward a vegetable-based diet also requires a better understanding of carotenoid accumulation in plants as, for the vegan population, plants are the only source of vitamin A precursors [5]. Therefore, understanding the relationship between the expression levels of carotenoid biosynthesis genes and final levels of β-carotene and lutein accumulation is important for plant breeding and for metabolic engineering approaches for increasing delivery of these health-promoting metabolites. There is a particular interest in identifying which of the genes involved in carotenoids biosynthesis are expressed early in plant development. Early gene expression profile could allow a robust prediction of carotenoids levels at the time of harvest and greatly accelerate screening for improved germplasm. Several studies have compared gene expression levels with carotenoid accumulation across plant life cycles, e.g., apple [11], tomato [12], cabbage [13], squash [14], carrot [15], chrysanthemum [16] and kiwifruit [17]. However, the networks underlying gene-to-metabolite are largely unknown, and the relationship between expression levels of carotenoid pathway genes and carotenoid levels in green leafy vegetables is not yet established.

In most plants, carotenoids usually accumulate in specialized plastids (chromoplasts) in organs such as roots, tubers, flowers and fruits and cause characteristic orange or yellowish coloration [18]. In photosynthetic tissues, carotenoid accumulation occurs in the chloroplasts, where they act as accessory pigments in light harvesting and protect against photooxidative damage [19,20]. In fact, lutein and β-carotene levels in green leafy vegetables such as lettuce, kale and sweet basil have been shown to significantly correlate with chlorophyll concentration and resulting green color intensity [21–23].

More than 700 types of carotenoids have been identified in photosynthetic organisms [4]. Carotenoids are involved in light harvesting and, as such, contribute to effective photosynthesis; they are also involved in non-photochemical quenching and protecting plant tissues from excess light or heat stress. In higher plants, carotenoids are synthesized through mevalonate pathway from Acetyl-Co-A in cytosols and non-mevalonate pathways that starts from pyruvate in chloroplasts. While these pathways are highly conserved, the regulation of carotenoid biosynthesis is tissue- and species-specific [2]. Both pathways result in C5-isopentenyl pyrophosphate (IPP), the starting point of carotenoids biosynthesis. The four molecules of IPP undergo condensation to form a single C20-geranylgeranyl diphosphate (GGPP) molecule [2,18,24]. GGPP is converted to phytoene by the phytoene synthase (PSY) and subsequently converted to lycopene via 9,15,9′-tris-cis-ζ-carotene by phytoene desaturase (PDS) and ζ-carotene desaturase (ZDS) [2]. The ends of the linear carotenoid lycopene can then be cyclized by lycopene β-cyclase (LCY-β1) and/or lycopene ε-cyclase (LCY-ε) and subsequently modified by hydroxylation [25]. The combined action of LCY-β1 and LCY-ε allows synthesis of α-carotene from lycopene, while LCY-β alone leads to β-carotene. After synthesis, α-carotene can be hydroxylated to form lutein by either α-ring or β-ring hydroxylase (CHY α, β); β-carotene can also be hydroxylated to zeaxanthin by CHYβ. Zeaxanthin is then transformed into violaxanthin via antheraxanthin by zeaxanthin epoxidase (ZEP) [2,26] (Supplementary Figure S1). In addition to the usual ε-β and β-β branches in lettuce and some other plants, the third ε-ε-branch from lycopene is identified (Supplementary Figure S1) [27,28]. This branch is activated by LCYE in the absence of active LCYB, and a number of unusual carotenes are synthesized from ε-carotene, including delta-carotene, epsilon-carotene and lactucaxanthin (epsilon,epsilon-carotene-3,3′-diol).

A range of molecular mechanisms regulate carotenoid biosynthesis and accumulation in plastids during plant development. Deoxyxylulose 5-phosphate synthase (DXS) and deoxyxylulose 5-phosphate reductoisomerase (DXR) catalyze the first two enzymatic steps of the methylerythritol 4-phosphate (MEP) pathway to supply the isoprene building-blocks of carotenoids and may be rate-limiting in different species. Indeed, the overexpression of DXS and DXR increases carotenoid and chlorophyll leaf content in *Daucus* and *Nicotiana* but has no affect in, e.g., *Lavandula* [29]. Differences in the transcript levels of the genes involved in carotenoid metabolism can be directly related to the level of carotenoid synthesis or degradation in many species, suggesting key roles for biosynthetic control at the transcriptional level [29,30]. For example, carotenoid accumulation in harvested ripe apples can be predicted during fruit development from increased levels of carotenoid isomerase (*CRTISO*) and *LCY-ε* transcripts [11]. Similar predictive relationships have been found in *Daucus* (carrot), in which increased transcript levels of *LCY-ε* could be correlated with final lutein concentration, while *ζ-carotene desaturase (ZDS)*, *ZDS1* and/or *ZDS2* transcript levels could be correlated with the final lycopene concentration [15].

The presence of chlorophyll-binding proteins and lipoproteins that sequester carotenoids in plastids can affect carotenoid accumulation [31]. Source–sink relationships affecting fluxes of metabolites through biosynthetic pathways also affect carotenoid accumulation in plants [32]. In fruits, the conversion of chloroplasts to chromoplasts during fruit development makes chromoplasts the major storage structures of carotenoids metabolites in ripe fruits [33]. However, in green leafy vegetables, such conversions do not occur, and such plants form carotenoid crystals to increase accumulation capacity [34,35]. A similar mechanism has been identified in cauliflower in which plastid differentiation is affected by a mutant gene *Or*, which causes β-carotene accumulation in plant curds and increases the sink capacity [36]. In addition, the degradation of carotenoids produced at the end of the pathway can affect carotenoid concentrations [37].

Lettuce (*Lactuca sativa* L., Asteraceae) is the most commercially important and widely consumed green salad crop in the world and has significant potential to provide carotenoids and other nutrients for human health [5]. However, lettuce cultivars display large variations in their nutritional composition. For example, the most widely consumed lettuce, the iceberg type, shows the least amount of antioxidant activity of the main lettuce cultivars [23,38,39]. Despite the importance of carotenoids for human health, the homeostasis of carotenoids during the lettuce leaf development remains surprisingly unexplored.

The aim of our research was to understand if carotenoids accumulation was differently regulated in plant with distinct architecture and leaf color. In this study, our objective was to investigate the role of 17 genes involved in carotenoid metabolism in lettuce leaves and to determine their expression in six commercial cultivars at different plant developmental stages. We studied differential gene expression changes and compared them to carotenoids (β-carotene and lutein) and chlorophyll metabolic changes in the inner and outer leaves throughout the plant life cycle. Through a gene-to-metabolic network analysis, we then tested the role of carotenoid biosynthetic and degradation pathways to understand β-carotene and lutein accumulation in lettuce plants for further use in lettuce breeding programs.

2. Results

2.1. Growth and Carotenoid Content Analysis of Commercial Lettuce Cultivars during Plant Development

The development of six contrasting commercial lettuce cultivars was followed from seedling stage (20 DAS) through mature stage (40 DAS) until commercial stage (Comm, 60 DAS)—that is, the usual stage at which lettuce is harvested for consumption. All cultivars had different growth rates, and the biggest difference between cultivars was already evident at 40 DAS, where Veronica and Grand Rapid cultivars had at least two-fold larger leaf areas of the outer leaves compared to Dark Land and Dragoon varieties, while Salinas 88 had the largest inner leaves (Figure 1).

As expected, the biggest difference between the six cultivars was visible at the commercial stage, when plants displayed extreme variation in leaf shape, thickness, leaf size and leaf color. At the commercial stage, leaf color ranged from light green for Grand Rapid and Veronica to dark green for Salinas 88 and Parris Island or dark green/purple for Dragoon and Dark Land (Figure 1).

Figure 1. Development of six lettuce cultivars growing under the same environmental conditions at seedling stage: 20 days after sowing (DAS), 40 DAS and at commercial size (60 DAS).

We quantified the β-carotene, lutein and chlorophyll a, chlorophyll b and total chlorophyll concentrations at seedling, 40 DAS and Comm stages in the inner leaves and at 40 DAS and Comm stage in the outer leaves of six commercial lettuce cultivars (Figure 2). All lettuce cultivars exhibited the lowest concentrations of β-carotene and lutein at the seedling stage (Figure 2A,B). The concentration of lutein (Figure 2A) had nearly doubled by 40 DAS in all cultivars, except in Veronica, which had the lowest lutein content (25 $\mu g\ g^{-1}$ FW). In the inner leaves at Comm stage, the lutein content was either reduced (e.g., Dark Land and Dragoon) or remained unchanged from 40 DAS (e.g., Grand Rapid and Veronica) (Figure 2A). At the same time, the lutein content in the outer leaves was slightly increased at commercial stage comparing to 40 DAS.

As with lutein, the β-carotene level was the lowest at the seedling stage (Figure 2A,B). However, β-carotene displayed the highest variability between the cultivars and leaves. For example, in the inner leaves of Dark Land β-carotene, content increased three-fold at 40 DAS (288.6 $\pm$ 19.7 $\mu g\ g^{-1}$ FW), and then at 60 DAS reduced to 110.4 $\pm$ 14.5 $\mu g\ g^{-1}$ FW, only slightly higher than at 20 DAS (65.2 $\pm$ 9.6 $\mu g\ g^{-1}$ FW). The β-carotene content in the outer leaves in Dark Land also reduced by 50% between 40 DAS and Comm stage. Interestingly, the inner leaves of Dark Land had higher β-carotene levels at 40 DAS compared to the outer leaves. In Veronica's inner leaves, β-carotene level was just slightly increased from 42.5 $\pm$ 6.4 $\mu g\ g^{-1}$ FW to 59.1 $\pm$ 3.85 $\mu g\ g^{-1}$ FW during the first 20 days of maturation, and β-carotene content had not changed by Comm stage.

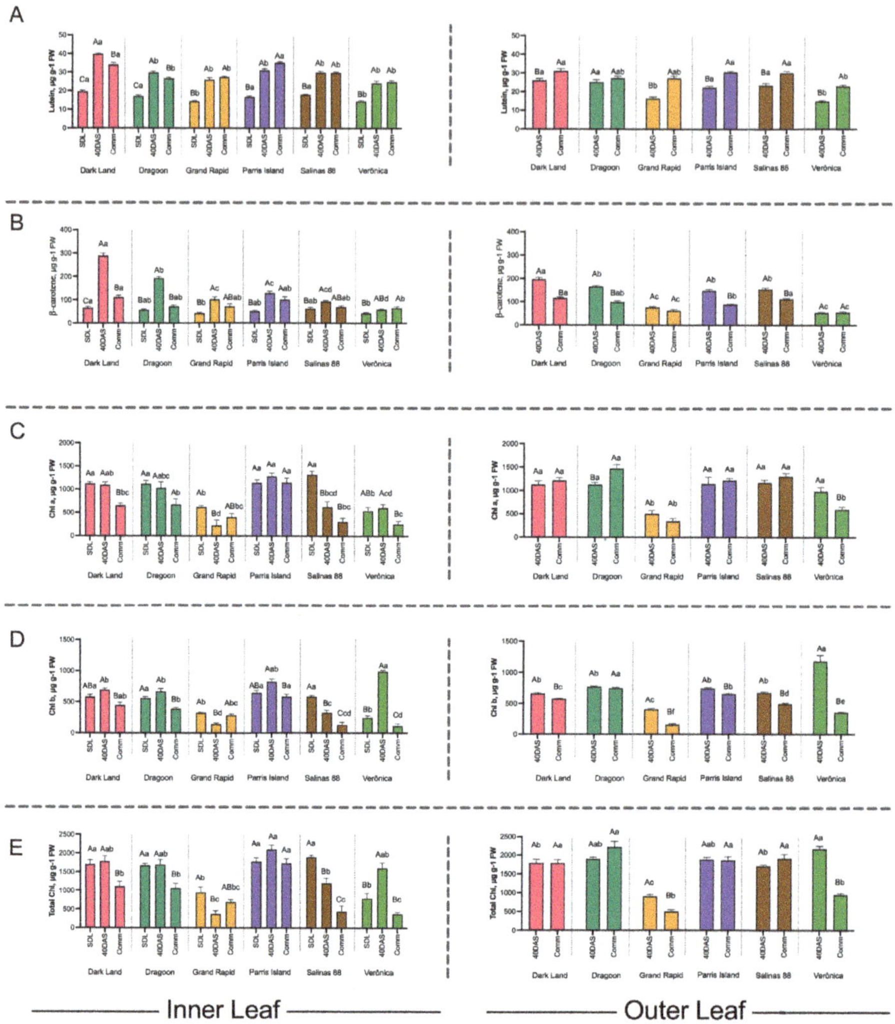

Figure 2. Analysis of chlorophyll and carotenoids in the inner and outer leaves of six lettuce cultivars. (**A**) lutein, (**B**) β-carotene, (**C**) Chla, (**D**) Chlb and (**E**) total chlorophyll content in the inner and outer leaves of six lettuce cultivars at seedling stage (SDL, 20 Days after sowing (Das)), 40 DAS and at commercial size (60 DAS). Error bars are standard deviation of the mean from three biological replicates, each being the pooled sample of at least three technical replicates, at $p = 0.05$ level. Bars with similar capital letters for each cultivar in each leaf position at different time points are not significantly different ($p \geq 0.05$), using one-way ANOVA analysis followed by Tukey pos hoc test. Bars with similar small letters at the same time point in each leaf position between cultivars are not significantly different ($p \geq 0.05$), using one-way ANOVA analysis followed by Tukey pos hoc test.

Grand Rapid and Salinas88 had 1.5–2-fold increase in β-carotene level at 40 DAS for inner leaves (Figure 2B). We observed that dark-leaved varieties had significant reductions in the β-carotene level (Figure 2B) at commercial stage in both inner and outer leaves (like Dark Land, Dragoon, Parris Island), while light-green cultivars Grand Rapid and Veronica had much smaller reductions in β-carotene compared to at 40 DAS. However, the overall β-carotene was the smallest among all cultivars at all developmental stages tested. These findings are critical for lettuce commercial stage definition and further lettuce selection if plants are to be a key source of provitamin A for human consumption [5].

Analysis of Chlorophyll a, b and total chlorophyll (Figure 2C–E) revealed that dark-leaved varieties consistently had the highest chlorophyll levels in the inner leaves up to 40 DAS, with significant reduction of chlorophyll content in the inner leaves at Comm stage in all except Parris Island, likely due to plant architecture. In contrast, chlorophyll in outer leaves remained steadily high or even increased slightly (1.3× in Dragoon) up to 60 DAS. Light-green cultivars, as expected, had the lowest chlorophyll content in the leaves, and chlorophyll content was significantly reduced by the Comm stage. In all genotypes, the concentration of chlorophyll a was higher than chlorophyll b at all leaf ages (Figure 2C,D). Only in Grand Rapids was total chlorophyll higher in inner than outer leaves (Figure 2E). The Pearson correlation between the normalized metabolites (Figure 3A) revealed very strong correlation between lutein and β-carotene concentrations ($R^2 = 0.86$, $p < 0.001$), with two distinct clusters present—cluster 1 for lutein, β-carotene and total carotenoids and cluster 2 for chlorophyll a, b and total chlorophyll (Figure 3A). The β-carotene had negative correlation with chlorophylls in leaves, while lutein concentration in lettuce had weak relationship with leaf color ($R^2 = -0.53$, $p > 0.05$, Figure 3A). Regression analysis confirmed that lutein concentration in leaves could predict β-carotene content, as suggested by the Pearson correlation ($R^2 = 0.497$, $p < 0.001$, Figure 3B, Table S2a), while chlorophyll a, b or total chlorophyll levels have no direct effect on β-carotene content in leaves ($p > 0.05$, Table S2a) if the model was excluding the plant developmental stage, leaf's location or variety. However, the full model, accounting for effect of variety, lettuce developmental stage and location of leaves used for the samples, had a better correlation coefficient ($R^2 = 0.77$, $p < 0.001$) and illuminated the importance of the Chlb ($p = 0.01$) and total chlorophyll as a predictor of β-carotene content in lettuce.

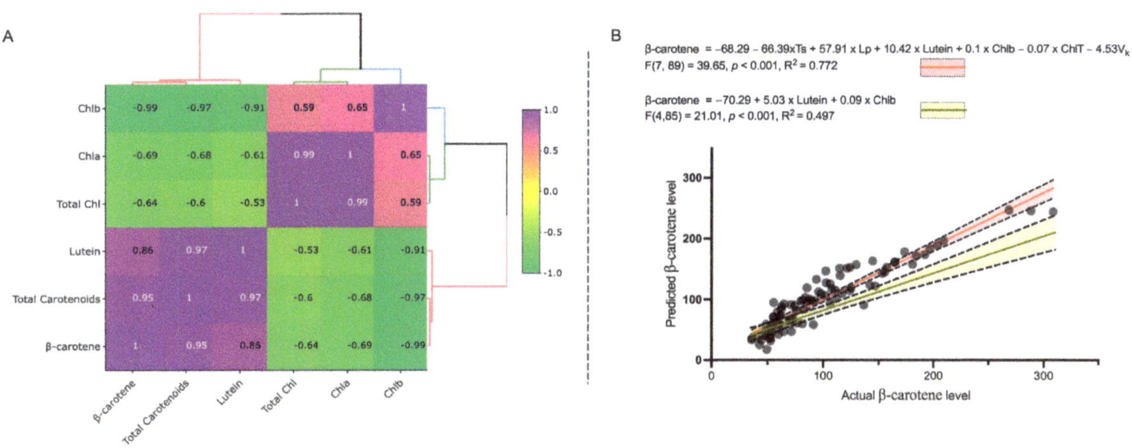

Figure 3. Pearson correlation on normalized metabolites in six lettuce varieties from the inner and outer leaves during three developmental stages revealed correlation between β-carotene and lutein and confirmed inverse correlations between carotenoids and chlorophyll contents in lettuce plants (**A**). Regression analysis showing that β-carotene content can be predicted by lutein content in plants (blue line) and strongly depend on plant genotype (Vk), plant developmental stages (Ts), sample location (Lp) and chlorophyll b level (**B**).

2.2. Relative Expression of Carotenoid Biosynthetic Pathway Genes during Lettuce Plant Development

To determine whether the expression level of regulatory carotenoid biosynthetic genes affects carotenoid accumulation [2,24] in different lettuce cultivars, we searched for candidate carotenoid biosynthetic genes in the lettuce genome project sequence (https://lgr.genomecenter.ucdavis.edu, last accessed on 17 June 2023). We identified 17 candidate genes (Table S1) encoding key enzymes of the core biosynthesis of carotenoids. In several cases, these genes were present as low copy number families: for example, only two copies of geranylgeranyl diphosphate (*GGPS* and *GGPPS*), phytoene synthase (*PSY1* and *PSY2*) and lycopene β-cyclase (*LCY-β1* and *LCY-β2*) were identified.

The expression patterns of different genes expression changed during plant development (Figures 4, S2 and S3).

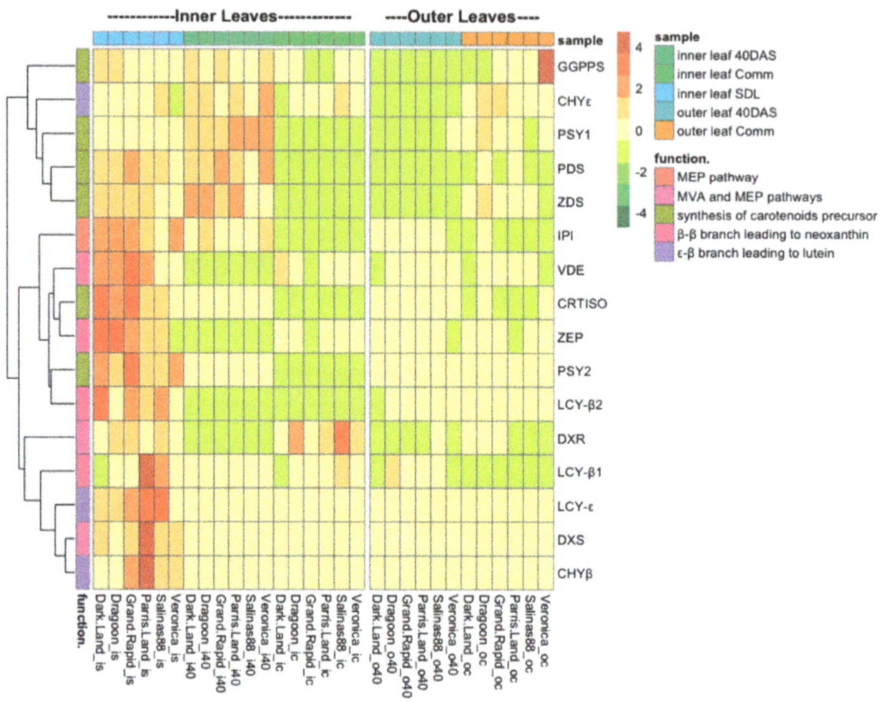

Figure 4. Analysis of carotenoids biosynthesis genes expression in the inner and outer leaves of six lettuce cultivars was measured at seedling stage (20 days after sowing (DAS)), 40 DAS, and commercial size (60 DAS). Gene expression was measured relative to *UBIQUITIN* (n = 3 biological replicates for each sample).

The heat maps generated from the gene analysis expression analysis revealed that transcription of the carotenoid genes is modulated in a largely uniform manner in response to developmental stages but did vary between leaf position (Figure 4). In general, the expression of most (13/17) carotenoid biosynthesis genes declines progressively throughout plants development. The highest fold change of expression levels between SDL and 40 DAS stages was noticed for *DXS* (358 ± 101-folds decline, depending on genotype, Figure S2); *LCY-B2* (110 ± 24 folds decline, depending on genotype, Figure S2); *LCY-ε* (232 ± 49 folds decline); and *CHY-β* (170 ± 54 folds decline, depending on genotype, Figure S2). Other genes involved in carotenoids biosynthesis, such as IPI, declined circa 2.4-fold, comparing expression at SDL to 40 DAS stages in the inner leaves, and the highest decline was noticed at the Comm stage (10.4 ± 2.7 folds decline, depending on genotype, Figure S2). The four exceptions are ε-ring hydroxylase (*CHY-ε*), phytoene desaturase (*PDS*), ζ-carotene desaturase (*ZDS*) and phytoene synthase (*PSY1*), whose expression levels remained the same at 40 DAS and were reduced only at the Comm stage (Figure S2). Notably, some genes had different patterns between the inner and outer leaves. For example, geranylgeranyl diphosphate (*GGPPS*), ε-ring hydroxylase (*CHY-ε*) and phytoene synthase (*PSY1*) genes were higher at the Comm stage than at 40 DAS in the outer leaves, unlike the inner leaves, where these genes were down-regulated in all cultivars (Figure 4).

To obtain a global perspective on the expression pattern of the gene involved in lutein, β-carotene and chlorophyll biosynthesis during plants development, we performed principal component analysis (PCA) on their normalized expression levels (Figure 5). The first three principal components (PCs) accounted for 75.71% of the total variance among the samples (Figure 5, left and right). Time of sampling had the highest effect on the expression profile in the inner and outer leaves. The strongest correlation was driven by the age of plants and sample source (inner or outer leaves).

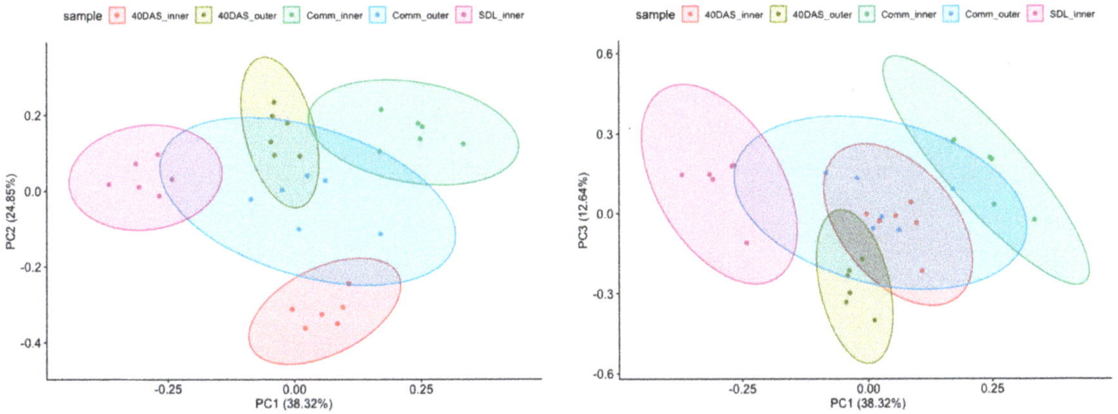

Figure 5. PCA analysis (PC1:PC2 and PC1:PC3) of gene expression at different developmental stages in inner and outer leaves of six lettuce cultivars.

We further analyzed the patterns of gene expression at each developmental stage (Supplementary Figure S4). Some interesting patterns were observed. For example, CRTISO, a key enzyme encoding lycopene synthesis, was co-expressed with many early genes in the carotenoids pathway when all samples were analyzed across all developmental stage, but when analyzed at the early developmental stage, a significant positive correlation between *CRTISTO* was observed only with the genes responsible for zeaxanthin degradation, *VDE* and *ZEP* (r = 0.89 and r = 0.90, respectively). Genes involved in β-carotene biosynthesis, *LCYβ1*, and *LCYβ2*, were co-expressed with genes involved in β-carotene conversion into abscisic acid. In addition, *LCYβ1* expression had significant positive correlation to *LCY-ε* (r = 0.87, $p < 0.05$), *CHY-ε* (r = 0.95, $p < 0.05$), *CHY-β* (r = 0.94, $p < 0.05$) and *ZEP* (r = 0.83, $p < 0.05$) in the inner leaves at the commercial stage, while in the outer leaves at the same developmental stage, *LCYβ1* expression had significant positive correlation only to *VDE* gene (r = 0.91, $p < 0.05$) (Supplementary Figure S4).

2.3. Regulation of Carotenoid Biosynthetic during Lettuce Plant Development

To elucidate the role transcriptional regulation plays in coordinating carotenoids biosynthesis, in particular lutein and β-carotene, and the synthesis of other functionally related compounds like chlorophyll a and chlorophyll b, we summarized the normalized gene expression of all lettuce cultivars and presented the metabolic map of gene expression pattern and fold changes of respective metabolites (Figure 6) and performed correlation analysis between chlorophylls and carotenoid synthesis through gene expression analysis and carotenoid and chlorophyll content (Supplementary Table S3). Deoxyxylulose 5-phosphate synthase (*DXS*) and deoxyxylulose 5-phosphate reductoisomerase (*DXR*) are the enzymes that catalyze the first steps of the pyruvate conversion to 1-deoxy-D-xylulose-5-phosphate and 2-C-methyl-D-erythriol-4-phosphate within the methylerythritol 4-phosphate (MEP) pathway to supply the isoprene building-blocks of carotenoids [29].

The transcript level of *DXS* was significantly higher in the inner leaves at seedling stage (20DAS) compared with other tissues of both stages, suggesting its relevant roles in photosynthetic young tissues. A single copy-downstream gene of *DXS*, *DXR*, was strongly expressed in the inner leaves at 20 DAS, but the highest expression level was noticed at the commercial stage, while in the outer leaves the *DXR* gene was downregulated at 40 DAS and at the commercial 60 DAS stage. The expression of genes encoding early pathway enzymes controlling the metabolite flux into the carotenoid pathway and chlorophyll synthesis, in particular isopentenyl pyrophosphate (IPP) isomerase (IPI) and geranylgeranyl pyrophosphate synthase (*GGPS*) which leads to gerangerandyl diphosphotase (GGPP) synthesis [40]) was also high at early developmental stages. Hence, the transcriptional regulation of the GGPPS gene serves as an important regulatory node in coordinating carotenoid and chlorophyll a and chlorophyll b biosynthesis.

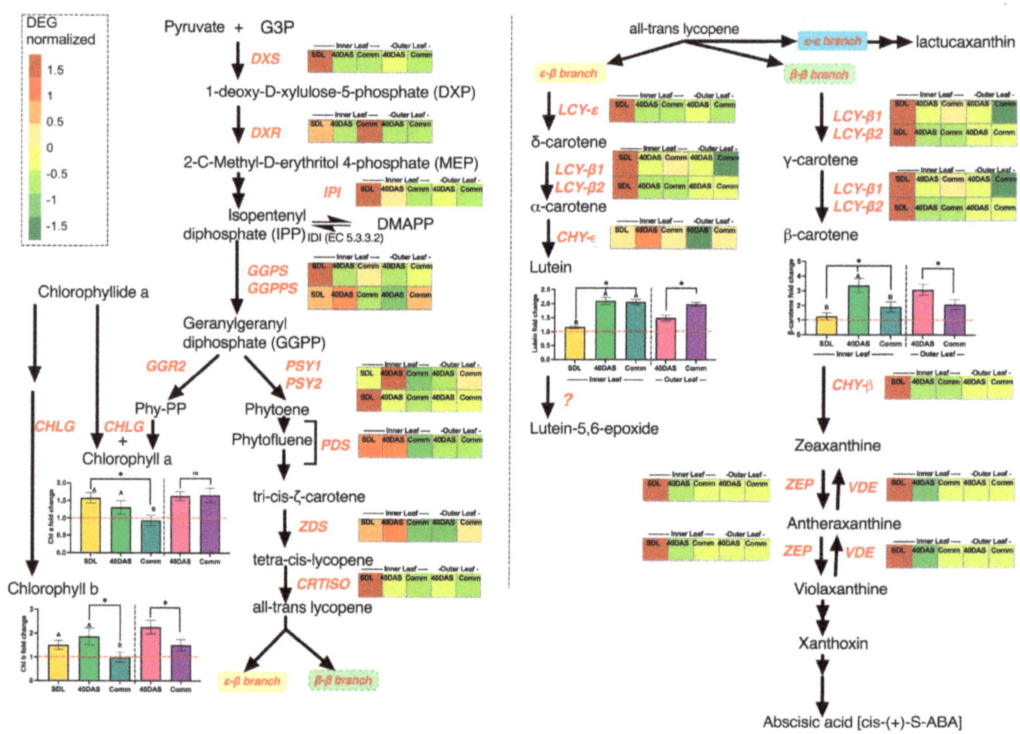

Figure 6. Metabolomic and transcriptomic pathways of carotenoid biosynthesis in *Lactuca sativa*, Asteraceae. Genes that encode pathway enzymes are shown in red italic letters. The heatmap is the normalized average of all six varieties at three developmental stages in inner and outer leaves. Abbreviations: GA3P, glyceraldehyde-3-phosphate; DXS, 1-deoxyxylulose 5-phosphate synthase; DOXP, D-1-deoxyxylulose-5-phosphate; DXR, 1-deoxyxylulose 5-phosphate reductoisomerase; MEP, 2-C-methyl-D-erythritol-2,4- cyclodiphosphate; IPP, isopentenyl diphosphate; IPI, IPP isomerase; GGPP, geranylgeranyl diphosphate; GGPS, GGPP synthase; PSY1, phytoene synthase; PDS, phytoene desaturase; ZDS, ζ-carotene desaturase; CRTISO, carotenoid isomerase; LCY-B, lycopene β-cyclase; LCY-E, lycopene ε-cyclase; CHYB, β-ring hydroxylase; CHYE, ε-ring hydroxylase; ZEP, zeaxanthin epoxidase; VDE, violaxanthin deepoxidase; CHLG, Chlorophyll synthase; GGR2, geranylgeranyl reductase2. Bars with similar capital letters at different time point means that samples are not significantly different ($p \geq 0.05$), using one-way ANOVA analysis followed by Tukey pos hoc test. * Means that samples are significantly different based on the *t*-test.

The peak in expression of *GGPPS* at 20 DAS and 40 DAS could explain the increased chlorophyll a and chlorophyll b changes in the inner leaves, although regulation of chlorophyll synthesis in the outer leaves at other levels of regulation such as post-transcriptional regulation should also be considered [24].

The phytoene synthase enzyme that converts gerangerandyl diphosphotase into phytoene (encoded by the two *PSY* genes in lettuce) is the first committed key-limiting step in carotenoid biosynthesis [2] and controls the expression of genes involved in carotenoids biosynthesis [41]. *PSY2* is highly expresses in the inner leaves at the seedlings stage of lettuce plants, while *PSY1* is highly expressed at 40 DAS and Comm stage in the inner and outer leaves, respectively.

In the subsequent steps of phytoene into lycopene transformation, plants employ two desaturases, phytoene desaturase (PDS), forming 11 and 11′ cis double bonds, and ζ-carotene desaturase (ZDS), forming 7 and 7′ cis double bonds [42]. The upregulation of any of the desaturation-related genes impacts the homeostasis between them, and the expression level is upregulated in adjustment [2,24]. Similarly, we observe that in lettuce, *PDS* and *ZDS* were highly upregulated in the inner leaves at the seedling stage and 40 DAS, while outer leaves had lower relative expression level. In the next stage the carotene isomerase (*CRTISO*), upregulated at the seedling stage, is employed to complete conversion into all-trans-lycopene containing 11 C-C bonds essential for light energy absorption and carotenoids functioning [43].

Cyclization of all-trans-lycopene bifurcates the pathway into two branches: the β-β branch leading to b-carotene, violaxanthin and, finally, abscisic acid; and the ε-β branch leading to lutein biosynthesis [25]. In lettuce and some other plant species, the cyclization could have the third ε-ε-branch, leading to lactucaxanthin biosynthesis (Supplementary Figure S1) [27,28]. The ε-ε-branch was out of the scope of this research, as we mainly focused on β-carotene and lutein biosynthesis and transcriptional regulation.

While key carotenoid biosynthesis genes (*DXS, DXR, CRTISO*) as well as first enzymes of ε−β branch were highly upregulated at 20 DAS, the lutein content was the lowest at this developmental stage across all genotypes, ranging from 13 to 21 $\mu g\ g^{-1}$ FW (Figure 2). The only enzyme of the ε−β branch wherein upregulation was corresponding a 2–2.1-fold lutein increase at 40 DAS was *CHY-ε*, the last enzymatic stage of lutein biosynthesis, wherein relative expression at 40 DAS was increased 1.95 ± 0.48-fold as compared to the 20 DAS relative expression level (Figure 6).

In the β-β branch, the enzyme lycopene β-cyclase (LCYB), encoded in lettuce by *LCY-β1* and *LCY-β2*, catalyzes the formation of two β-rings at both ends of the lycopene molecule, resulting in β-carotene production. Again, the highest expression of *LCY-β1* and *LCY-β2* was observed at 20 DAS (seedling stage), while β-carotene accumulation was on average 3-fold increased (Figure 6) at 40 DAS (up to 288.6 ± 19.7 $\mu g\ g^{-1}$ FW, Figure 2A) compared to at 20 DAS (42.5 ± 6.4 $\mu g\ g^{-1}$ FW to 65.2 ± 9.6 $\mu g\ g^{-1}$ FW, depending on genotype, Figure 2A). At the Comm stage, it had fallen to just half its 40 DAS levels, a trend observed in both inner and outer leaves.

Next, we performed a co-expression correlation analysis using *LCY-β1* and *LCY-β2* genes as drivers to determine the level of co-expression that genes encoding lycopene β-cyclase share with all other analyzed genes (Supplementary Table S4). The highest significant co-expression pattern was observed for the β-ring hydroxylase (*CHY-β*) gene ($p < 0.01$, Table S3), encoding the enzyme involved in β-carotene conversion to zeaxanthine, and for zeaxanthin epoxidase (*ZEP*)/violaxanthin deepoxidase (*VDE*) genes, regulating zeaxanthine biosynthesis [2,25,44].

The second pathway after bifurcation, the ε-β branch is controlled by LCYE (Lycopene Epsilon Cyclase, encoded by single gene *LCY-ε*) and Lycopene Beta Cyclase [25]. The final step of ε- and β-ring hydroxylation of α-carotene are catalyzed by CHYE, ε-ring hydroxylase, yielding lutein. The expression correlation analysis using ε-ring hydroxylase (*CHY-ε*) as the driver gene revealed a significant co-expression pattern between *CHY-ε* and the *LCY-ε* and *LCY-β2* genes ($p < 001$). *CHY-ε* was up-regulated in the inner leaves at all developmental stages and in the outer leaves at the commercial stage, which corresponded to the increased of lutein accumulation at 40 DAS and 60 DAS.

We noted that *CRTISO*, *CHY-β* and *VDE/ZEP* were co-expressed if *CHY-ε* was used as the driver (Supplementary Table S5), which could indicate that both pathways were controlled by levels of their common precursor (all-trans-lycopene), and upregulation of one branch would lead to reduced biosynthesis of compounds in the second branch.

In summary, the co-expression analysis suggested that to ensure high carotenoids content in leaves, the metabolites sink from β-carotene and lutein needs to be regulated. For example, by preventing zeaxanthin accumulation (e.g., via *CHY-β* gene silencing or overexpression of *VDE*), we could prolong accumulation of β-carotene and lutein in the leaves and, as a result, to delay the senescence symptoms appearance in lettuce leaves.

3. Discussion

3.1. Changes in Carotenoid Accumulation Related to Lettuce Genotype and Developmental Stages

The regulation of carotenoid accumulation occurs at multiple levels: transcriptional, post-transcriptional, post-translational and storage/degradation (see review in [19]). Here, we focused only on the transcriptional regulation of carotenoid biosynthesis in six lettuce cultivars during three developmental stages, since if transcriptional regulation on carotenoids biosynthesis is only part of the whole picture, it should not be neglected. In leafy vegetables, the biosynthesis and accumulation of carotenoids occurs during plant cultivation. Carotenoids are accessory pigments, are involved in light harvesting and are essential for effective photosynthesis. The steady-state concentrations of carotenoids in leaves are considered to result from a balance of their biosynthesis and degradation [45], and changes in carotenoid content are usually associated with photosystem imbalance [46]. Carotenoid content can change in response to the type and source of light and external factors like temperature [2]. Green lettuce cultivated in greenhouses supplemented with UV-A and UV-A+UV-B has higher β-carotene and lutein concentrations compared to those cultivated without UV- radiations [47,48], as found in our experiments. In this study, metabolite data obtained by reverse HPLC was combined with gene expression data generated by qRT-PCR analysis to characterize carotenoids biosynthesis during lettuce plant development across cultivars which display contrasting green color intensity.

To better understand the relationship between expression levels of carotenoid biosynthesis pathway genes and carotenoid levels, we investigated the expression levels of gene-encoding enzymes involved in the control of carotenoid accumulation in inner and outer leaves of the lettuce plants during different developmental stages, after identifying these from the lettuce genome. Lutein and β-carotene levels were measured in different-age leaves for all tested cultivars. The levels of these carotenoids at commercial size were comparable with carotenoid levels detected in other green leafy vegetables, such as basil, kale and Brassica oleracea cultivars [21,22,49]. In a previous study [23], levels of β-carotene and lutein were quantified in three of the lettuce genotypes used in our study. We report similar levels to those detected by Mou [23], except for the cultivar Salinas 88, which we found to contain almost ten times greater amounts of both carotenoids in leaves at similar growth stages, which could be attributed to our soil mineral composition, light intensity and the use of outer leaves which had higher green color intensity than the leaves used previously [23].

The lettuce cultivars investigated displayed a significant variation in chlorophyll, β-carotene and lutein levels. It was previously reported that green color intensity could serve as a proxy indicator of β-carotene concentration in lettuce [23]. The chlorophyll a is synthesized through geranylgeranyl diphosphate (GGPP) (Supplementary Figure S1), and mutations in GGPPs lead to alteration of both chlorophyll and carotenoids content in plants [19]. The Pearson correlation analysis (Supplementary Table S3) indicated that the *PSY1* gene in inner leaves at the commercial stage is positively correlated with total chlorophyll in outer leaves ($r = 0.85$, $p < 0.05$), which suggests an indirect relation to β-carotene. In our study, we also observed negative correlation between chlorophyll levels and β-carotene and lutein contents (Figure 3B). The lutein concentrations had weak negative correlation with total chlorophyll content in leaves ($R^2 = -0.53$, $p > 0.05$, Figure 3A), meaning that lutein content could not be predicted by color intensity levels, appealing though this might be for consumers. For example, while 'Dragoon' and 'Salinas 88' had different leaf color between inner and outer leaves, these cultivars had the same concentration of lutein in the inner and outer leaves, indicating that that lutein concentrations are unrelated to leaf color intensity.

During senescence, the chlorophylls are degraded (Figure 2), and chlorophyll a is decomposed into phytol and pheophorbide a [19]. In most lettuce cultivars, we found carotenoid and lutein concentrations to peak at 40 DAS (Figure 2), while at the usual commercial stage of harvesting, 60 DAS, when chlorophyll was already degraded, carotenoid levels had also started declining, leading to accumulation of abscisic acid (Figure 6). A similar pattern has been observed previously in *Arabidopsis thaliana* leaves, where a 50% increase in carotenoid levels was observed in newly-formed adult leaves compared to older ones [50], and therefore, carotenoid accumulation was associated with the developmental stage.

Leaf maturation is a multicomponent process that includes morphological and cellular changes, like the increase in chloroplast size and chloroplast density per cells, and is tightly regulated at the molecular level. For all lettuce cultivars tested, chlorophyll and carotenoid concentrations varied according to plant development stage. Thus, chlorophyll and chlorophyll b were the highest at the seedling and 40 DAS stages, followed by reduction in all cultivars (Figure 2) when plants reached the commercial size. Although all photosynthetic pigments will eventually degrade, chlorophyll was lost more rapidly than β-carotene, while lutein remained relatively stable (Figure 2) like in senescent Arabidopsis and tobacco leaves [48,50,51]. The role of ABA in carotenoids degradation is supported by co-expression of *CHY-β* gene, encoding the enzyme involved in β-carotene conversion to zeaxanthin with zeaxanthin epoxidase (ZEP) gene, essential for ABA biosynthesis (Supplementary Table S4). The transition of chloroplasts into chromoplasts results in an enhanced storage capacity for β-carotene [52] and accumulation of carotenoids. In all lettuce cultivars, chlorophyll concentrations were lower in the inner leaves, and the peak in lutein and β-carotene accumulation at 40 DAS vs 20 DAS could be consequence of this transition. In the outer leaves, this reduced chlorophyll content was noticeable only in Grand Rapid and Veronica, which both have similar plant architecture. In both cases, reduction in chlorophyll a and chlorophyll b was associated with lutein and β-carotene accumulation (Figure 2). The reduction of β-carotene content at 60 DAS as compared to 40 DAS, and similar or higher levels of lutein at 40 and 60 DAS in the inner and outer leaves could be explained by the cleavage of β-carotene branch carotenoids to produce strigolactones and ABA, which further promote leaf senescence [24,53].

3.2. Relationship between Biosynthetic Gene Expression and Carotenoid Concentration

Upon exposure to light, both carotenoids and chlorophyll are synthesized, and co-expression of chlorophyll- and carotenoid-related genes is evident in many plants [24,40]. The light-green leaf color in the Verônica and Grand Rapids cultivars was consistent with the lower amounts of carotenoids detected, while the dark green color observed in 'Dark Land' and 'Parris Island' was consistent with higher concentrations of detected carotenoids. The relationship between color intensity and carotenoid levels was also evident for 'Dragoon' and 'Salinas 88', which displayed significant differences in carotenoid concentrations in outer and inner leaves. This correlation between carotenoid levels and green color intensity is likely related to the protective role of carotenoids in the photosynthetic matrix, since carotenoids are essential to protect chlorophyll from degradation [48,54,55].

The rate-limiting enzyme, DXS, is essential for controlling carotenoid biosynthesis flux [29]. The negative correlation between DXS levels and total chlorophyll is consistent with reduced levels of expression at later stages for a gene which is involved in the first step of the biosynthetic pathway (Figure 6, Supplementary Table S3). Indeed, DXS expression levels were significantly reduced in all genotypes compared to other tested genes, suggesting a significant reduction in pigment synthesis in older tissues (Figure 6). Similar DXS reduction during developmental stages was observed in red pepper fruits [56].

The expression of *PSY1* and *PSY2*, which encode the PSY enzymes that catalyze the first committed step in carotenoid biosynthesis, was also found to be positively correlated with chlorophyll and carotenoid accumulation (Supplemental Tables S4 and S5). A similar relationship between carotenoid synthesis and *PSY1* expression has been shown in *Brassica napus* and in *Arabidopsis* [57,58] and *PSY1* expression in ripe tomato fruits [59].

A correlation of *CRTISO* and *LCY-β2* transcript levels with chlorophyll and carotenoids was not observed during the early stages of development, but it was detected at the commercial size stage (Figure 6, Supplementary Table S3). A similar pattern was observed in apple fruits, where the expression of some genes predominates after pollination but is reduced before the ripening stage [11]. The differential expression of *CRTISO* between inner and outer leaves at the commercial stage suggests that this gene could be associated with a 'pause' in carotenoid synthesis. In inner leaves, the *CRTISO* gene showed higher levels of expression compared to outer leaves while the carotenoid synthesis was found to be constant, following the same pattern of accumulation between cultivars (Figure 2, Supplementary Figure S2). In contrast, in outer commercial leaves, the *CRTISO* expression was different, with higher transcript levels detected in dark green cultivars. This observation was confirmed by the positive and negative correlations between the *CRTISO* gene expression levels and carotenoids accumulation in inner and outer leaves, which could help to explain the lower carotenoid levels detected in the plants at the commercial stage (Figure 2, Supplementary Figures S2 and S3 and Supplementary Table S3).

The difference in carotenoid accumulation between lettuce cultivars could also potentially be explained by degradative enzyme action. Some early pathway genes displayed a reduction in total transcript levels at the commercial size stage (Figure 4, Supplementary Figures S2 and S3). An increase in total transcript levels of VDE observed at 60 DAS (when cultivars had reduced carotenoid concentration compared to 40 DAS) could suggest that carotenoid degradation had occurred in the internal leaves, resulting in total carotenoid concentration being reduced at 60 DAS. Our results also indicate a reduction in β-carotene levels in the inner leaves of 'Salinas 88' and 'Dragoon' when they reached the commercial size. The production of carotenoids after synthesis of phytoene in these cultivars may be affected by cleavage enzymes during carotenoid synthesis. A similar mechanism has been suggested for chrysanthemum, in which white cultivars had similar expression levels of gene-encoding intermediate enzymes of the pathway as in a yellow cultivar but increased expression of carotenoid cleavage genes [16].

Total carotenoid levels in plant tissues result from an interplay between carotenoid biosynthesis, sequestration, storage and degradation. Key genes which could affect carotenoid synthesis have been identified. Transcript levels of *LCY-ε* can be directly related to increased lutein concentrations in several species [11,15,16]. In our experiments, the pattern of transcript level changes was similar among all cultivars, where critical genes had similar levels of expression. Hence, to explain the different levels of carotenoids and chlorophyll in lettuce cultivars, additional mechanisms of regulation beyond transcript levels should be considered.

In expression correlation analysis using *CHY-ε* as the driver gene for lutein biosynthesis, we found significant correlation between *CHY-ε* and *LCY-ε* expression, leading to lutein accumulation (Supplementary Tables S3 and S5). We also found co-expression between *LCY-ε* and *DXS* genes which regulate the carotenoids biosynthetic pathway. This could suggest that the synthesis of lutein is predominantly controlled by the expression of early pathway genes and that a later degradation process occurs during plant development. The *LCY-β1* and *LCY-β2* genes, encoding the LCYB enzyme which participates in controlling the flux of metabolites through the $\varepsilon-\beta$ and $\beta-\beta$ branches [58,60,61], were also highly expressed at early stages of development. The lack of correlation between LCYB expression and β-carotene accumulation points towards the existence of an upper threshold in *LCY-β1* and *LCY-β2* transcript levels and the accumulation of β-carotene [25]. The downregulation of LCY-β2, and reduced expression of *LCY-β1* at both 40 and 60 DAS in the inner leaves as compared to the seedling stage, opposite to lutein and β-carotene accumulation, suggests different levels of carotenoid regulation at the later stages of plant development—for example, carotenoid biosynthesis via the retrograde signaling pathway from plastids to the nucleus [52]. The co-expression with *CHY-β* gene suggests that flux toward ABA biosynthesis is an important regulatory mechanism that controls β-carotene biosynthesis and degradation in adult leaves.

To summarize, in our research, we compared gene expression patterns to carotenoids (β-carotene and lutein) and chlorophyll metabolic changes in the inner and outer leaves throughout the plant life cycle to understand if carotenoids accumulation is differently regulated in plants with distinct architecture and leaf color. We did not identify master regulators responsible for divergent expression of carotenoid biosynthetic genes leading to variety led changes in carotenoid contents. However, our data prompt to a clear natural variation in carotenoid biosynthesis pathway, likely controlled at least partly at the gene expression level. A major finding in our paper is the identification of three genes (*GGPPS*, *PSY1* and *LCY-β1*) whose transcript levels can be used to predict carotenoid levels at the commercial harvested stage for lettuce. Our results also suggested that plants selected based on the delayed leaf senescence phenotype will also have a higher concentration of β-carotene and lutein for extended period.

4. Materials and Methods

4.1. Plant Material and Growing Conditions

Six commercial lettuce cultivars were chosen for analysis based on differences in plant architecture and color intensity: Dark Land Cos MT (romaine type, dark green), Dragoon (mini romaine, dark green), Grand Rapids (batavia type, light green), Parris Island (romaine type, medium green), Salinas 88 (crisphead, medium-green intensity) and Verônica (batavia, light green) (Supplementary Table S6). Plants were grown at the University of Galway, Ireland, in a plant growth chamber under fluorescent lighting with light intensities of 300 $\mu mol \cdot m^{-2} \cdot s^{-1}$ under long-day conditions (16 h light–8 h dark). Lettuce seeds were sown on soil in 200 mL pots and transplanted to three-liter pots at the seedling size (20 days after sowing (DAS)). The experiments were conducted in a complete randomized design with three replications and three plants per genotype per replication. All plants were grown at the same time and in the same chamber, with the same light intensity and temperature to minimize environmental variability. Leaf samples of each cultivar were harvested at three different stages, corresponding to seedling size (transplanting time), 40 DAS and at commercial size (60 DAS). Samples at each plant stage represent an independent group of plants. At the 40 and 60 DAS stages, the outer leaves (first external leaves) and inner leaves (a leaf from the seventh inner layer of leaves from the outside of the head) were sampled. In each lettuce plant, sampling was done on three different leaves around the plant in the same location in each outer or inner leaf, corresponding to a total of three pooled samples per plant, and nine pooled samples per sample. The harvested samples from each sampling time and leaf stage were frozen immediately and stored at −80 °C until analysis.

4.2. Carotenoid Extraction and HPLC Analyses

Carotenoid pigments were extracted following procedures described by Norris [62] with some modifications. Leaf tissue (0.3 g) was ground under liquid nitrogen in a porcelain mortar and transferred to a 2 mL centrifuge tube with one glass bead. Then, 200 μL of 80% *v*/*v* acetone was added before adding ethyl acetate (200 μL), and the tubes were agitated at 30,000 rpm for 1 min in a tissue-lyser (RETSCH MM200—Qiagen, Manchester, UK). Water (140 μL) was added, and the mixture was agitated again at 30,000 rpm for 1 min and then centrifuged at 15,800× *g* for 5 min in a microcentrifuge (Heraeus Fresco 17—ThermoScientific, Ireland). The upper phase, containing carotenoids, was then transferred into a new tube. The samples were extracted at least three more times, adding 200 μL of ethyl acetate, agitating, centrifuging at 15,800× *g* for 5 min and removing the upper phase until the precipitate did not have any visible green color. The combined ethyl acetate phases were vacuum dried in a centrifugal evaporator (miVac—GeneVac SP Scientific, UK). The dried samples were subsequently redissolved in 1.5 mL of 0.8% of BHT/acetone [17] and analyzed by the reverse phase of high performance liquid chromatography (HPLC). The HPLC system (Alliance, Waters Co., Milford, Mass.) consisted of a separation unit (model 2695), YMC 4.6 × 10 mm C30 guard cartridge and YMC RP C30 column (3 μm, 250 × 4.6 mm—YMC, Wilmington, North Carolina, USA). The column temperature was 25 °C; samples were kept in a 4 °C sample cooler, and a 50 μL aliquot was injected into a 1 mL min^{-1} flow rate. The elution was performed using a mobile phase comprising solvent A (MeOH), solvent B [H_2O/MeOH, 20:80) containing 0.2% *w*/*v* ammonium acetate] and solvent C (tert-butyl methyl ether). The elution gradient was a reduced-time version of that described by [11]. The gradient started with 95% A/5% B for 2 min, decreasing to 80% A/5% B/15% C between 2 and 10 min, decreasing to 30% A/5% B/65% C by 15 min, decreasing to 25% A/5% B/70% C at 20 min, and returning to 95% A/5% B at 25 min. The β-carotene and lutein picks were identified by comparing the retention time (RT) and absorption spectra of individual peaks within each sample with the β-carotene and lutein standards RT (Sigma-Aldrich, Arklow, Ireland). To ensure that RT was not changed/masked by other plants metabolites, β-carotene and lutein standards were also spiked into plant samples followed by RT comparison to standards containing no plant extracts (Supplementary Figure S5). Plant samples extraction losses were determined from the initial concentration of the internal standard trans-β-Apo-Carotenal (Sigma-Aldrich, Ireland). The linear portions of the standard curves were used to convert the integrated areas of both β-carotene and lutein as μg per g of fresh weight.

4.3. Chlorophyll Extraction and Quantification

Chlorophyll pigments were extracted by crushing the samples in a tissue-lyser (RETSCH MM200—Qiagen, UK) in tubes with one glass bead for 1 min at 30,000 rpm. Then, 2 mL of 96% *v*/*v* ethanol was added, and the samples were agitated again for 1 min at 30,000 rpm and left for 24 h at 4 °C in a 2 mL centrifuge tube. Next, the absorbance of the samples was measured using spectrophotometer (Nanophotometer—Inplen, Germany) at 649 and 665 nm wavelengths. The chlorophyll a (*Chla*) and *b* (*Chlb*) concentrations were calculated by the following formulas:

$$Chl_a = 13.95A_{665} - 6.88A_{649} \quad (1)$$

$$Chl_b = 24.96A_{649} - 7.32A_{665} \quad (2)$$

The total chlorophyll was calculated by the sum of *Chla* and *Chlb*, and the values were determined as μg g^{-1} of fresh leaves.

4.4. RNA Extraction and cDNA Synthesis

Total RNA was extracted from leaf samples using the Isolate II RNA Mini Kit (Bioline, UK). Samples were crushed with glass beads in a tissue-lyser (RETSCH MM200—Qiagen, UK) for 1 min at 30,000 rpm. Lysis buffer was added to the ground tissue followed by vigorous vortexing. The lysate was loaded in the filter tube column and centrifuged for 1 min at 11,000× *g* (Heraeus Fresco 17—Thermo Scientific, Dublin, Ireland) to isolate impure particles. In the filtered samples, 70% *v*/*v* ethanol was added, followed by vortexing and loading in a second filter tube column. The samples were centrifuged for 30 s at 11,000× *g* and washed two times with wash buffer and dried. Total RNA was eluted with RNase-free water, followed by DNase treatment according to the manual (AMPD1, Sigma-Aldrich, Ireland). cDNA was synthesized from total RNA (0.5–1 μg) using a SensiFAST cDNA Synthesis kit (Bioline, UK) according to the manufacturer's protocol. The reaction components were 5× TransAmp Buffer and Reverse Transcriptase. The reaction was incubated at 25 °C for 10 min (primer annealing), 42 °C for 15 min (reverse transcription) and 85 °C for 5 min (inactivation).

4.5. Quantitative Real-Time PCR Analysis

Lettuce genes involved in the carotenoid biosynthetic pathway were identified through searches of the Expressed Sequence Tag (EST) database deposited in NCBI (National Center for Biotechnology Information) with tBlastn where known carotenoid biosynthetic proteins from chrysanthemum (*Chrysanthemum indicum L.*, Asteraceae), apple (*Malus domestica Borkh.*, Rosaceae) and Arabidopsis (*Arabidopisis thaliana (L.)* Heynh., Brassicaceae) were used as queries. Primers were designed for the 17 target carotenoid biosynthesis genes and for a lettuce UBIQUTIN housekeeping gene (LsUBI; Supplementary Table S2) using Quantprime (http://www.quantprime.de/main.php?page=home, last accessed on 17 June 2023). cDNA was generated from the same samples that had been used for carotenoid and chlorophyll quantification, and expression levels of the 17 carotenoid biosynthesis genes were determined by Quantitative real-time PCR (qRT-PCR) using a CFX96 Real-Time system (BioRad, Watford, UK). The SYBR Green master mix was used following manufacturer's protocol with minimum modifications. cDNA templates were diluted 1:4 times and used in a 5 μL final volume reaction (1 μL of cDNA template, 0.25 μL of each primer, 1 μL of water and 1 μL of SYBR green). For each sample, three technical replicates were prepared with three negative controls per plate. The PCR reaction conditions were 95 °C for 10 min (preincubation), followed by 40 amplification cycles (95 °C for 15 s, 60 °C for 30 s and 72 °C for 30 s). A melting curve analysis with continuous fluorescence measurement during the 65–95 °C melt was generated after the amplification reactions. Data were analyzed using BioRad CFX manager software (BioRad, UK). The expression level of each gene was normalized to the lettuce UBIQUTIN gene. Standard deviation was calculated using the three biological repetitions for each sample.

4.6. Bioinformatic Identification of Candidate Genes

Candidate genes were identified by tBLASTn using reference sequences from chrysanthemum (which is in the same family, Asteraceae, as lettuce), apple and A. thaliana as queries against lettuce EST databases (NCBI) and genome project sequence (https://lgr.genomecenter.ucdavis.edu/Links.php, last accessed on 17 June 2023).

4.7. Statistical Analysis

The transcriptome and metabolites levels (carotenoids and chlorophylls) were compared by one-way ANOVA followed by Tukey HSD post hoc test assuming normal distribution and variance homogeneity. If the assumption did not apply, a Kruskal–Wallis one-way ANOVA on ranks was performed. Significant differences were considered at $p \leq 0.05$ and were indicated by different letters or asterisks. Outlier identification was performed in RStudio (2023.03.0+386 "Cherry Blossom" Release for MacOS) using Grubbs test (package "outliers"). Metabolite levels are presented as mean concentrations ± standard error unless otherwise stated. Fold changes of carotenoids and chlorophyll change are normalized to the corresponding levels in the Grand Rapid variety. Linear regression analysis was performed in SPSS 28.0 to predict the effect of chlorophyll and lutein on β-carotene content in leaves. Pearson correlation analysis to identify correlations between chlorophyll and carotenoids levels was performed using the heatmaply package in R. Principle Component Analysis (PCA) was conducted in RStudio (package ggfortify) by using ESTs or normalized carotenoids levels as variables. The heatmap analysis was performed with z-normalized data using pheatmap package in RStudio. The variables were clustered according to the inner or outer leaves samples, time of sampling and genotype. The heatmap data is presented as row normalized. A Pearson correlation analysis between gene expression level was performed in SAS Statistical Analysis Software STAT 14.1 (www.sas.com, last accessed on 17 June 2023) and tested for statistical significance using "proc corr".

5. Conclusions

To date, carotenoid biosynthesis and degradation pathways have been mostly elucidated using *Arabidopsis thaliana* and other models. In addition to specific carotenoid degradation enzyme activities, e.g., Carotenoid Cleavage Dioxygenases, carotenoids can be degraded non-enzymatically due to high levels of lipid peroxidation and oxidative stress in senescing tissues [24,63]. Simultaneous detection of gene expression levels and carotenoid levels across contrasting color (and architecture) lettuce genotypes has allowed us to reveal the relationship between gene expression and carotenoid accumulation in inner and outer leaves at different stages of the lettuce life cycle. Our study demonstrates that carotenoids and chlorophyll accumulation levels vary during development and leaf age of lettuce plants. Our results also suggest that expression levels of genes encoding key enzymes of the carotenoid biosynthetic pathway may cause these differences and that strategies to manipulate carotenoids genes and carotenoid content in plants must be evaluated case by case. Our study provides a basis for further determination of the control of carotenoid biosynthesis in a crop species so that nutritionally improved lettuce varieties can be developed.

Supplementary Materials: The following supporting information can be downloaded at: https://www.mdpi.com/article/10.3390/ijms241210310/s1.

Author Contributions: Conceptualization and methodology, C.L.d.O., L.A.A.G., G.B. and C.S.; investigation, C.L.d.O.; bioinformatic assistance, M.d.S.G.; gene expression and biochemical analyses, A.F., A.A.E.-F., R.S., P.C.M. and G.B.; formal analysis, C.L.d.O.; data curation, C.L.d.O. and G.B.; visualization, G.B. and C.L.d.O.; writing—original draft preparation, C.L.d.O. and G.B.; supervision, L.A.A.G. and C.S.; writing—review and editing G.B., P.C.M. and C.S.; funding acquisition, C.L.d.O., L.A.A.G. and C.S. All authors have read and agreed to the published version of the manuscript.

Funding: This research was funded by Science Foundation Ireland, SFI 19//FFP6711 (to G.B. and C.S), and Coordenação de Pessoal de Nível Superior of Brazil Ministry of Education (CAPES—BRAZIL [BEX 7368/13-5 to C.L.O.]).

Institutional Review Board Statement: Not applicable.

Informed Consent Statement: Not applicable.

Data Availability Statement: The data available upon request.

Acknowledgments: C.O. and L.G. acknowledge International Strategic Cooperation Award (ISCA) Programme supporting the Research Brazil Ireland Initiative (http://rbi.ie/, last accessed on 17 June 2023). Authors grateful to Gerard Fahy of University of Galway for technical support.

Conflicts of Interest: The authors declare no conflict of interest. The funders had no role in the design of the study; in the collection, analyses, or interpretation of data; in the writing of the manuscript; or in the decision to publish the results.

References

1. Young, A.J.; Pallett, K.E. Carotenoids. In *Antioxidants in Higher Plants*; CRC Press: Boca Raton, FL, USA, 2017; pp. 59–89.
2. Harbart, V.; Frede, K.; Fitzner, M.; Baldermann, S. Regulation of carotenoid and flavonoid biosynthetic pathways in *Lactuca sativa var capitate* L. in protected cultivation. *Front. Plant Sci.* **2023**, *14*, 1124750. [CrossRef]
3. Fraser, P.D.; Bramley, P.M. The biosynthesis and nutritional uses of carotenoids. *Prog. Lipid Res.* **2004**, *43*, 228–265. [CrossRef]
4. Saini, R.K.; Nile, S.H.; Park, S.W. Carotenoids from fruits and vegetables: Chemistry, analysis, occurrence, bioavailability and biological activities. *Food Res. Int.* **2015**, *76*, 735–750. [CrossRef]
5. Harari, A.; Melnikov, N.; Kandel Kfir, M.; Kamari, Y.; Mahler, L.; Ben-Amotz, A.; Harats, D.; Cohen, H.; Shaish, A. Dietary β-Carotene Rescues Vitamin A Deficiency and Inhibits Atherogenesis in Apolipoprotein E-Deficient Mice. *Nutrients* **2020**, *12*, 1625. [CrossRef] [PubMed]
6. de Oliveira, C.L.; Brychkova, G.; Esteves-Ferreira, A.A.; McKeown, P.; de Souza Gomes, M.; Maluf, W.R.; Gomes, L.A.A.; Spillane, C. Thermal disruption of the food matrix of biofortified lettuce varieties modifies absorption of carotenoids by Caco-2 cells. *Food Chem.* **2020**, *308*, 125443. [CrossRef]
7. West, K.P.J.; Darnton-Hill, I. Vitamin A deficiency. In *Nutrition and Health Indeveloping Countries*; Humana Press: Totowa, NJ, USA, 2001; pp. 267–306.

61. Welsch, R.; Arango, J.; Bar, C.; Salazar, B.; Al-Babili, S.; Beltran, J.; Chavarriaga, P.; Ceballos, H.; Tohme, J.; Beyer, P. Provitamin A accumulation in cassava (*Manihot esculenta*) roots driven by a single nucleotide polymorphism in a phytoene synthase gene. *Plant Cell* **2010**, *22*, 3348–3356. [CrossRef]
62. Norris, S.R.; Barrette, T.R.; DellaPenna, D. Genetic dissection of carotenoid synthesis in arabidopsis defines plastoquinone as an essential component of phytoene desaturation. *Plant Cell* **1995**, *7*, 2139–2149. [PubMed]
63. Gonzalez-Jorge, S.; Ha, S.-H.; Magallanes-Lundback, M.; Gilliland, L.U.; Zhou, A.; Lipka, A.E.; Nguyen, Y.-N.; Angelovici, R.; Lin, H.; Cepela, J. Carotenoid cleavage dioxygenase4 is a negative regulator of β-carotene content in Arabidopsis seeds. *Plant Cell* **2013**, *25*, 4812–4826. [CrossRef]

International Journal of
Molecular Sciences

MDPI

Article

The Content of Total Carotenoids, Vitamin C and Antioxidant Properties of 65 Potato Cultivars Characterised under the European Project ECOBREED

Beata Tatarowska *, Dorota Milczarek and Jarosław Plich

Plant Breeding and Acclimatization Institute—National Research Institute in Radzików, Młochów Division, Department of Potato Genetics and Parental Lines, Platanowa Str. 19, 05-831 Młochów, Poland; d.milczarek@ihar.edu.pl (D.M.); j.plich@ihar.edu.pl (J.P.)

* Correspondence: b.tatarowska@ihar.edu.pl; Tel.: +48-22-729-92-48; Fax: +48-22-729-92-47

Abstract: The aim of this study was to determine the effect of cultivars on the concentration of antioxidant compounds: total carotenoid content (TC) and vitamin C (VC), and their correlation with the total antioxidant activity (TAA) in 65 potato cultivars (*Solanum tuberosum*) from 10 countries. The TC content revealed a highly significant effect of the year (Y), cultivar (C) and flesh colour (FC). The TC ranged from 101.5 μg 100 g^{-1} DM (in cv. Kelly) to 715 μg 100 g^{-1} DM (in cv. Mayan Gold). The TC values were weakly correlated with years and higher in yellow-fleshed potatoes than in white-fleshed potatoes (319.9 vs. 175.6 μg 100 g^{-1} DM, respectively). The VC content ranged from 1.0 mg 100 g^{-1} FM (in cv. Bzura) to 14.8 mg 100 g^{-1} FM (in cv. Twinner). The content of VC were higher in yellow-fleshed (6.5 mg 100 g^{-1} FM) than in white-fleshed potatoes (5.8 mg 100 g^{-1} FM). The highest TAA were observed in cvs. Colleen, Basa, Triplo, Gatsby, Ditta, Twinner, Riviera, Michalina, Damaris, Belmonda, Ambo, Savinja, 12-LHI-6. For these cultivars, the FRAP values were 0.53 μmol TE 100 mg^{-1} DM and DPPH 0.55 μmol TE 100 mg^{-1} DM. The lowest TAA were observed in cvs.: Owacja, Mayan Gold, Kokra, Magnolia and Kelly. For them, the FRAP and DPPH values were slightly above 0.2 μmol TE 100 mg^{-1} DM. It was shown that the concentration of TC in potato tubers has an impact on TAA.

Keywords: potato; health compounds; carotenoids; vitamin C; total antioxidant activity

1. Introduction

The potato is the fifth most important staple food crop in the world, which supplies energy and some nutritionally relevant ingredients [1,2]. The nutritional value of potato tubers is mainly characterised by the presence of essential amino acids, high contents of starch and dietary fibre, as well as a low concentration of fats. Potato tubers also contain important levels of bioactive compounds and antioxidants, including phenolic acids, carotenoids and flavonoids. In the human body, these bioactive compounds may act as very strong antioxidants. In addition to scavenging free radicals, multiple activities of antioxidants include inactivating metal catalysts by chelation, reducing hydroperoxides to stable hydroxyl derivatives, and interacting synergistically with other reducing compounds [3,4]. Therefore, a higher consumption of potato tubers may increase the level in blood and tissues and acts against oxidative stress, which is responsible for damage to lipids, proteins, enzymes and DNA, resulting in chronic diseases such as cancer or cardiovascular disease (CVD) [5–7]. These compounds and their antioxidative potential, which describes potentially health-promoting properties of potato tubers, are determined by genetics as well as environmental factors. Currently, the market's demand for food rich in antioxidant compounds is increasing. Plant breeding has invested significant resources in the selection of potato cultivars with a higher content of carotenoids, flavonoids, vitamin C and, consequently, a higher antioxidant potential [8–10]. One of the methods that allows us to improve the composition of these compounds is breeding focused on biofortification (increased nutritional security), which has the task of enriching the nutritional value of the product by supplementing it with bioavailable nutrients.

In this research, 65 potato cultivars from the work collection created within the European project ECOBREED were used. Cultivars which were selected for the collection were characterised by high or increased resistance to *Phytophthora infestans* and were commercially available in the European seed market. These cultivars come from 10 countries and are used to identify the most important characteristics for organic farming systems. The aim of this research was to evaluate the variability of the bioactive compounds (total carotenoids and vitamin C) in 65 potato cultivars. In this study, the total antioxidant activity of potato tubers depends on the content of individual phytocompounds which depend on the total antioxidant activity of the tubers. In addition, cultivars were screened for the presence of dominant allele 3 in the *Chy2* locus.

2. Results and Discussion

Potato has been identified by the Food and Agriculture Organization of the United Nations (FAO) as a basic and sustainable food for the growing world population [11]. Therefore, it is very important to understand its contribution to both our daily and long-term health. It is very important to develop cultivars with increased antioxidant capacity as 'functional foods' and encourage potato consumers to buy such cultivars, rich in carotenoids, vitamin C and other pro-health compounds.

Potato cultivars show great variability in terms of carotenoid accumulation in tubers. The concentration of carotenoids in potato tubers is affected by several factors such as genotype, agronomic factors, post-harvest storage, cooking and processing conditions [12–16].

In our study, 65 cultivars were evaluated YI. Based on the values obtained for YI, cultivars were divided into two groups: white and yellow flesh. In total, 18 cultivars were classified as white-fleshed and 47 as yellow-fleshed. The YI of the evaluated white-fleshed potato cultivars ranged from 35.5 to 48.8 and in yellow-fleshed cultivars ranged from 50.3 to 80.4 (Table 1). The values of the YI range from 35.5 for cv Kelly to 0.4 to cv. Mayan Gold (Table 1). Also, the 65 potato cultivars differed significantly in terms of their total carotenoid content. The three-year mean TC content ranged from 101.5 μg 100 g^{-1} DM (in cv. Kelly) to 715 μg 100 g^{-1} DM (in cv. Mayan Gold) (Table 1). The examined cultivar Mayan Gold is a diploid that has *S. phureja* in its ancestry. This potato species is reported to be a good source of high total carotenoid content. The results of ANOVA showed that TC is also significantly influenced by the factor year (Y) and the correlations between the TC values between individual years were rather weak ($r_{2019\times2020}$ = 0.43*; $r_{2019\times2021}$ = 0.26*; $r_{2020\times2021}$ = 0.31*; significant at $p < 0.05$) (Table 2). Differences in rainfall intensity and distribution were probably important factors contributing to TC content in potato tubers in individual years (Figure S1). In the group with the highest TC, there were 11 cultivars: Otolia, Bionta, Anuschka, Belmonda, Carolus, Gardena, Salome, Belana, Twinner, Caprice, and Mayan Gold. The mean TC values from three years research for these cultivars range from 422.0 to 715 μg 100 g^{-1} DM (Table 1). In the group with the lowest of TC 13 cultivars were included: Kelly, Yona, Savinja, Nofy, Basa, Valor, Botond, Sarpo Mira, Riviera, Vipava, Tinca, Bzura and Erica. The mean values of these cultivars from three years of research range from 101.5 to 168.6 μg 100 g^{-1} DM (Table 1). The highest mean value of TC was recorded in 2020 (Table 1). The mean values of TC were significantly higher in potatoes with yellow-flesh (319.9 μg 100 g^{-1} DM) than in white-fleshed potatoes (175.6 μg 100 g^{-1} DM) (Table 3; Figure 1). Iwanzik et al. [17] reported a range from 27 to 74 μg 100 g^{-1} FW for TC in white-fleshed potato cultivars. In yellow-fleshed potatoes, TC is usually higher, up to 560 μg 100 g^{-1} FW [18–20]. In diploid clones of *Solanum* species, reported levels of TC are much higher and reached more than 2000 μg 100 g^{-1} FW [21]. A very high TC of up to 3850 μg 100 g^{-1} FW was found in cultivars from the *Andigenum* group [22]. Othman [23] evaluated 32 potato cultivars grown in New Zealand. Agria, a dark-yellow-fleshed cultivar, was found to have the highest TC (169.57 μg/g DW), the lowest TC was found in the white-fleshed cultivar Moonlight (1.18 µg/g DW). He also reported that genotype, location and their interaction were significant. Tatarowska et al. [10] reported a TC range from 5.57 to 20.20 mg kg^{-1} FW and proved that several factors, location, year, genotype, and their interactions influence on carotenoid accumulation.

Table 1. Mean values of yellow index (YI), total carotenoid content (TC), presence loci of marker *Chy2*, vitamin C (VC) and total antioxidant capacity (TAA) for 65 potato cultivars.

Cultivar	YI [(1)]	FC [(2)]	TC in µg 100 g^{-1} DM [(3)]			Mean Values TC Years 2019–2021	*Chy2* Loci [(4)]	VC [(5)] mg 100 g^{-1} FM 2020	TAA DPPH [(6)]	TAA FRAP [(7)]
			2019	2020	2021				µmol TE 100 mg $^{-1}$ DM 2020	
Ambo	40.8	w	119.3	332.7	166.0	206.0 AB	0	5.7	0.53	0.48
Balatoni Rozsa	47.2	w	272.1	188.8	186.0	215.6 AB	1	2.8	0.28	0.24
Botond	46.3	w	91.8	237.3	93.0	140.7 B	1	5.2	0.36	0.30
Bzura	48.8	w	211.5	232.5	52.0	165.3 AB	0	1.0	0.22	0.23
Cara	42.4	w	130.6	517.0	80.0	242.5 AB	1	6.4	0.43	0.41
Colleen	48.4	w	143.5	316.5	80.0	180.0 AB	1	11.5	0.57	0.62
Denar	44.3	w	269.7	290.7	180.0	246.8 AB	1	1.2	0.42	0.36
Gatsby	45.2	w	156.5	264.8	207.0	209.4 AB	0	4.0	0.56	0.57
Kelly	35.5	w	103.1	72.4	129.0	101.5 B	0	12.0	0.13	0.18
Kokra	45.0	w	470.1	150.0	52.0	224.1 AB	0	8.3	0.17	0.30
Owacja	47.1	w	137.1	224.4	173.0	178.2 AB	1	3.9	0.19	0.23
Premiere	46.9	w	172.6	250.3	185.0	202.6 AB	1	5.6	0.22	0.31
Sarpo Mira	47.0	w	237.3	138.7	52.0	142.7 B	1	3.9	0.22	0.26
Savinja	42.9	w	67.6	172.6	127.0	122.4 B	1	6.2	0.52	0.48
Valor	39.8	w	141.9	146.8	113.0	133.9 B	0	10.3	0.32	0.33
Vipava	35.8	w	345.6	91.8	49.0	162.2 AB	0	7.5	0.40	0.36
White Lady	45.8	w	219.5	151.6	140.0	170.4 AB	0	2.1	0.31	0.32
Yona	44.7	w	141.9	156.5	49.0	115.8 B	1	7.6	0.34	0.37
12-LHI-6	57.8	y	428.1	287.4	87.0	267.5 AB	1	12.9	0.5	0.48
Agria	65.7	y	436.2	437.8	145.0	339.7 AB	1	11.8	0.41	0.46
Alouette	59.4	y	156.5	373.1	415.0	314.9 AB	1	5.1	0.31	0.31
Anuschka	71.4	y	339.2	673.9	312.0	441.7 AB	1	10.5	0.45	0.41
Basa	55.8	y	84.5	248.6	60.0	131.1 B	1	3.6	0.61	0.61
Belana	70.4	y	405.5	497.6	821.0	574.7 AB	0	4.9	0.43	0.40
Belmonda	70.0	y	303.6	555.8	508.0	455.8 AB	1	4.5	0.53	0.49
Bionta	66.3	y	444.3	332.7	517.0	431.3 AB	1	3.4	0.40	0.40
Caprice	65.3	y	505.7	512.2	1000.6	672.8 AB	1	3.4	0.41	0.39
Capucine	68.3	y	390.9	492.8	64.0	315.9 AB	1	6.0	0.20	0.25
Carolus	58.6	y	358.6	314.9	863.0	512.2 AB	1	10.5	0.38	0.35
Casablanca	50.3	y	114.4	169.4	290.0	191.3 AB	1	9.4	0.29	0.36
Charlotte	58.5	y	151.6	313.3	249.0	238.0 AB	1	3.5	0.34	0.41
Colomba	51.3	y	373.1	319.8	93.0	262.0 AB	1	2.3	0.39	0.41
Damaris	53.5	y	298.8	276.1	112.0	229.0 AB	1	3.9	0.50	0.50
Delila	52.3	y	348.9	185.6	133.0	222.5 AB	1	2.4	0.29	0.31
Ditta	62.6	y	184.0	449.1	225.0	286.0 AB	1	4.3	0.57	0.55
Edony	61.2	y	127.4	303.6	88.0	173.0 AB	1	9.6	0.41	0.38

Table 1. *Cont.*

Cultivar	YI [(1)]	FC [(2)]	TC in μg 100 g^{-1} DM [(3)]			Mean Values TC Years 2019–2021	*Chy2* Loci [(4)]	VC [(5)] mg 100 g^{-1} FM 2020	TAA DPPH [(6)]	TAA FRAP [(7)]
			2019	2020	2021				μmol TE 100 mg^{-1} DM 2020	
Elfe	68.8	y	184.8	483.1	328.0	332.0 AB	1	7.9	0.47	0.46
Erika	59.0	y	130.6	318.2	57.0	168.6 AB	1	3.2	0.41	0.41
Fidelia	65.1	y	281.0	347.3	161.0	263.1 AB	1	6.8	0.34	0.35
Fortus	59.0	y	295.5	305.2	120.0	240.2 AB	1	7.6	0.38	0.37
Gardena	59.3	y	212.3	316.5	1049.0	525.9 AB	1	3.8	0.42	0.36
Goldmarie	64.5	y	307.7	473.4	129.0	303.3 AB	1	4.3	0.39	0.39
Granola	63.3	y	378.0	450.7	93.0	307.2 AB	1	1.2	0.30	0.31
Karlena	62.4	y	344.0	284.2	133.0	253.7 AB	1	11.9	0.30	0.31
Levante	59.4	y	96.7	281.0	501.0	292.9 AB	0	12.3	0.36	0.38
Lilly	67.7	y	399.8	528.3	57.0	328.4 AB	1	5.4	0.33	0.34
Lord	53.7	y	227.6	344.0	272.0	281.2 AB	1	2.3	0.34	0.30
Magnolia	54.9	y	187.2	264.8	119.0	190.3 AB	1	2.7	0.14	0.15
Mayan Gold	80.4	y	778.2	639.9	727.0	715.0 A	1	7.9	0.18	0.24
Michalina	53.7	y	205.0	534.8	113.0	284.3 AB	1	2.6	0.52	0.50
Noblesse	63.3	y	399.8	365.0	186.0	317.0 AB	1	8.2	0.28	0.31
Nofy	56.5	y	65.9	182.3	143.0	130.4 B	0	10.7	0.24	0.30
Omega	66.4	y	240.6	314.9	78.0	211.2 AB	1	10.0	0.44	0.37
Otolia	58.4	y	641.5	431.3	193.0	422.0 AB	0	2.6	0.36	0.32
Riviera	52.7	y	146.8	219.5	101.0	155.8 AB	1	2.6	0.55	0.50
Salome	63.2	y	596.3	407.1	658.0	553.8 AB	1	7.8	0.40	0.37
Sarpo Shona	50.3	y	182.3	245.4	104.0	177.3 AB	0	3.5	0.22	0.26
Slavnik	56.0	y	284.2	289.1	63.0	212.1 AB	1	3.3	0.40	0.27
Tajfun	54.9	y	546.1	476.6	20.0	347.6 AB	1	6.7	0.35	0.34
Tinca	58.5	y	171.8	237.3	85.0	164.7 AB	0	12.4	0.45	0.43
Triplo	64.4	y	323.0	329.5	96.0	249.5 AB	1	3.4	0.63	0.59
Twinner	65.3	y	260.0	392.5	1213.0	621.8 AB	1	14.8	0.53	0.51
Twister	59.7	y	465.3	298.8	126.0	296.7 AB	1	9.7	0.34	0.32
Voyager	55.4	y	395.0	363.4	60.0	272.8 AB	1	10.3	0.32	0.34
Wega	67.0	y	187.2	704.6	175.0	355.6 AB	0	7.7	0.31	0.34
Mean value	56.2		274.1	330.9	234.7	279.9	-	12.9	0.50	0.50

YI [(1)]—yellow index, FC [(2)]—flesh colour: y—yellow-fleshed; w—white-fleshed; TC [(3)]—total carotenoid content; marker presence [(4)]: 1—*Chy2* positive; 0—*Chy2* negative; VC [(5)]—vitamin C; TAA [(6)]—total antioxidant capacity measured by DPPH; TAA [(7)]—total antioxidant capacity measured by FRAP; A, B, AB—homogenous groups.

Table 2. Sources of variation and one- and two-way ANOVA results for TC, VC, DPPH and FRAP in potato tubers.

Sources of Variation	ANOVA				
	Sum of Squares	Degrees of Freedom	Mean Square	F Statistic	Significance
			ANOVA results for TC		
Cultivar (C)	3.704.438	64	57.882	1.9978	***
Year (Y)	304.012	2	152.006	4.0722	**
(C) × (Y)	3.462.533	128	27.051		ns
			ANOVA results for VC		
Cultivar (C)	1285.732	64	20.090	2.4049	***
			ANOVA results for DPPH		
Cultivar (C)	1.48101	64	0.02314	2.496	***
			ANOVA results for FRAP		
Cultivar (C)	1.08066	64	0.01689	2.499	***

ns = not significant; ** significant at $p < 0.01$; *** significant at $p < 0.001$.

Table 3. Student's *t*-test for the two independent groups of potato flesh colour (white and yellow) relative to TC, VC, DPPH and FRAP.

Variable	Student's *t*-Test				
	Mean Yellow	Mean White	*t*	df	Significance
Total carotenoid (TC)	319.86 μg 100 g^{-1} DM	175.56 μg 100 g^{-1} DM	4.854	193	***
Vitamin C (VC)	6.49 mg 100 g^{-1} FM	5.84 mg 100 g^{-1} FM	0.670	63	ns
DPPH	0.38 μmol TE 100 mg $^{-1}$ DM	0.34 μmol TE 100 mg $^{-1}$ DM	1.276	63	ns
FRAP	0.38 μmol TE 100 mg $^{-1}$ DM	0.35 μmol TE 100 mg $^{-1}$ DM	1.010	63	ns

ns = not significant; *** significant at $p < 0.001$.

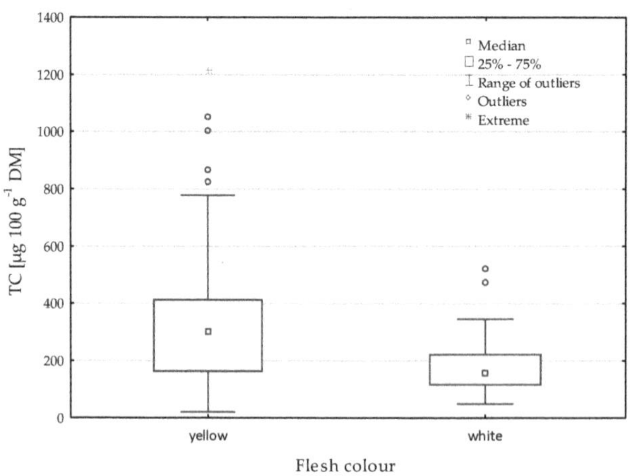

Figure 1. Mean values of total carotenoid content (TC) in two groups of potato with white and yellow flesh.

Two main loci have been shown to control potato tuber flesh colour. The Yellow (Y) locus was shown to co-segregate with the *β-Carotene Hydroxylase 2* (*CHY2*) gene [24]. A dominant *Y* allele of this locus was associated with a high flesh carotenoid content and showed increased expression of the *CHY2* transcript [25,26]. The Orange (Or) locus was associated with high amounts of flesh zeaxanthin [27]. The *Or* locus was found to co-segregate with the *Zeaxanthin Epoxidase* (*ZEP*) gene. All orange genotypes carry only a recessive *zep* allele (*zep1*) characterised by the insertion of a non-LTR retrotransposon into intron 1 and by reduced steady-state levels of the *ZEP* transcript.

Brown et al. [24] examined 21 yellow fleshed and 28 white fleshed potato cultivars and breeding clones for the presence of a unique allele of *bch* associated with yellow flesh colour and obtained the expected for gene product from all yellow-fleshed cultivars and breeding clones tested. The same primers amplified no product from any of the white-fleshed clones tested. In the research by Sulli et al. [28], the dominant *CHY2* allele 3, previously found to be a major determinant of carotenoid accumulation, was present only in yellow or orange-fleshed clones. Goo et al. [29] also observed a higher expression level of *Chy2* associated with yellow flesh colour in a small number of potato cultivars. In our investigation, the dominant *Chy2* allele 3 was amplified in 50 of the 65 cultivars tested (Table 1). In Figure 2, the amplification of CAPS marker linked to *Chy2* gene allele 3 is presented. Cultivars with amplified diagnostic fragment were both yellow and white fleshed. Cultivars without this fragment were also yellow and white fleshed. However, this fragment was absent in more than 44% of white-fleshed cultivars and only in ~15% of yellow-fleshed cultivars. The examined cultivars with the marker had higher YI and TC than the cultivars without the marker, but the differences were statistically significant in the case of YI but not TC (Table 4). In conclusion, the examined marker of *β-Carotene Hydroxylase 2* can help in selection of potato clones/cultivars with high carotenoid content, but its usefulness is restricted.

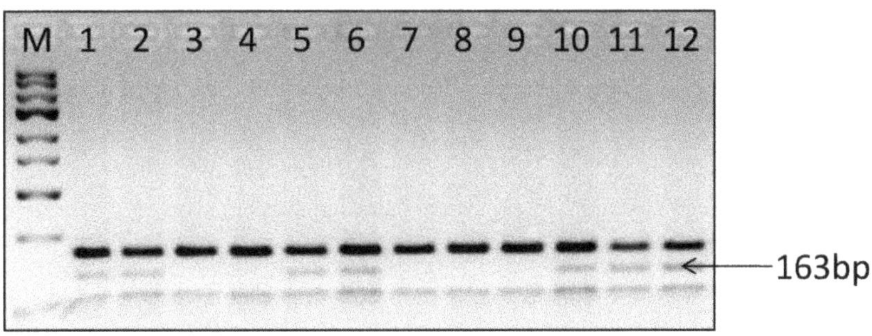

Figure 2. Amplification of the CAPS marker linked to the *Chy2* gene allele 3. Cultivars having *Chy2* gene allele 3 are shown in lanes 1, 2, 5, 6, 10, 11 and 12, respectively. Lanes 3, 4, 7, 8 and 9 show that cultivars lack the *Chy2* gene allele 3. Lane M contains the 100 bp DNA ladder.

Table 4. Student's *t*-test for the two independent groups of potato with or without the *Chy2* allele 3 marker relative to YI and TC.

Variable	Student's *t*-Test				
	Mean *Chy2*+	Mean *Chy2*−	*t*	df	Significance
Yellow index (YI)	57.97	50.47	2.813	63	**
Total carotenoid (TC)	294.07 µg 100 g^{-1} DM	232.68 µg 100 g^{-1} DM	1.852	193	ns

ns = not significant; ** significant at $p < 0.01$.

One of the powerful antioxidants in plants is vitamin C. In potato tubers, it occurs as L-ascorbic acid and dehydroascorbic acid in the acid oxidation position [30]. Potatoes are one of the most important sources of vitamin C, but its value in human nutrition is often underestimated [31]. Vitamin C is the major naturally occurring inhibitor of enzymatic browning of potatoes [32]. In the human body, it acts as an antioxidant [11] and plays an important role in the protection against oxidative stress [33]. The VC content in potato tubers influences many factors, such as cultivar, maturity, year, method and growing conditions and location [16,34–36]. Various studies have measured large differences between potato cultivars in their VC content from 7.54 to 46.0 mg 100 g^{-1} FM [34,37–40]. Our studies showed a great variation between potato cultivars' VC content. The mean VC values for 65 cultivars are presented in Table 1. The average VC content in 65 cultivars ranged from 1.0 to 14.8 mg 100 g^{-1} FM. The highest level of VC was noted for cultivars: Twinner (14.8 mg 100 g^{-1} FM), 12-LHI-6 (12.9 mg 100 g^{-1} FM), Tinca (12.4 mg 100 g^{-1} FM), Levante (12.3 mg 100 g^{-1} FM), Kelly (12.0 mg 100 g^{-1} FM), Karlena (11.9 mg 100 g^{-1} FM), Agria (11.8 mg 100 g^{-1} FM) and Colleen (11.6 mg 100 g^{-1} FM) (Table 1). In Figure 3 are presented mean values of VC in two groups of flesh colour. Mean values of VC were higher in potatoes with yellow-flesh (6.5 mg 100 g^{-1} FM) than in potatoes with white-flesh (5.8 mg 100 g^{-1} FM), but this difference was not statistically significant (Figure 3; Table 3). One-way ANOVA analysis showed a significant influence of the cultivar on the content of VC in potato tubers (Table 2).

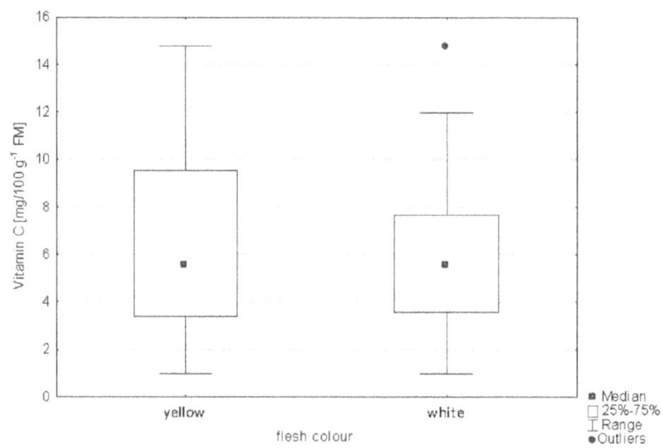

Figure 3. Content of vitamin C in two groups of potatoes with white and yellow flesh.

Similar to our research, Sawicka et al. [41] proved that in the formation of VC in potato tubers, the greatest impact is the genotypic variability. According to Brown [42], VC content on the level 20 mg/100 g^{-1} FM can raise up to 13% in the TAA of tubers. Several authors also confirm the influence of earliness on the VC content in potato tubers. Trawczyński et al. [43] and Hrabowska et al. [44] showed that a higher VC was noted for late cv. compared to early cv. According to Grudzińska and Mańkowski [5], the flesh colour significantly influenced the VC content in potato tubers. The authors observed significant differences in the VC content between tubers with yellow flesh and cream flesh. Hamouz et al. [35] determined that VC content was 2.9-times higher in red and purple potatoes than in potatoes with yellow and white flesh. Valcalcer et al. [16] proved that the highest VC content was in yellow flesh cv. and the lowest in cream flesh cv. By contrast, Hejtmánková et al. [45] showed that white and yellow potatoes did not differ in terms of VC content, similar to our research. The high diversity of VC in potato tubers shows that to conduct breeding towards the production of cultivars with high VC content is justified. Biofortification using breeding programs aimed at increasing the VC content in potato tubers is fully justified.

The concentration of bioactive compounds in potato tubers is determined by both genetic potential and environmental factors. The purpose of these experiments is, among others, to determine the influence of genotype on the TAA of potato tubers. The antioxidant capacity of 65 potato cultivars from nine countries dedicated to organic farming was determined. Two different methods of measuring antioxidative capacity, i.e., DPPH and FRAP, were used. On the basis of these analyses conducted, it can be said that the evaluated cultivars are different in the TAA of potato tubers. The full results of the two methods and all the cultivars are presented in Table 1. For the value of FRAP and DPPH, one-way ANOVA revealed highly significant effect for cultivar (C). Higher antioxidant activity was observed in 10 cultivars: Colleen, Basa, Triplo, Gatsby, Ditta, Twinner, Riviera, Michalina, Damaris, Belmonda, Ambo, Savinja, 12-LHI-6 (Table 1). For these cultivars, the mean FRAP values were 0.53 μmol TE 100 mg^{-1} DM and the mean DPPH values were 0.55 μmol TE 100 mg^{-1} DM. Tubers of these potato cultivars have many benefits to prevent oxidative stress and potential as sources of natural antioxidant can be further used in breeding work. A lower antioxidant activity was observed in cultivars: Owacja, Mayan Gold, Kokra, Magnolia and Kelly (Table 1). For them, the FRAP and DPPH values were slightly above 0.20 μmol TE 100 mg^{-1} DM. Strong positive correlations (r = 0.917*; significant at $p < 0.05$) were found between FRAP and DPPH (Figure 4). These methods can be used interchangeably in the assessment of the TAA in potato tubers. In potato cultivars with yellow-fleshed tubers higher antioxidant activity was observed (mean DPPH = 0.39 μmol TE 100 mg^{-1} DM; mean FRAP = 0.39 μmol TE 100 mg^{-1} DM) than in white-fleshed potatoes (mean DPPH and FRAP = 0.33 μmol TE 100 mg^{-1} DM) (Figure 5), but these differences were not statistically significant (Table 3).

In the literature, can be find research showing a large variation of TAA among potato cultivars/clones, and many other factors having an impact on their level in potato tubers. For example, Wichrowska [46] reported a significantly differentiated TAA of potato tubers cv Satina measured using the FRAP method. The highest TAA was characteristic of tubers from the plots fertilized with manure, a full dose of mineral fertilization and bio-fertiliser. FRAP values ranged from 6.98 to 9.91 mmol Fe^{2+} kg^{-1} DM. Hu et al. [47] reported a significant effect of cultivar and year on TAA. In the FRAP assay, antioxidant activity ranged from 6 to 64 μmol AAE g^{-1} DW and in ORAC ranged from 42 to 168 μmol TE g^{-1} DW. Lee et al. [48] reported that the TAA was significantly higher in purple-coloured potato cultivars than in the white- and yellow-coloured potatoes. In their research, they showed that cvs. Hongyoung and Jayoung had the highest antioxidant activity. Antioxidant effect was highly correlated with polyphenol content. Kim et al. [49] reported that the effective compounds that contribute to TAA were phenolics acids. Fidrianny et al. [50] proved that the phenolic and flavonoids compounds of potato tubers are very strong antioxidants. Seijo-Rodríguez et al. [51] stated that due to the high correlation between the phenolic content and antioxidant activity, the phenolic content in potato tubers can be used as an indicator of the antioxidant activity of a tuber. The Keutgen et al. [8] reported that antioxidative capacity measured by FRAP was correlated with chlorogenic acid content (r = 0.590**; significant at $p < 0.01$) and glutathione fractions, especially with the reduced form (GSH, r = 0.692**; significant at $p < 0.01$). Hejtmánková et al. [45] did not detect a significant correlation between content of VC and antioxidant activity ($R^2 = 0.08$) in unpeeled raw potato. Grudzińska and Mańkowski [5] also indicated the lack relationship between TAA and VC in unpeeled potatoes ($R^2 = 0.17$).

In our research, we also wanted to indicate which compounds in potato tubers have the greatest impact on TAA. The FRAP and DPPH assays indicate that TC in potato tubers contributes to their antioxidant activities. The correlation between TC and TAA measured by FRAP and DPPH was significant (r = 0.275* and r = 0.283*, respectively; significant at $p < 0.05$) (Figure 6). On the other hand, the concentration of VC in the potato tubers had no effect on TAA. The correlation between VC and TAA measured by FRAP was very low and nonsignificant (r = 0.113) and DPPH (r = 0.024) (Figure 7). Our research also reported that cultivars have significantly influence on TAA (Table 2), but not on flesh colour (Table 3).

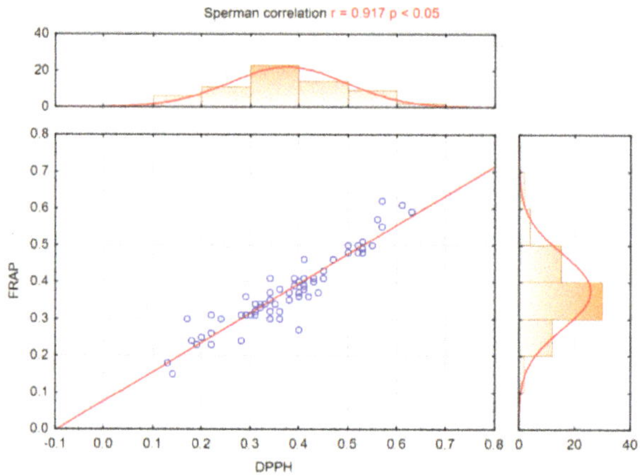

Figure 4. Correlation between two methods, FRAP and DPPH, used to measure TAA in tubers of 65 potato cultivars.

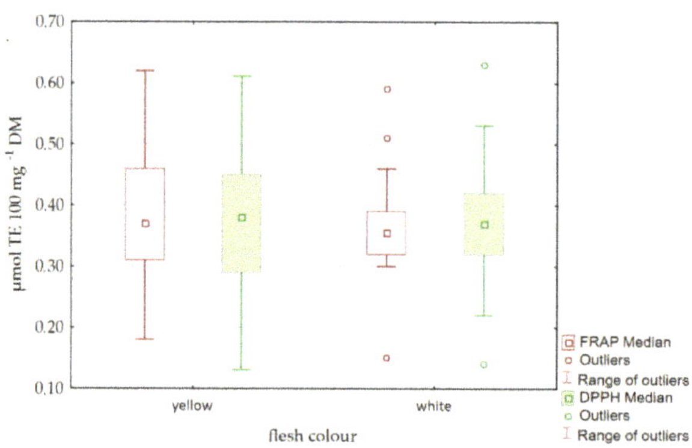

Figure 5. TAA of potato tubers measured by DPPH and FRAP in two groups of potato with white and yellow flesh.

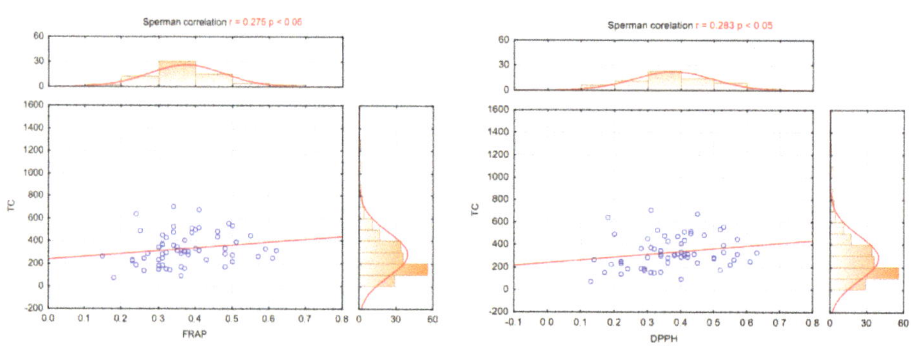

Figure 6. The relationship between TAA measured by FRAP and DPPH and the content of TC in tubers of 65 potato cultivars.

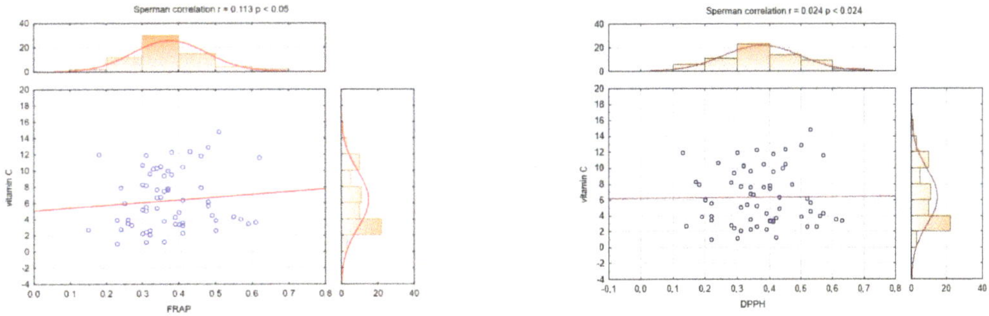

Figure 7. The relationship between TAA measured by FRAP and DPPH and the vitamin C content in tubers of 65 potato cultivars.

3. Materials and Methods

3.1. Plant Material

Plant material consisted of 65 potato cultivars from 10 countries: from Poland—Bzura, Michalina, Denar, Lord, Gardena, Magnolia, Owacja, Tajfun; from Denmark—Tinca, 12-LHI-6; from Germany—Caprice, Damaris, Fidelia, Goldmarie, Karlena, Salome, Wega, Anuschka, Belana, Elfe, Otolia, Agria, Omega, Belmonda, Lilly, Granola; from Netherlands—Alouette, Carolus, Erika, Levante, Nofy, Premiere, Riviera, Twinner, Twister, Colomba, Fortus, Noblesse, Triplo, Voyager; from Austria—Bionta, Ditta; from Slovenia cvs.—Kokra, Savinja, Slavnik, Vipava; from United Kingdom—Gatsby, Casablanca, Valor, Mayan Gold, Cara, Sarpo Mira, Sarpo Shona; from Ireland—Ambo, Colleen; from Hungary—Balatoni Rozsa, Basa, Botond, White Lady—and from France—Capucine, Charlotte, Delila, Edony, Kelly, Yona. These cultivars came from the Working Collection created within the European project ECOBREED.

3.2. Field Experiments

All these cultivars were multiplied under field conditions. The trials were carried out for three years in the central part of Poland, in Młochów. In these experiments, a low-input system was applied (in Supplementary Materials Table S1). Pesticides were applied during the vegetation period, i.e., copper fungicides and pyrethrin (plant extract) against late blight and Novodor against Colorado beetle. All application times and dose were established in accordance with the recommendations on the labels of each plant protection product. The experimental design was a randomised complete block. In each of the two blocks (replications), the cultivars were planted in 60 hill plots. The spacing of the plants in the row was 70 cm × 40 cm. Tubers were planted at the end of April and harvested after about 130 days. The monthly average temperature and precipitation were collected from a weather station located near the experimental field. The weather conditions for the growing seasons 2019, 2020 and 2021 are shown in Supplementary Materials Figure S1.

3.3. Total Carotenoid Content (TC)

TC was estimated for tubers obtained from field experiments. For each potato cultivar and each repetition, three tubers were collected, cut into small cubes, frozen in liquid nitrogen, lyophilised and milled. The total carotenoid content was measured according to Milczarek and Tatarowska [52]. All analyses were performed using three technical repetitions. Analyses were conducted via three-year experiments.

3.4. Yellowness Index (YI)

The intensity of potato flesh colour was assessed using CR-400 Colorimeter (Minolta, Osaka, Japan). YI was calculated according to the formula:

$$YI_{E313} = \frac{100(C_x X - C_z Z)}{Y}$$

where C_x and C_z are illuminant and observer-specific constants; X, Y, Z—trichromatic values [53].

3.5. Genotyping of the Chy2 Loci

All tested cultivars were screened for the presence of dominant allele 3 at the *CHY2* locus. The tests were performed with the use of a DNA marker developed by Wolters et al. [26]. DNA was isolated from young leaves of all tested cultivars using a Genomic Mini AX Plant (A&A Biotechnology, Gdańsk, Poland). PCR amplification was performed with PCR Mix Plus Green, a high specificity ready-to-use master mix for PCR (A&A Biotechnology, Gdańsk, Poland), and 300 nM of each primer in a final volume of 25 µL. Standard amplification conditions were as follows: initial denaturation for 4 min at 94 °C followed by 30 cycles of 30 s denaturation at 94 °C, 30 s annealing at 55 °C and 1 min elongation at 72 °C. Reactions were completed with an elongation step of 7 min at 72 °C. The PCR with primers CHY2ex4F (5′-CCATAGACCAAGAGAAGGACC-3′) and Beta-R822 (5′-GAAAGTAAGGCACGTTGGCAAT-3′) amplified product of size 308-bp. PCR products were digested with *Alu*I restriction enzyme (Thermo Fisher Scientific, Dreieich, Germany) according to the manufacturer instructions. The CAPS markers of the *Chy2* gene were separated with a 2% ethidium bromide-stained agarose gel. In the case where the dominant *Chy2* allele 3 is present, a diagnostic fragment of 163 bp was observed, whereas all the other alleles at the same locus produce only a specific fragment of 237 bp.

3.6. Vitamin C (VC)

Vitamin C was estimated for tubers obtained from field experiments. For each potato cultivar and each repetition, three tubers were collected. Potato tubers were peeled and then rubbed on a plastic grater. About 6 g of sample was weighed into a 50 mL test tube. It was supplemented with an extraction solution (metaphosphoric acid, acetic acid, pH 2) at a weight of 25 g, which prevented the breakdown of vitamin C. The sample was shaken on a Thermomixer compact (Eppendorf AG, Hamburg, Germany) for 15 min at 4 °C and after this was spun for 10 min at 4 °C and 604× *g* in Centrifuge 5417R (Eppendorf AG, Hamburg, Germany). After centrifugation, the clear solution was filtered through a 0.40 µm PTFE syringe filter. The injection volume was 2 µL. The mobile phase was water: acetonitrile in relative 2:98, flow 1.5 mL min^{-1}. Analyses were performed using Column Luna 3u HILIC 150 × 4.6 mm (Phenomenex, Torrance, CA, USA). Vitamin C concentrations were evaluated using ultra-high efficiency HPLC liquid chromatography kit (LC-20AD, DGU-20A 3, SIL-20AC HT, CTO-10AS) with a UV detector (SPD-20A) from Shimadzu (Kyoto, Japan).

3.7. Total Antioxidant Activity (TAA)

3.7.1. Preparation of Samples for Analysis

The antioxidant potential for tubers obtained from field experiments was evaluated. For each potato cultivar and each repetition, three tubers were collected. They had been cut into small cubes and frozen in liquid nitrogen, lyophilised and milled. Samples (0.5 g) were extracted using ethanol (80%) and placed in an ultrasonic bath for 20 min. The extracts were transferred to falcon tubes (per 15 mL) and centrifuged at a speed of approximately 537× *g* for 10 min (centrifuge Heroeus multifuge 3 S-R, Thermo Fisher Scientific, Cleveland, OH, USA).

3.7.2. Measurements DPPH

Measurements were made using the spectrophotometric method. Up to 1900 μL of DPPH radical solution in methanol with a concentration of 0.05 mg ml^{-1} with 100 μL of sample added. The mixture was prepared in this way and the end reaction was set aside for 30 min. Then, the absorbance was measured at the length of $\lambda = 515$ nm. The measurements were repeated 3 times for each sample. The results are calculated based on the calibration curve for Trolox (6-hydroxy-2,5,7,8-tetramethylchroman-2-carboxylic acid).

3.7.3. Measurements FRAP

The measurements were made spectrophotometrically with a UV-Vis HITACHI U-1900 spectrophotometer (Tokyo, Japan). The absorbance was measured at 593 nm whereby 0.05 mL of the sample solution were added to 1 mL of FRAP reagent. The mixture prepared this way was incubated for 4 min at 37 °C. The FRAP results were calculated from the calibration curve for the Fe (II) standard solution. The FRAP unit was assumed to determine the ability to reduce 1 mole of iron (III) to iron (II). The results were calculated for the standard substance (Trolox) and presented as (μmol TE 100 mg^{-1} DM).

3.8. Statistical Analysis

The mean values for TC, VC and TAA were compared by using one- and two-way ANOVA to assess the effects of cultivar, year, and their interactions, and Tukey's multiple range test was used to identify significant differences between means. To compare mean values for groups of clones with or without markers and white and yellow fleshed, Student's *t*-test with unequal variances was applied. The Pearson correlation coefficient was used to estimate the relationships between the study variables. All statistical analyses were performed with the Statistica software (Dell, Round Rock, TX, USA), version 13.

4. Conclusions

Our results permit optimisation of the vitamin C and carotenoid content of the diet by using potato cvs., such as Twinner, Tinca, Levante, Kelly, Karlena, Agria and Colleen with high-vitamin C, and by using potato cvs., such as Otolia, Bionta, Anuschka, Belmonda, Carolus, Gardena, Salome, Belana, Twinner, Caprice, and Mayan Gold with high carotenoids content. Obtained results showed how breeding efforts have influenced the biofortification of these important phytonutrients in the potato tubers cultivars registered in European markets.

Supplementary Materials: The following supporting information can be downloaded at: https://www.mdpi.com/article/10.3390/ijms241411716/s1.

Author Contributions: Conceptualization, B.T.; methodology, B.T. and D.M.; validation, B.T., J.P. and D.M.; formal analysis, B.T.; resources, B.T., J.P. and D.M.; writing—original draft preparation, B.T.; writing—review and editing, B.T., J.P. and D.M.; visualization, B.T.; project administration, B.T. All authors have read and agreed to the published version of the manuscript.

Funding: This work was financed by the European Union Horizon 2020 project ECOBREED (771367).

Institutional Review Board Statement: Not applicable.

Informed Consent Statement: Not applicable.

Data Availability Statement: Supplementary data (climatic conditions, list of cultivars, results of chemical analyses) are available at https://doi.org/10.5281/zenodo.8012892.

Conflicts of Interest: The authors declare no conflict of interest.

References

1. FAOSTAT. 2018. Available online: http://www.fao.org/faostat/en/#data (accessed on 8 March 2018).
2. Hellmann, H.; Goyer, A.; Navarre, D.A. Antioxidants in potatoes: A functional view on one of the major food crops worldwide. *Molecules* **2021**, *26*, 2446. [CrossRef] [PubMed]

3. Lovat, C.; Nassar, A.M.; Kubow, S.; Li, X.Q.; Donnelly, D.J. Metabolic biosynthesis of potato (*Solanum tuberosum* L.) antioxidants and implications for human health. *Crit. Rev. Food Sci. Nutr.* **2016**, *56*, 2278–2303. [CrossRef] [PubMed]
4. Samtiya, M.; Aluko, R.E.; Dhewa, T.; Moreno-Rojas, J.M. Potential health benefits of plant food-derived bioactive components: An overview. *Foods* **2021**, *10*, 839. [CrossRef] [PubMed]
5. Grudzińska, M.; Mańkowski, D. Bioactive compounds in yellow, light yellow and cream-coloured potato tubers after short-term storage and boiling. *Ital. J. Food Sci.* **2020**, *32*, 4.
6. Gumul, D.; Ziobro, R.; Noga, M.; Sabat, R. Characterization of five potato cultivars according to their nutritional and pro-health components. *Acta Sci. Pol. Technol. Aliment.* **2011**, *10*, 77–81. [PubMed]
7. Tsao, R. Phytochemical profiles of potato and their roles in human health and wellness. *Food. Spec. Issue* **2009**, *3*, 125–135.
8. Keutgen, A.J.; Wszelaczyńska, E.; Pobereżny, J.; Przewodowska, A.; Przewodowski, W.; Milczarek, D.; Tatarowska, B.; Keutgen, N. Antioxidant properties of potato tubers (*Solanum tuberosum* L.) as a consequence of genetic potential and growing conditions. *PLoS ONE* **2019**, *14*, e0222976. [CrossRef]
9. Marone, D.; Mastrangelo, A.M.; Borrelli, G.M.; Mores, A.; Laidò, G.; Russo, M.A.; Ficco, D. Specialized metabolites: Physiological and biochemical role in stress resistance strategies to improve their accumulation and new applications in crop breeding and management. *Plant Physiol. Biochem.* **2022**, *172*, 48–55. [CrossRef]
10. Tatarowska, B.; Milczarek, D.; Wszelaczyńska, E.; Pobereżny, J.; Keutgen, N.; Keutgen, A.J.; Flis, B. Carotenoids variability of potato tubers in relation to genotype, growing location and year. *Am. J. Potato Res.* **2019**, *96*, 493–504. [CrossRef]
11. FAO. International Year of the Potato—Potato, Food and Nutrition. 2008. Available online: http://www.potato2008.org/en/potato/IYP-6en.pdf (accessed on 13 January 2022).
12. Ezekiel, R.; Singh, N.; Sharma, S.; Kaur, A. Beneficial phytochemicals n potato—A review. *Food Res. Int.* **2013**, *50*, 487–496. [CrossRef]
13. Hamouz, K.; Pazderu, K.; Lachman, J.; Čepl, J.; Kotíková, Z. Effect of cultivar, flesh colour, locality and year on carotenoid content in potato tubers. *Plant Soil Environ.* **2016**, *62*, 86–91. [CrossRef]
14. Hejtmánková, K.; Kotíková, Z.; Hamouz, K.; Pivec, V.; Vacek, J.; Lachman, J. Influence of flesh colour, year and growing area on carotenoid and anthocyanin content in potato tubers. *J. Food Compos. Anal.* **2013**, *32*, 20–27. [CrossRef]
15. Kotíková, Z.; Hejtmankova, A.; Lachman, J.; Hamouz, K.; Trnkova, E.; Dvorak, P. Effect of selected factors on total carotenoid content in potato tubers (*Solanum tuberosum* L.). *Plant Soil Environ.* **2007**, *53*, 355. [CrossRef]
16. Valcarcel, J.; Reilly, K.; Gaffney, M.; O'Brien, N. Total Carotenoids and L-Ascorbic Acid Content in 60 Varieties of Potato (*Solanum tuberosum* L.) Grown in Ireland. *Potato Res.* **2015**, *58*, 29–41. [CrossRef]
17. Iwanzik, W.; Tevini, M.; Stute, R.; Hilbert, R. Carotenoid content and composition of various German potato varieties and their importance for tuber flesh colour. *Potato Res.* **1983**, *26*, 149–162. [CrossRef]
18. Andre, C.M.; Oufir, M.; Guignard, C.; Hoffmann, L.; Hausman, J.F.; Evers, D.; Larondelle, Y. Antioxidant profiling of native Andean potato tubers (*Solanum tuberosum* L.) reveals cultivars with high levels of β-carotene, α-tocopherol, chlorogenic acid and petanin. *J. Agric. Food Chem.* **2007**, *55*, 10839–10849. [CrossRef]
19. Breithaupt, D.E.; Bamedi, A. Carotenoids and carotenoid esters in Potatoes (*Solanum tuberosum* L.): New Insights into an Ancient Vegetable. *J. Agric. Food Chem.* **2002**, *50*, 7175–7181. [CrossRef]
20. Haynes, K.G.; Clevidence, B.A.; Rao, D.; Vinyard, B.T.; White, J.M. Genotype × environment interactions for potato tuber carotenoid content. *J. Am. Soc. Hortic. Sci.* **2010**, *135*, 250–258. [CrossRef]
21. Bonierbale, M.; Grüneberg, W.; Amoros, W.; Burgos, G.; Salas, E.; Porras, E.; Zum Felde, T. Total and individual carotenoid profiles in *Solanum phureja* cultivated potatoes: II. Development and application of near-infrared reflectance spectroscopy (NIRS) calibrations for germplasm characterization. *J. Food Compos. Anal.* **2009**, *22*, 509–516. [CrossRef]
22. Andre, C.M.; Ghislain, M.; Bertin, P.; Oufir, M.; del Rosario Herrera, M.; Hoffmann, L.; Evers, D. Andean potato cultivars (*Solanum tuberosum* L.) as a source of antioxidant and mineral micronutrients. *J. Agric. Food Chem.* **2007**, *55*, 366–378. [CrossRef]
23. Othman, R. Biochemistry and Genetics of Carotenoid Composition in Potato Tubers. Ph.D. Thesis, Lincoln University, Lincoln, New Zealand, 2009; pp. 38–48.
24. Brown, C.R.; Kim, T.S.; Ganga, Z.; Haynes, K.; de Jong, D.; Jahn, M.; Paran, I.; de Jong, W. Segregation of total carotenoid in high level potato germplasm and its relationship to beta-carotene hydroxylase polymorphism. *Am. J. Potato Res.* **2006**, *83*, 365–372. [CrossRef]
25. Kloosterman, B.; Oortwijn, M.; Uitdewilligen, J.; America, T.; de Vos, R.; Visser, R.G.; Bachem, C.W. From QTL to candidate gene: Genetical genomics of simple and complex traits in potato using a pooling strategy. *BMC Genom.* **2010**, *11*, 158. [CrossRef] [PubMed]
26. Wolters, A.M.A.; Uitdewilligen, J.G.; Kloosterman, B.A.; Hutten, R.C.; Visser, R.G.; van Eck, H.J. Identification of alleles of carotenoid pathway genes important for zeaxanthin accumulation in potato tubers. *Plant Mol. Biol.* **2010**, *73*, 659–671. [CrossRef] [PubMed]
27. Brown, C.R.; Edwards, C.G.; Yang, C.P.; Dean, B.B. Orange flesh trait in potato: Inheritance and carotenoid content. *J. Am. Soc. Hortic. Sci.* **1993**, *118*, 145–150. [CrossRef]
28. Sulli, M.; Mandolino, G.; Sturaro, M.; Onofri, C.; Diretto, G.; Parisi, B.; Giuliano, G. Molecular and biochemical characterization of a potato collection with contrasting tuber carotenoid content. *PLoS ONE* **2017**, *12*, e0184143. [CrossRef]
29. Goo, Y.M.; Kim, T.W.; Ha, S.H.; Back, K.W.; Bae, J.M.; Shin, Y.W.; Lee, C.H.; Ahn, M.J.; Lee, S.W. Expression profiles of genes involved in the carotenoid biosynthetic pathway in yellow-fleshed potato cultivars (*Solanum tuberosum* L.) from South Korea. *J. Plant Biol.* **2009**, *52*, 49–55. [CrossRef]

30. Vreugdenhil, D.; Bradshaw, J.; Gebhardt, C.; Govers, F.; Taylor, M.A.; MacKerron, D.K.; Ross, H.A. *Potato Biology and Biotechnology. Advances and Perspectives*; Elsevier: Amsterdam, The Netherlands, 2007; 857p, ISBN 978-0-444-51018-1.
31. Dale, M.F.; Griffiths, W.; Todd, D. Effects of genotype, environment, and postharvest storage on the total ascorbate content of potato (*Solanum tuberosum*) tubers. *J. Agric. Food Chem.* **2003**, *51*, 244–248. [CrossRef]
32. Almeida, M.E.M.; Nogueira, J.N. The control of polyphenol oxidase activity in fruits and vegetables: A study of the interaction between the chemical compounds used and heat treatment. *Plant Foods Hum. Nutr.* **1995**, *47*, 245–256. [CrossRef]
33. Bates, C. Bioavailability of vitamin C. *Eur. J. Clin. Nutr.* **1997**, *1*, 28–33.
34. Burgos, G.; Auqui, S.; Amoros, W.; Salas, E.; Bonierbale, M. Ascorbic acid concentration of native Andean potato varieties as affected by environment, cooking and storage. *J. Food Compos. Anal.* **2009**, *22*, 533–538. [CrossRef]
35. Hamouz, K.; Lachman, J.; Dvořák, P.; Hejtmánková, K.; Čepl, J. Antioxidant activity in yellow and purple-fleshed potatoes cultivated in different climatic conditions. *Zesz. Probl. Postępów Nauk Rol.* **2008**, *530*, 241–247.
36. Murnice, I.; Karklina, D.; Galoburda, R.; Santare, R.; Santare, D.; Skrabule, I.; Costa, H.S. Nutritional composition of freshly harvested and stored Latvian potato (*Solanum tuberosum* L.) varieties depending on traditional cooking methods. *J. Food Comp. Anal.* **2011**, *24*, 699–710. [CrossRef]
37. Hamouz, K.; Lachman, J.; Dvořák, P.; Orsák, M.; Hejtmánková, K.; Čížek, M. Effect of selected factors on the content of ascorbic acid in potatoes with different tuber flesh colour. *Plant Soil Environ.* **2009**, *55*, 281–287. [CrossRef]
38. Han, J.S.; Kozukue, N.; Young, K.S.; Lee, K.R.; Friedman, M. Distribution of ascorbic acid in potato tubers and in home-processed and commercial potato foods. *J. Agric. Food Chem.* **2004**, *52*, 6516–6521. [CrossRef] [PubMed]
39. Jimenez, M.E.; Rossi, A.M.; Samman, N.C. Phenotypic, agronomic and nutritional characteristics of seven varieties of Andean potatoes. *J. Food Compos. Anal.* **2009**, *22*, 613–616. [CrossRef]
40. Nordbotten, A.; Loken, B.; Rimestad, H. Sampling of potatoes to determine representative values for nutrient content in a national food composition table. *J. Food Comp. Anal.* **2000**, *13*, 369–377. [CrossRef]
41. Sawicka, B.; Michałek, W.; Pszczółkowski, P. The relationship of potato tubers chemical composition with selected physiological indicators. *Zemdirbyste-Agriculture* **2014**, *102*, 41–50. [CrossRef]
42. Brown, C.R. Antioxidants in potato. *Am. J. Potato Res.* **2005**, *82*, 163–172. [CrossRef]
43. Trawczyński, C. Relationship between vitamin C and nitrate content in potato tubers depending on the maturity groups of cultivars. *J. Elem.* **2021**, *26*, 971–983. [CrossRef]
44. Hrabovská, D.; Heldák, J.; Volnová, B. Changes in the content of vitamin C in potato tubers depending on variety. *J. Microbiol. Biotechnol. Food Sci.* **2013**, 2052–2058.
45. Hejtmánková, K.; Pivec, V.; Trnková, E.; Hamouz, K.; Lachman, J. Quality of coloured varieties of potatoes. *Czech J. Food Sci.* **2009**, *27*, 310–313. [CrossRef]
46. Wichrowska, D. Antioxidant capacity and Nutritional Value of potato tubers (*Solanum tuberosum* L.) as a Dependence of Growing Conditions and Long-Term Storage. *Agriculture* **2021**, *12*, 21. [CrossRef]
47. Hu, C.; Tsao, R.; Liu, R.; Alan Sullivan, J.; McDonald, M.R. Influence of cultivar and year on phytochemical and antioxidant activity of potato (*Solanum tuberosum* L.) in Ontario. *Can. J. Plant Sci.* **2012**, *92*, 485–493. [CrossRef]
48. Lee, S.H.; Oh, S.H.; Hwang, I.G.; Kim, H.Y.; Woo, K.S.; Woo, S.H.; Jeong, H.S. Antioxidant contents and antioxidant activities of white and colored potatoes (*Solanum tuberosum* L.). *Prev. Nutr. Food Sci.* **2016**, *21*, 110. [CrossRef] [PubMed]
49. Kim, J.; Soh, S.Y.; Baer, H.; Nam, S.Y. Antioxidant and phenolic contents in potatoes (*Solanum tuberosum* L.) and micropropagated potatoes. *Appl. Biol. Chem.* **2019**, *62*, 17. [CrossRef]
50. Fidrianny, I.; Suhendy, H.; Insanu, M. Correlation of phytochemical content with antioxidant potential of various sweet potato (*Ipomoea batatas*) in West Java, Indonesia. *Asian Pac. J. Trop. Biomed.* **2018**, *8*, 25. [CrossRef]
51. Seijo-Rodríguez, A.; Escuredo, O.; Rodríguez-Flores, M.S.; Seijo-Coello, M.C. Assessment of Antioxidant Potential of Potato Varieties and the Relationship to Chemical and Colourimetric Measurements. *Am. J. Potato Res.* **2018**, *95*, 71–78. [CrossRef]
52. Milczarek, D.; Tatarowska, B. Evaluation of potato cultivars and breeding lines for carotenoids content in tubers. *Plant Breed. Seed Sci.* **2017**, *75*, 11–16. [CrossRef]
53. ASTM (American Society for Testing and Materials). *E 313–05 Standard Practice for Calculating Yellowness and Whiteness Indices from Instrumentally Measured Color Coordinates*; ASTM International: West Conshohocken, PA, USA, 2005.

International Journal of
Molecular Sciences

MDPI

Article

Staphylococcus aureus Modulates Carotenoid and Phospholipid Content in Response to Oxygen-Restricted Growth Conditions, Triggering Changes in Membrane Biophysical Properties

Laura Zamudio-Chávez [1], Elizabeth Suesca [1], Gerson-Dirceu López [2], Chiara Carazzone [3], Marcela Manrique-Moreno [4] and Chad Leidy [1,*]

1 Biophysics Group, Physics Department, Universidad de los Andes, Bogotá 111211, Colombia; ld.zamudio@uniandes.edu.co (L.Z.-C.); e.suesca87@uniandes.edu.co (E.S.)
2 PhysCheMath Research Group, Chemistry Department, Universidad de América, Bogotá 111211, Colombia; gerson.lopez@profesores.uamerica.edu.co
3 Laboratory of Advanced Analytical Techniques in Natural Products (LATNAP), Chemistry Department, Universidad de los Andes, Bogotá 111211, Colombia; c.carazzone@uniandes.edu.co
4 Chemistry Institute, Faculty of Exact and Natural Sciences, University of Antioquia, Medellin 050010, Colombia; marcela.manrique@udea.edu.co
* Correspondence: cleidy@uniandes.edu.co; Tel.: +57-1-3394949 (ext. 2741)

Abstract: *Staphylococcus aureus* membranes contain carotenoids formed during the biosynthesis of staphyloxanthin. These carotenoids are considered virulence factors due to their activity as scavengers of reactive oxygen species and as inhibitors of antimicrobial peptides. Here, we show that the growth of *S. aureus* under oxygen-restricting conditions downregulates carotenoid biosynthesis and modifies phospholipid content in biofilms and planktonic cells analyzed using LC-MS. At oxygen-restrictive levels, the staphyloxanthin precursor 4,4-diapophytofluene accumulates, indicating that the dehydrogenation reaction catalyzed by 4,4′-diapophytoene desaturases (CrtN) is inhibited. An increase in lysyl-phosphatidylglycerol is observed under oxygen-restrictive conditions in planktonic cells, and high levels of cardiolipin are detected in biofilms compared to planktonic cells. Under oxygen-restriction conditions, the biophysical parameters of *S. aureus* membranes show an increase in lipid headgroup spacing, as measured with Laurdan GP, and decreased bilayer core order, as measured with DPH anisotropy. An increase in the liquid–crystalline to gel phase melting temperature, as measured with FTIR, is also observed. *S. aureus* membranes are therefore less condensed under oxygen-restriction conditions at 37 °C. However, the lack of carotenoids leads to a highly ordered gel phase at low temperatures, around 15 °C. Carotenoids are therefore likely to be low in *S. aureus* found in tissues with low oxygen levels, such as abscesses, leading to altered membrane biophysical properties.

Keywords: *Staphylococcus aureus*; biofilms; staphyloxanthin; LC-DAD-APCI-MS/MS; phospholipids; LC-ESI-MS/MS; fluorescence spectroscopy; infrared spectroscopy

1. Introduction

Staphylococcus aureus (*S. aureus*) is an opportunistic pathogen; a Gram-positive, facultative anaerobic bacterium, naturally present in human skin and mucous membranes [1,2], which can also invade and reside in the interior of host cells, including human keratinocytes [3]. *S. aureus* is responsible for multiple hospital-acquired diseases such as pneumonia, different kinds of human skin infections, osteomyelitis, endocarditis and sepsis, presenting an important health risk worldwide [1]. Due to the emergence of multiple-drug resistant strains [4], understanding the physiology of this bacterium is crucial in the development of complementary treatment strategies.

S. aureus has evolved the ability to modify the lipid composition of its membranes in response to environmental stresses such as oxidative stress, osmotic stress, and the presence of antimicrobial agents [5–8], as well as during changes in *S. aureus* growth conditions [6,8]. Modifications in phospholipid headgroup and acyl chain composition or carotenoid content influence the biophysical properties of *S. aureus* membranes [6,9], leading to changes in membrane rigidity and lipid packing [6,8,10,11].

The biosynthesis of carotenoids in *S. aureus* has been identified as an important virulence factor leading to resistance to antimicrobial peptide activity and to reactive oxygen species (ROS) generated during the immune response. Carotenoids also act as scavengers of free radicals (hydroxyl OH) generated during oxidative stress or during cellular respiration [12]. The production of different carotenoid species in *S. aureus* is mainly associated with the biosynthesis of staphyloxanthin (STX). The STX biosynthetic pathway begins with the dimerization of two farnesyl diphosphates to form 4,4-diapophytofluene (4,4′-DPE) [8]. This is followed by a series of dehydrogenase reactions catalyzed by 4,4′-diapophytoene desaturases (CrtN). Oxidation reactions then lead to the formation of 4,4′-diaponeurosporenoic acid (4,4′-DNPA). The carboxylic acid is then functionalized with a glucose, through CrtQ, and a fatty acid with a varying acyl chain is finally attached, through CrtO, to form STX [8,13]. Variations in oxygen levels have been shown to modulate this biosynthetic path [14]. Due to a series of conjugated double bonds, the rigid structure of the diaponeurosporenoic acid leads to increased membrane rigidity, which confers on *S. aureus* a higher resistance to antimicrobial peptides [15]. This is in addition to its protective role as a free radical scavenger [8,13,16–18]

The main phospholipid components in *S. aureus* membranes are the anionic phospholipids phosphatidyl-glycerol (PG) and cardiolipin (CL), as well as the cationic phospholipid lysyl-phosphatidylglycerol (Lys-PG), which are present in the membrane with varying acyl chain lengths [6,10,19]. Phospholipid acyl chains are mainly saturated or branched in *iso-* and *anteiso-* conformations, where the branched/saturated ratio helps modulate the level of lipid packing in the membrane [6,20]. CL is an anionic tetra-acyl chained glycerophospholipid, which has been shown to increase lipid packing levels [21], modulate membrane curvature [22], and confer lipid bilayer membranes with increased mechanical resistance towards antimicrobial peptides [21,23]. The MprF-mediated biosynthesis of the cationic lipid Lysyl-PG modulates the surface charge of the membrane, decreasing the anionic surface charge, and leading to a reduction in the affinity of antimicrobial peptides, therefore inhibiting their lytic activity [9].

S. aureus can invade different tissues in the host, where oxygen levels vary widely [24], showing the capacity to live at low oxygen levels [25] or even under fully anaerobic conditions in certain tissues such as abscesses [26]. For this reason, *S. aureus*, as a facultative anaerobic bacterium, relies on switching its metabolism from aerobic respiration, in the presence of excess oxygen, to a combination of fermentation and anaerobic respiration, using alternative electron acceptors such as nitrate, at low oxygen levels [24]. This switch in metabolism at oxygen-restrictive levels leads to lower ATP production and slower growth. One of the main roles that carotenoids play is as scavengers of free radicals, which are produced in high quantities at high oxygen levels. A two-component AirSR oxygen response regulator in *S. aureus* has been reported to positively regulate the STX biosynthetic pathway operon *crtOPQMN* in the presence of increased oxygen levels, showing the importance of STX biosynthesis during oxidative stress [14]. The downregulation of STX production at low oxygen levels may help reduce metabolic costs under anaerobic respiration or fermentation, since the need to scavenge oxygen free-radical species will be reduced.

In the present study, *S. aureus* was grown either as biofilms or under planktonic conditions, exploring how the membrane lipid composition was modulated under restrictive and non-restrictive oxygen levels. Since carotenoid production in *S. aureus* is closely tied to available oxygen levels, we mainly focused on changes in carotenoid content as determined via liquid chromatography coupled to mass spectrometry (LC-MS), reporting on overall levels of carotenoids and the proportion of STX precursors for the different growth conditions. We also report on changes in phospholipid headgroup composition, showing shifts in the proportion of PG, CL, and Lysyl-PG phospholipids under restrictive oxygen growth conditions. These shifts in lipid composition alter the biophysical properties of *S. aureus* membranes. Changes in lipid headgroup spacing measured with Laurdan Generalized Polarization (Laurdan GP), bilayer core dynamics measured with DPH anisotropy, and overall lipid phase behavior measured with Fourier transform infrared (FTIR) spectroscopy are reported. Scanning electron microscopy images are used to evaluate changes in biofilm and planktonic average cell size and changes in the production of exopolysaccharides in biofilm cells. Oxygen levels vary greatly for different body parts and tissues [27]. For this reason, understanding how *S. aureus* lipid composition is affected by oxygen, and how bacterial membrane biophysical properties are affected by these shifts, may be relevant in understanding the susceptibility of *S. aureus* in different tissues to treatments involving oxidative stress or antimicrobial peptide treatments.

2. Results

2.1. Cell Morphologies in Biofilm and Planktonic Cells and Production of Exopolysaccharides in Biofilms

Images obtained with scanning electron microcopy (SEM) are shown in Figure 1 for planktonic cells and biofilms under different aeration conditions. In Figure 1c,d, the formation of biofilm structures, characterized by an increase in exopolysaccharide structures, is evident in the images for both aerated and non-aerated conditions. However, exopolysaccharides are more prevalent for aerated biofilms (Figure 1c). Biofilms can be found in different stages of development [28,29], which are the initial attachment where the planktonic cells attach to surface due to electrostatic interaction, the growth phase where cells will begin to divide and accumulate to obtain a mature biofilm, and finally the detachment stage where cells are dispersed over the catheter. The biofilm images shown in Figure 1c,d are likely in different growth stages due to the amount of biomass that is observed in the sample, where the non-aerated samples showed more biomass.

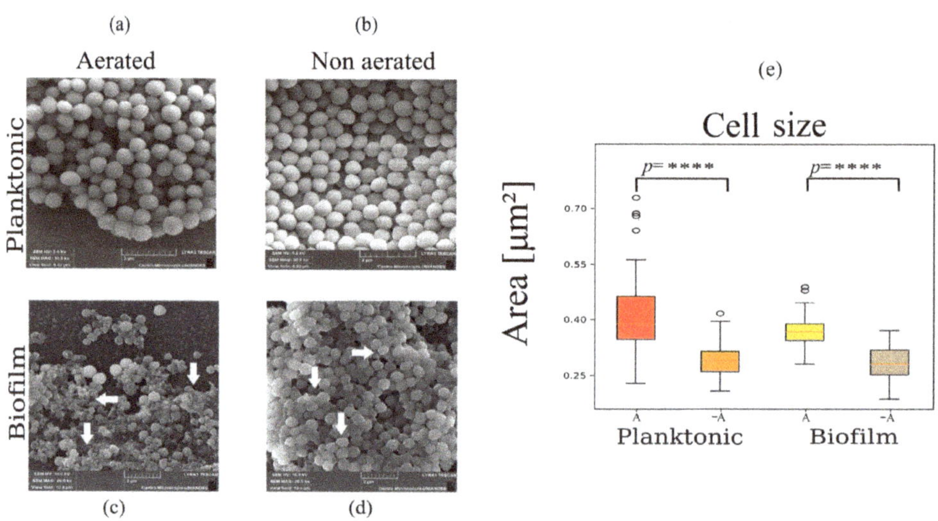

Figure 1. Scanning electron microscopy images show changes in cell morphologies for planktonic (**a**,**b**) and biofilm cells (**c**,**d**) grown under different aeration conditions. Aerated biofilms show variations in cell morphologies, the production of exopolysaccharides, and the presence of damaged cells, as pointed out by the arrows in (**c**). Non-aerated biofilms show tightly packed cells covered by a smooth exopolysaccharide structure as pointed out by the arrows in (**d**). (**e**) Boxplots showing the mean area of the cells, where open circles show the dispersal of the data points. These data show significant difference in the size between the different growth conditions (**** $p < 0.0001$).

Analysis of cell size, measured as area per cell, showed a statistically significant reduction in cell size under non-aerated conditions for planktonic and biofilm growth states (Figure 1e and Table 1). This is likely induced by a switch from aerobic to anaerobic respiration and upregulation of fermentation which has a lower ATP yield [24].

Table 1. Mean area of *S. aureus* cells measured using scanning electron microcopy images for the different growth conditions. Non-aerated cells are observed to be smaller than aerated cells for biofilms and planktonic growth conditions.

Growth Condition	Average Cell Area [$\mu m^2 \pm$S.D.]
Aerated biofilms	0.41 ± 0.12
Non-aerated biofilms	0.29 ± 0.05
Aerated planktonic cells	0.37 ± 0.04
Non-aerated planktonic cells	0.28 ± 0.04

Additionally, exopolysaccharide structures differed between aerated and non-aerated samples for biofilms. In the case of biofilms fed with aerated media, images show dispersed cells embedded in a rugged and extensive exopolysaccharide structure (Figure 1c), while under oxygen-restricting conditions, biofilms showed densely packed cells closely surrounded by a smooth exopolysaccharide structure (Figure 1d).

2.2. *Biofilms Pigmentation Levels Vary When Exposed to Oxygen Stress*

For planktonic cells, significant differences in pigmentation were detected by modifying agitation levels (Figure 2). Total lipid extractions were obtained from these samples. The oxygen-restricting culture always showed a lower OD than the aerated culture which agrees with a lower cell count, as expected under non-aerated conditions. In the case of biofilms, a variation in the experimental set up, introducing aeration of the feeding media (Figure 2a), induced changes in cell pigmentation and biomass, showing a remarkable loss in pigmentation and reduction of biofilm mass under oxygen-restricting conditions (Figure 2b,c). Total lipids were extracted from the samples following the protocol in [6,8]. Lipid yields for biofilm samples were significantly lower due to the high amount of protein in the samples.

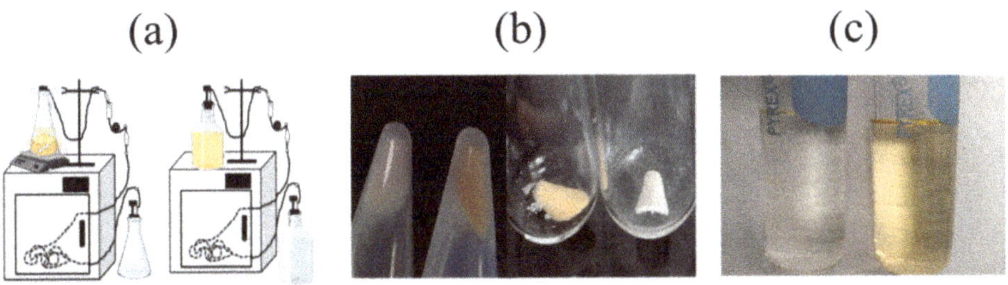

Figure 2. Biofilms lose pigmentation levels in response to oxygen restriction. (**a**) Experimental setup for biofilm formation in the presence (left) and absence (right) of aeration. Incubation temperature is maintained at 37 °C. (**b**) Biofilm obtained from the catheter before (left) and after (right) lyophilization, showing variations in the pigmentation levels. (**c**) Lipid sample after extraction; the lighter-colored sample is the biofilm grown under low aeration conditions (left). Biofilms grown in the presence of higher aeration (right) levels show intense pigmentation.

2.3. *Carotenoid and Phospholipid Content and Composition*

Carotenoid content and composition was analyzed with liquid chromatography-diode array detection-atmospheric pressure chemical ionization-tandem mass spectrometry (LC-DAD-APCI-MS/MS) for all growth conditions, and the results are presented in Figure 3 and Table 2. The three chemical species that were evaluated were 4,4′-DPE, a precursor of STX that results from the dimerization of two farnesyl diphosphates; 4,4′-DNPA, which presents an extensive conjugated double bond structure, but is not functionalized with glucose; and the different STX species varying in the chain length of the fatty acid group attached to the glucose (Figure 3a) [8]. We observed a significant reduction in the overall amounts of STX species produced in the cells reported as μg/g dry cell weight (orange pallet in Figure 3b) under oxygen-restricting conditions. The precursor species 4,4′-DPE (Figure 3a,c) greatly accumulates in planktonic cells under oxygen-restricting conditions (blue pallet in Figure 3b), indicating that the dehydrogenation reaction catalyzed by 4,4′-diapophytoene desaturases (CrtN) is inhibited when oxygen is restricted. Low concentrations of 4,4′-DNPA and STX (Figure 3b,d) were also found for these conditions. Accumulation of 4,4′-DPE was also observed in non-aerated biofilms, but to a smaller degree. 4,4′-DNPA appears to accumulate more prevalently in biofilms at the expense of STX compared to planktonic cells (yellow in Figure 3b). This may be a result of reduced levels of available glucose in the biofilm growth phase.

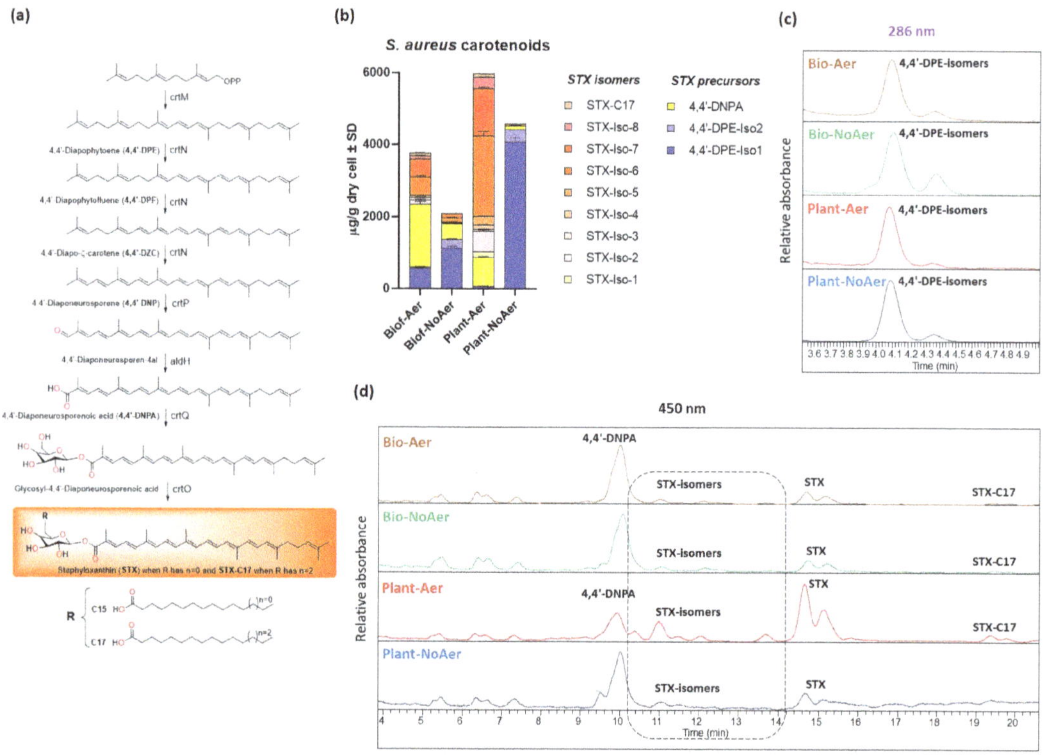

Figure 3. Carotenoids analyzed using LC-DAD-APCI-MS/MS from biofilms and planktonic *S. aureus* cells. (**a**) Biosynthetic pathway for carotenoids in *S. aureus* (**b**) Carotenoid concentration in μg/g dry *S. aureus* cell. (**c**) Chromatogram of 4,4′-Diapophytoene isomers at 286 nm. (**d**) Chromatogram of carotenoids at 450 nm.

Figure 4 and Table S1 shows an analysis of the main phospholipid components of *S. aureus*, PG, CL, and Lysyl-PG for biofilms and planktonic cells. The chemical structure of the three phospholipid species is shown in Figure 4a. PG is an anionic phospholipid with the negative charge in the phosphate group at physiological pH with a glycerol attached at the headgroup side of the molecule. CL has two anionic phosphates attached through a glycerol group, and a total of four acyl chains.

Table 2. Concentrations of carotenoids and carotenoids precursor in biofilms and planktonic *S. aureus* cells.

No.	Compound	Biof-Aero	Biof-NoAero	Plant-Aer	Plant-NoAer
		(μg/g Dry Cell Weight ± SD)			
1	4,4-DPE-Iso1	556.72 ± 22.29	1129.74 ± 56.27	67.36 ± 2.51	4069.07 ± 125.34
2	4,4-DPE-Iso2	42.78 ± 3.45	252.12 ± 7.31	n.d.	346.36 ± 3.85
3	4,4-DNPA	1742.18 ± 88.96	437.63 ± 19.86	802.07 ± 20.87	111.24 ± 11.81
4	STX-Iso-1	n.d.	n.d.	n.d.	146.78 ± 6.91
5	STX-Iso-2	132.17 ± 11.52	22.06 ± 2.29	591.55 ± 32.91	n.d.
6	STX-Iso-3	64.23 ± 4.48	17.76 ± 2.16	49.72 ± 4.87	n.d.
7	STX-Iso-4	42.20 ± 2.83	n.d.	121.79 ± 4.80	n.d.
8	STX-Iso-5	n.d.	n.d.	243.92 ± 6.30	n.d.
9	STX-Iso-6	520.01 ± 36.17	119.57 ± 9.81 [a]	2232.06 ± 112.15	41.83 ± 3.73 [a]
10	STX-Iso-7	491.00 ± 15.36	109.58 ± 4.50	1316.68 ± 34.64	n.d.
11	STX-Iso-8	93.78 ± 4.66	n.d.	307.91 ± 5.40	n.d.
12	STX-C17	80.20 ± 5.40 [b]	n.d.	92.29 ± 4.72 [b]	n.d.
	Total carotenoids precursor	599.50 ± 36.34	1376.86 ± 61.70	67.36 ± 2.51	4415.43 ± 263.23
	Total carotenoids	3168.78 ± 57.73	706.59 ± 15.69	5758.00 ± 72.50	299.85 ± 53.38

Equal letters in the same rows are not significantly different, according to the ANOVA test (Tukey, $p < 0.05$).
n.d. = not detected.

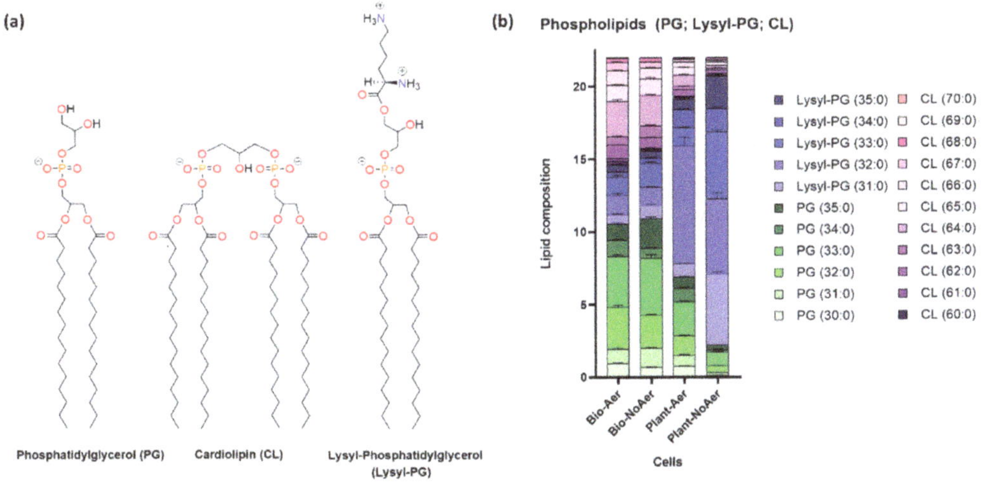

Figure 4. (**a**) Molecular structures of main phospholipid components. (**b**) Lipid analysis by LC-ESI-MS/MS shows shifts in the proportions of PG, CL, and Lysyl-PG for the different growth conditions. An increase in the proportion of Lys-PG is observed for planktonic *S. aureus* cells under oxygen-restricting growth conditions.

When comparing biofilms with planktonic cells (Figure 4b), the results show an increase in the proportion of CL in the membranes of biofilm cells compared to planktonic cells. Phospholipid composition in biofilms is observed to be independent of aeration levels (Figure 4b). In contrast, planktonic cells show an increase in the proportion of Lysyl-PG under oxygen-restricting growth conditions compared to aerated planktonic cells. The mass spectra of PG, Lysyl-PG, CL, and STX are presented in Figure S1.

2.4. Biophysical Properties of Membranes

The biophysical properties of the bilayers under different growth conditions were probed using a variety of techniques in order to obtain an integrated view of the effect of carotenoid production on membrane properties. This is performed as a function of temperature to assess the gel to liquid-crystalline phase transitions that are present in *S. aureus* membranes. At lower temperatures, the lipid bilayer is normally in the gel phase regime, which presents increased packing and rigidity [30]. At higher temperatures, the membrane enters the liquid-crystalline phase regime, which has higher levels of acyl chain disorder and decreased headgroup packing (among other parameters that change). The liquid-crystalline phase is considered to be the physiologically functional phase that allows the bacteria to undergo metabolism and cell division.

Laurdan GP fluorescence spectroscopy was used to detect changes in lipid headgroup spacing for the different samples as a function of temperature. Laurdan is a fluorescent molecule with an emission spectrum that is sensitive to the polarity of the surrounding environment, resulting in a red-shifted emission spectrum when the polarity increases. This shift was measured by calculating the Laurdan GP parameter (see methods). The fluorescence moiety of Laurdan positions itself at the headgroup/tail region of the bilayer. A reduction in GP is associated with increased hydration of this region due to an increase in headgroup spacing. For this reason, Laurdan probes lipid headgroup spacing, which decreases when lipid packing increases.

All measurements showed an increase in headgroup spacing as temperature increases, evidenced by a reduction in Laurdan GP with temperature (Figure 5a,b). This is consistent with the membranes entering the liquid-crystalline phase, which is characterized by increased headgroup spacing. In Figure 4c,d, the Laurdan GP values are compared in the gel phase regime (15 °C) and the liquid-crystalline phase regime (37 °C).

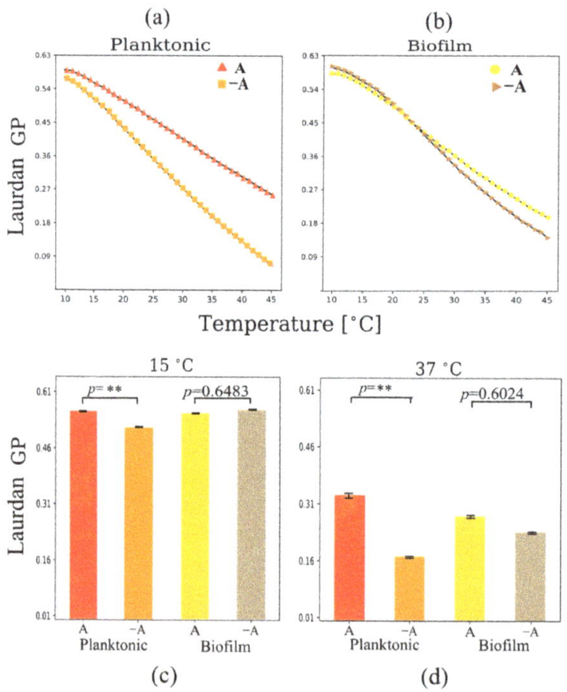

Figure 5. Laurdan GP polarization as a function of temperature for multilamellar vesicles from lipid extracts of (**a**) planktonic and (**b**) biofilm systems. Changes in Laurdan GP at (**c**) 15 °C and (**d**) 37 °C under planktonic and biofilm growth conditions in the presence and absence of aeration. Planktonic cells show a reduction in headgroup spacing under aerobic conditions. Measurements were performed in triplicate. Error bars in (**a**,**b**) are smaller than the symbols. ** $p < 0.01$.

In relation to aeration conditions, both biofilm and planktonic cells showed higher levels of Laurdan GP under aerated conditions at physiological temperatures (37 °C), indicating an increase in lipid packing related to a reduction in headgroup spacing as carotenoid concentration increases (Figure 5c,d). However, biofilms do not show notable differences in GP. The data indicate that non-aerated biofilms show high levels of lipid packing even in the absence of carotenoids. Only planktonic cells show significant differences in headgroup spacing in the presence of carotenoids, and these differences are relatively small, indicating that carotenoid content does not affect significantly headgroup spacing.

DPH fluorescence anisotropy was used to explore the level of disorder at the bilayer core for the different growth conditions [31]. When DPH is excited with polarized light, the molecule also emits polarized light. However, DPH light polarization is lost after excitation if the tumbling rate of the molecule is high compared to the excitation lifetime. A reduction in DPH fluorescence polarization anisotropy indicated high DPH tumbling rates, and therefore high levels of disorder at the bilayer hydrophobic core, where DPH is positioned. DPH anisotropy results also showed a decrease with temperature for all samples, which is consistent with the increased level of acyl-chain disorder in the liquid-crystalline phase compared to the gel phase (Figure 6a,b). Again, a cooperative melting event is not evident from the DPH thermograms.

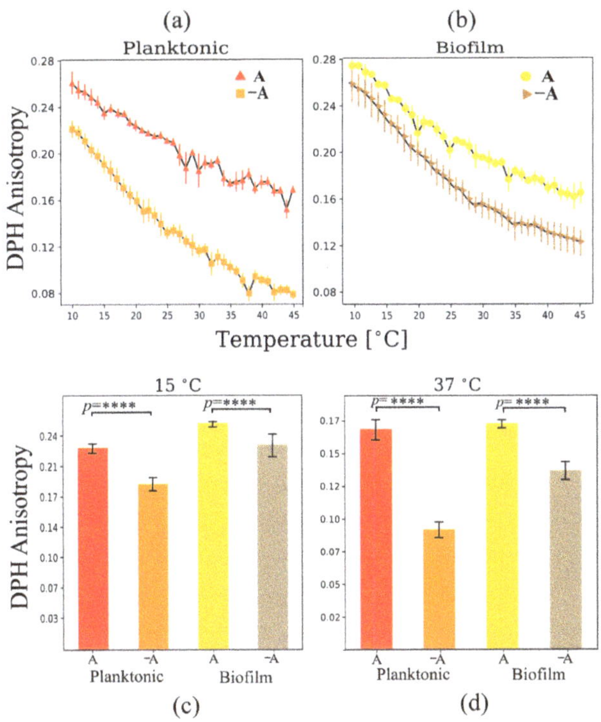

Figure 6. DPH anisotropy as a function of the temperature for multilamellar vesicles from lipid extracts of planktonic and biofilm systems for (**a**) planktonic and (**b**) biofilm growth states. Changes in DPH anisotropy at (**c**) 15 °C and (**d**) 37 °C under planktonic and biofilm growth conditions in the presence and absence of aeration. The anaerobic samples showed lower levels of lipid packing in the liquid-crystalline phase temperature regime. **** $p < 0.0001$.

In contrast with Laurdan GP, DPH anisotropy changes significantly in the presence of carotenoids induced by aeration for both planktonic and biofilm samples. The increase in order at the membrane core occurs both in the gel phase regime at 15 °C and in the liquid-crystalline regime at 37 °C (Figure 6c,d). This is consistent with previous results that show that the presence of carotenoids increases DPH anisotropy levels at 37 °C [6]. In general, the presence of carotenoids increases membrane rigidity at the bilayer core [6,8,32].

Fourier transform infrared spectroscopy (FTIR) is used to monitor the symmetric stretching vibration ($vsCH_2$) of the acyl chains in the bacterial membrane [33]. The wavenumber presents a sharp increase when the membrane transitions from the gel phase to the liquid-crystalline phase, as the number of +gauche and -gauche rotomers increases in relation to the trans rotomers when the acyl chains become disordered. An increase in the number of gauche rotomers compared to the trans configuration results in an increase in the $vsCH_2$. Therefore, this vibrational parameter is an indicator of the level of acyl-chain disorder in the bilayer membrane. FTIR is also very useful in detecting the cooperative melting event in membranes, as the $vsCH_2$ vibration undergoes a cooperative shift at the melting temperature [34].

Overall, an increase in the $vsCH_2$ stretch wavenumber is observed with increasing temperature for all samples (Figure 7a,b). Both planktonic and biofilm cells show cooperative events, where the wavenumber changes drastically. In contrast with the DPH fluorescence anisotropy results, the level of chain disorder does not increase significantly at 37 °C in the presence of carotenoids under aerated conditions (See Figure 7a,b at 37 °C). This indicates that, although the lipid bilayer core mobility decreases drastically with increased carotenoid content, as indicated by DPH fluorescence anisotropy (Figure 6d), the lipid acyl chains remain relatively disordered.

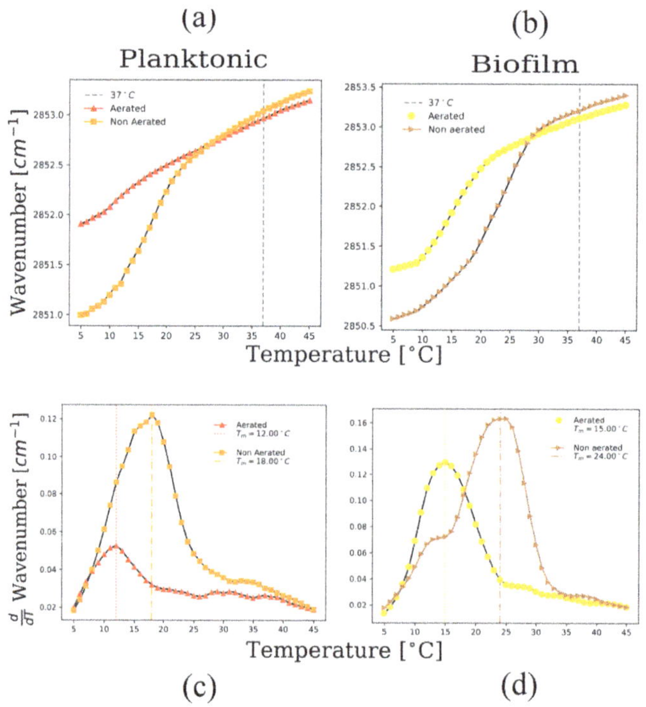

Figure 7. FTIR spectroscopy as a function of temperature for multilamellar vesicles from lipid extracts of planktonic and biofilm systems. Peak position of the symmetric CH_2 stretching vibration band of the acyl chains as a function of temperature for (**a**) biofilms, and (**b**) planktonic cells under different aeration levels. In (**c**,**d**) the first derivative of the wavenumber as a function of temperature is plotted for the different systems, showing in general that the liquid-crystalline to gel phase transition temperature decreases under aerated conditions.

In the gel phase regime, acyl chain disorder is significantly higher in the presence of carotenoids (Figure 7a,b). The presence of carotenoids induced by aeration keeps the membranes highly disordered in the gel phase regime. This is likely caused by the diaponeurosporenoic chain in STX, as well as the free diaponeurosporenoic acid that disrupts lipid packing in the gel phase regime [7,31]. The destabilization of the gel phase in the presence of carotenoids leads to a strong decrease in the gel to liquid-crystalline phase temperature (Figure 7c,d) for both planktonic and biofilm growth states under aerated conditions. The results point to an unusual result. The presence of carotenoids appears to stabilize the liquid-crystalline phase even though this phase has increased rigidity. This is explained by the effects of carotenoids in the gel phase regime, where a very large increase in acyl chain disorder occurs at 15 °C. This disordering effect in the gel phase is the dominant phenomena which therefore leads to a depression of the phase transition temperature.

3. Discussion

3.1. Regulation of Lipid Composition in S. aureus Biofilms and Planktonic Cells

S. aureus biofilms play an important role in chronic infections, in particular hospital-acquired *S. aureus* infections related to the use of contaminated catheters or implants. Biofilms are resilient to different treatments and stress factors, making them difficult to treat effectively [35,36]. It is known that the lipid composition of *S. aureus* membranes can influence its ability to resist the lytic activity of antimicrobial peptides. In particular, the increased level of lipid packing induced by the presence of carotenoids in the membrane leads to increase resilience to antimicrobial peptides such as the human neutrophil defensin-1 [15]. Cardiolipin has also been shown to increase resilience to cationic antimicrobial peptide activity [21], and acidic lipopeptides such as daptomycin [17], due to its role as an inhibitor of pore formation [23]. Carotenoids are also critical in protecting the cell from oxidative stress due to the emergence of reactive oxygen species generated as an immune response [14,37]. Understanding how variations in the main lipid components of *S. aureus* membranes are modulated in response to environmental factors may be useful in identifying under what circumstances will *S. aureus* be susceptible or resistant to these defense mechanisms.

In a previous publication, catheter-induced *S. aureus* biofilm formation was shown to strongly influence lipid composition [6], strongly inhibiting the biosynthesis of carotenoids. The present work shows that the lack of aeration in the medium is responsible for inhibiting carotenoid production in *S. aureus* biofilms, and that low aeration levels also inhibit carotenoid production in planktonic cells. This regulation is in line with a recently reported AirSR oxygen detection system in *S. aureus*, which regulates oxygen-sensing and redox-signaling, and which upregulated the biosynthesis of STX in response to oxidative stress [14]. In this previous publication, the AirSR signaling system was shown to positively regulate the STX biosynthetic pathways via operon crtOPQMN. Figure 3b shows that oxygen-restricting growth conditions lead to the accumulation of the precursor 4,4′-DPE, which indicates that CrtN is downregulated at low oxygen levels. It is also worthwhile noting that, overall, there is a higher accumulation of STX biosynthesis species in planktonic cells compared to biofilms, indicating higher levels of lipid metabolism in the planktonic growth condition (Figure 3b). In addition, the non-glycosylated 4,4′-DNPA species accumulates in biofilms indicating that the completion of the biosynthesis of STX may be due to the lack of accessible glucose in biofilms.

Regarding phospholipid composition, the present results show that cardiolipin appears in high proportions in both aerated and non-aerated biofilms (Figure 4b). This molecule is known to increase mechanical strength and induce an increase in membrane rigidity [21,23]. Its increased presence in *S. aureus* biofilms is likely to increase the resistance of biofilm cells to osmotic stress [38], and acute acid stress [39], in addition to its role as inhibitor of antimicrobial peptide activity [21,40]. In contrast to carotenoids, phospholipid composition does not appear to be very sensitive to aeration levels in biofilms. Instead, planktonic cells do show a pronounced increase in Lysyl-PG under oxygen-restricting growth conditions. Lysyl-PG is a cationic lipid that changes the surface electric potential of *S. aureus* membranes. It has been shown to act as a defense mechanism of cationic antimicrobial peptides by reducing the binding affinity of the peptides to the membrane surface [41].

3.2. Variation in the Biophysical Properties of S. aureus Membranes

A variety of physical techniques were used to assess how the biophysical properties of *S. aureus* membranes change in biofilms and planktonic cells under oxygen-restricting or aerated growth conditions. The physical aspects that were explored were lipid headgroup spacing as measured with Laurdan fluorescence GP, mobility in the membrane core as measured with DPH fluorescence anisotropy, and lipid acyl chain order as measured with FTIR spectroscopy. Thermotropic measurements were performed in order to observe the behavior of the membrane in the gel phase regime at low temperatures and the liquid-crystalline regime at high temperatures, and also to assess gel to liquid-crystalline phase transition temperatures, which are important indicators of the membrane physical state [6,8,10,11]. Downward shifts in the phase transition temperature indicate that the liquid-crystalline phase of the membrane is stabilized with respect to the gel phase.

Under aerated growth conditions, headgroup spacing was observed to decrease in planktonic cells (Figure 4). indicating a more condensed membrane. However, the changes in headgroup spacing were not significant in biofilms when varying aeration levels (Figure 4c,d). In contrast, the level of lipid bilayer mobility at the core decreases significantly for both biofilms and planktonic cells in the presence of carotenoids (Figure 5). These results are consistent with previous publications [31,42]. It is interesting to compare these results with those found via FTIR. At 37 °C, the acyl chain order does not vary significantly in the presence of carotenoids. This is in sharp contrast with the DPH fluorescence anisotropy results. A possible explanation for this is that the presence of carotenoids somehow maintains acyl chain disorder but keeps low internal mobility. This could be explained by the fact that the 4,4′-DNPA present in STX spans a distance longer than the thickness of the bilayer. This may imply that the molecule may anchor the leaflets without perturbing the acyl chain order very much but reducing internal mobility. An angled orientation of polar xanthophylls to fit into the membrane geometry has been reported [43]. This angled configuration may lead to disorder in the lipid acyl chains, while increasing the structure of the lipid bilayer core.

It is important to notice that in the gel phase regime, acyl chain order is significantly affected by the level of aeration (Figure 6), showing a sharp decrease in acyl chain order at 15 °C (Figure 6 and Table 3). This coupled to a sharp decrease in the gel to liquid-crystalline phase transition temperature (Table 3), indicating that the gel phase is destabilized when carotenoids are abundant. This result is important due to the fact that the more disordered liquid-crystalline phase is considered the physiologically relevant phase due to its increased mobility. By making the gel phase more disordered, carotenoid production may help the bacteria maintain a physiologically active membrane at low temperatures [8,10]. A recent study by Seel et al. showed that carotenoids regulate membrane fluidity in Staphylococcus xylosus, in particular greatly affecting the fluidity of the membrane at low temperatures [32]. The authors argue that this may be a mechanism for thermoregulation of membrane properties, helping the bacterium to maintain highly fluid membranes at low temperature growth conditions. Table 3 presents an integration of all the biophysical observations using the different techniques described here. The variations of biophysical parameters are relevant in the susceptibility of bacteria to membrane active antimicrobial agents. For example, the activity of phospholipase A_2 type IIA has been shown to be closely linked to lipid phase behavior and lipid packing in *S. aureus* [11], and the activity of antimicrobial peptides has been shown to be sensitive to changes in lipid packing induced by the presence of STX [6,15].

Table 3. Biophysical parameter measurements for planktonic and biofilm *S. aureus* total lipid extracts under aerated and non-aerated conditions.

Biophysical	Planktonic		Biofilm	
Membrane Parameters	Aerated	Non-Aerated	Aerated	Non-Aerated
10 °C	0.539 ± 0.001	0.498 ± 0.002	0.535 ± 0.001	0.545 ± 0.001
Laurdan GP				
37 °C	0.327 ± 0.006	0.168 ± 0.001	0.274 ± 0.004	0.233 ± 0.002
10 °C	0.231 ± 0.006	0.189 ± 0.008	0.259 ± 0.004	0.234 ± 0.014
DPH Anisotropy				
37 °C	0.178 ± 0.008	0.096 ± 0.006	0.181 ± 0.003	0.143 ± 0.007
10 °C	2851.36	2850.74	2852.08	2851.20
CH_2 Assy. Strech				
37 °C	2853.11	2853.21	2852.97	2853.04
T_m [°C]	12	18	15	24

It is clear when comparing Figure 6c,d that biofilm membranes in general have higher phase transition temperatures than planktonic cells. Comparing the reported melting temperatures in Table 3 for the different growth conditions, an increase of 3 °C in the T_m is observed in aerated biofilms compared to aerated planktonic cells, and an increase of 6 °C is observed in non-aerated biofilms compared to non-aerated planktonic cells. Pure cardiolipin has phase transition temperatures that are significantly higher than PG lipids, having equal chain lengths [44]. For example, while DMPC has a phase transition temperature at 24 °C, CL, which also 14 carbon saturated acyl chains, has a phase transition temperature around 47 °C. Since the proportion of cardiolipin increases significantly for *S. aureus* grown as biofilms (Figure 4b), this increase in CL likely explains the upward shift in T_m. This implies that biofilm cells have more rigid lipid bilayers, compared to planktonic cells. This could be correlated with the higher rigidity of biofilm structures related to the presence of exopolysaccharides and the harsher osmotic conditions that can be found in biofilms [45].

4. Materials and Methods

4.1. Materials

The fluorescent probes LAURDAN (6-Dodecanoy-l2-dimethylaminonaphthalene) and DPH (diphenylhexatriene) were purchased from Molecular Probes (Eugene, OR, USA). HPLC-grade methanol (MeOH), chloroform (CHCl3), acetonitrile, 2-propanol, and ethyl acetate were purchased from Honeywell (Charlotte, NC, USA), while methyl tertbutyl ether (MTBE) was purchased from J.T. Baker (Palo Alto, CA, USA). LC-MS-grade ammonium acetate was purchased from Honeywell Fluka (St. Louis, MO, USA). Powder components for Luria-Bertani medium, and butylated hydroxytoluene (BHT) were purchased from Sigma-Aldrich (St. Louis, MO, USA). HPLC-grade water was obtained from a water purification system, Heal Force Smart-Mini (Shanghai, China). β-carotene (99%) with Lot number S18A010 was purchased from Alfa Aesar (Madrid, Spain).

4.2. Bacterial Cultures

The *S. aureus* clinical isolate strain SA401 was used in this work [8]. The isolate was recovered from the stock that is kept stored at −80 °C. In sterile conditions, an amount of the inoculum was streaked over LB agar plates. The plate was incubated at 37 °C for at least 16 h to obtain visible single colonies. After the colonies were visible, a single colony was taken and added to 10 mL of LB medium with constant agitation at 250 rpm and 37 °C for 18 h. Then, an aliquot of 10 μL was seeded in 150 mL of fresh LB in aerobic conditions or in 300 mL in oxygen-restricted conditions. The new growth was incubated for 24 h into a shaker at 250 or 150 rpm for aerobic or oxygen-restricted conditions, respectively. This study also had the objective of comparing biofilms with planktonic cells grown under different agitation conditions (A, −A).

4.3. Gravity Fed System for Biofilm Growth under Aerobic and Anaerobic Conditions

For biofilm growth, three clinical catheters by Baxter (MRC0002mp) were used, two Schott Duran bottle glass of 1 L, two Duran Erlenmeyer flask of 2 L, 10 mL syringes, microfluidics connectors, a magnetic stir bar, and a magnetic stir plate. Connecting three clinical catheters with the microfluidic connectors, the Schott bottle with 1L of LB medium was placed higher than the empty bottle. Both bottles have a rubber stopper to avoid contamination; the flow was then started. Then, 10 mL of growth liquid was introduced via syringe into the injection port of the catheter and the flow was stopped for at least 2 h to promote cell adhesion; the catheter must be incubated at 37 °C. Afterwards, the flow was restarted at a rate of 1 drop/min and incubated for 48 h. The difference between aerated and non-aerated conditions were controlled through stirring of the feeding medium. For this, two Erlenmeyer flasks of 2 L each were used, and a sterile magnet was introduced to the Erlenmeyer flask with fresh medium, and it was placed on the magnetic stirrer under constant agitation (350 rpm).

4.4. Separation of Total Lipids

For lipid extraction, 100 mg of lyophilized cells were previously weighed, macerated, and added in glass tubes with 10 glass beads. They were dissolved in a solution of MeOH:$CHCl_3$ (2:1); 10 mL of $CHCl_3$ and 20 mL of MeOH, and vortex-mixed for 5 min. After this time, 10 mL of $CHCl_3$ was added to make the ratio 1:1 and vortex-mixed for 1 min. Finally, 10 mL of saline solution (1.7 M NaCl) was added and then cells were vortexed every 15 min for 4 h to obtain a homogeneous solution. The sample was centrifuged at $750\times g$ at 4 °C for 15 min. After centrifugation, the phase separation was achieved, and it was possible to observe three different organic phases; the thick layer of proteins was removed using a syringe and the $CHCl_3$-containing lipids were gently aspirated and added into a new glass vial. The lipids were dried with nitrogen, lyophilized, and weighted.

4.5. Scanning Electron Microscopy (SEM) of Biofilms and Planktonic Cells

After 48 h, a section of the catheter was cut and split open by cutting laterally; for planktonic cells, 1 mL of overnight was centrifuged to obtain the pellet. The samples were fixed for 24 h with glutararaldehyde 2.5%. Therefore, the samples were first rinsed to remove the salts with HPLC-water, then they were submerged in a series of mixtures of 70%, 95%, and 100% ethanol for 30 min each, and finally, the samples were covered with gold. ImageJ software (1.53t) was used to measure cell areas; for this, the bar included in the picture of 2 μm was useful for the transformation into pixels, and in this way, it was possible to measure the area, assuming a circle for the bacteria figure. A total of 20 cells of 5 pictures for each sample (biofilm A, −A and planktonic A, −A) were measured.

4.6. Fluorescence Spectroscopy

To perform measurements, it was necessary to form multilamellar vesicles. From bacterial lipid extracts, 1 mg were dissolved in MeOH:$CHCl_3$ (1:9, v/v) with LAURDAN or DPH at a proportion 1:150 (probe to lipid ratio). The solution was dried under nitrogen flow and lyophilized for at least 10 h to remove residual $CHCl_3$. After this time, the samples were rehydrated with 1 mL of buffer, and five cycles of vortex and heat were repeated to obtain multilamellar vesicles. During the preparation, an additional step was added due to the difficulty in detaching the sample; thus, cycles of tip sonication were used to help homogenize the sample. All measurements were taken from 10 to 45 °C using 30 μM of lipid in 1 mL of buffer in a cuvette. Fluorescence spectroscopy was performed using a PC-1 ISS photon-counting spectrofluorometer that has a temperature controller and continuous magnetic stirring to maintain homogeneity in the samples.

Laurdan has a low solubility in water; it is used as a sensitive probe to study the level of hydration of the headgroup acyl chain interphase of the membrane. Laurdan properties have been described using the generalized polarization GP that is calculated using [46] the following equation:

$$GP = \frac{I_B - I_R}{I_B + I_R} \quad (1)$$

where I_B (intensity at 440 nm) and I_R (intensity at 500 nm) are the measured fluorescence intensities. When Laurdan is in the membrane, it exhibits a red shift of the emission spectrum as the membrane changes from gel to liquid-crystalline phase; the factor GP can be used to measure this shift. For the measurements, the samples were excited at 350 nm and the emission was recorded at 440 and 500 nm. On the other hand, using DPH, we can measure the anisotropy which provides information on the dynamic behavior of the lipid bilayer core. For DPH, it is necessary to use polarized light, and this will result in a selective excitation of those molecules whose absorption transition dipole is parallel to the electric vector; samples were excited at 350 nm and the fluorescence intensity was monitored at 428 nm; anisotropy is defined by [46]

$$R = \frac{I_{\parallel} - I_{\perp}}{I_{\parallel} + 2I_{\perp}} \quad (2)$$

where is $I_{\parallel}$ the vertically polarized fluorescence intensity and $I_{\perp}$ the perpendicularly polarized fluorescence intensity.

4.7. Fourier Transform Infrared (FTIR) Spectroscopy

The experiments were prepared on a BioATR II cell integrated to a Tensor II spectrometer (Bruker Optics, Ettlingen, Germany) with a liquid nitrogen MCT detector using a spectral resolution of 4 cm^{-1} and 120 scans per spectrum. The temperature range was set by a computer-controlled circulating water bath Huber Ministat 125 (Huber, Offenburg, Germany). All the experiments were carried out with a heating rate of 1 °C/min and a stabilization time of 120 s at each temperature. First, the background was taken using buffer (HEPES 20 mM, 500 mM NaCl, and 1 mM EDTA) from 10 to 45 °C. Subsequently, the biofilms extracts were deposited on the silicon crystal (0.3 mg) to prepare, in situ on the BioATR II cell, the solid-supported lipid bilayers, where they were hydrated with 20 µL of buffer at 37 °C for 10 min and measurements were recorded at each temperature of the ramp. Finally, to determine the position of the vibrational band in the range of the second derivative of the spectra, all the absorbance spectra were cut in the 2970–2820 cm^{-1} range, shifted to a zero baseline, and the peak picking function included in OPUS software 7.5. The results were plotted as a function of the temperature using the OriginPro 8.0 software (OriginLab Corporation, Northampton, MA, USA). To determine the melting temperature T_m of the lipids, the curve was fitted according to the Boltzmann model to calculate the inflection point of the obtained thermal transition curves using Python (3.11.3).

4.8. Targeted Lipidomics Analysis by LC-MS/MS

In this research, three families of lipids from *S. aureus* PG, Lysyl-PG, and CL were analyzed with LC-MS/MS. A Dionex UltiMate 3000 Ultra-high-performance liquid chromatographer (UHPLC) equipped with an online degasser, binary pump, autosampler, and a thermostated column compartment coupled with an LCQ Fleet Ion Trap Mass Spectrometer through an ESI source operated in positive and negative mode was employed (Thermo Scientific, San Jose, CA, USA). Raw lipids data were acquired and processed using the Xcalibur 4.3 software (Thermo Scientific, San Jose, CA, USA). The dried extracts from *S. aureus* and standards were dissolved at 10 ppm in $CHCl_3$:MeOH (1:2) and analyzed using 10 µL injection volume of the samples kept at 15 °C (in the autosampler) on a C18 column at 50 °C (Phenomenex Kinetex C18, 2.1 × 50 mm, 2.6 µm particle size) protected with a SecurityGuard Cartridge Phenomenex C18 (4 × 2 mm, 3 µm particle size) pre-column. The chromatographic separation was carried out, employing as solvent A acetonitrile:ammonium formate 10 mM (6:4) and 2-propanol:acetonitrile (9:1) as solvent

B. The gradient elution, at a flow of 300 μL/min, was as follows: a 45% B hold for the first 2 min, then from 45 to 65% B in 18 min, successively from 65 to 100% B in 11 min, maintaining isocratic at 100% B for 4 min, then from 100 to 45% B in 2 min and maintained at 45% B for 5 min (total running time: 40 min), an optimization of previously published conditions [47]. The parameters of the ESI source were as follows for the positive ionization mode ESI(+): sheath gas flow rate at 15 (arbitrary units); spray voltage at 4.50 kV; capillary temp at 330 °C; capillary voltage at 48 V; tube lens at 105 V. For ESI(−), conditions were as follows: sheath gas flow rate at 15 (arbitrary units); spray voltage at 5.50 kV; capillary temp at 330 °C; capillary voltage at −47 V; and tube lens at −85.6 V. Mass spectra were acquired in full ion scanning over a mass range *m*/*z* 110–2000, and tandem mass analysis, using automated data-dependent MS/MS with a 30% collision energy, was used to obtain the corresponding fragment ions with an isolation amplitude of 3 *m*/*z*.

4.9. Carotenoids Analysis by LC-DAD-APCI-MS/MS

The identification and quantification of carotenoids were performed using a method previously published, with some modifications [8]. Dried extracts were resuspended at 10 ppm in MeOH:$CHCl_3$ (2:1) and analyzed with liquid chromatography (LC) using an UHPLC Dionex UltiMate 3000 equipped with diode-array detector (DAD) coupled to a LCQ Fleet Ion Trap Mass Spectrometer (MS) employing an atmospheric pressure chemical ionization (APCI) source operated in negative mode. Raw data were acquired and processed using Xcalibur 4.3 software. Both equipment and software were obtained from the brand Thermo Scientific, San Jose, CA, USA. The chromatographic separation was performed at room temperature with 10 μL injection volume of samples kept at 5 °C on a YMC-C30 column (150 × 4.6 mm i.d., 3 μm particle size; YMC America, Inc., Devens, MA, USA) protected with a SecurityGuard Cartridge Phenomenex C18 (4 × 2 mm, 3 μm particle size) pre-column. The mobile phases contained 400 mg/L of ammonium acetate dissolved in a mixture of methanol:methyl tert-butyl ether:water (80:18:2 *v*/*v*/*v*, for solution A and 8:89:3 *v*/*v*/*v*, for solution B). The elution gradient at a constant flow rate of 450 μL/min was as follows: 5% B for the first 3 min, then from 5 to 10% B in 6 min, from 10 to 25% B in 10 min, from 25 to 40% B in 4 min, maintaining isocratic elution at 40% B for 4 min, then from 40 to 100% B in 4 min and constant 100% B for 3 min. Reconditioning from 100 to 5% B in 2 min and maintaining isocratic conditions for additional 4 min, with a total running time of 40 min per sample. The DAD was performed over the entire UV–vis range (240–600 nm), and the characteristic absorbances of the carotenoids were extracted at 450 nm and STX precursor at 286 nm. The APCI source was operated with the following parameters: vaporizer temperature, 300 °C; discharge current, 15.0 μA; capillary voltage, −25.0 V; tube lens, −80.0 V; capillary temperature, −350 °C; envelope gas flow, 50 arbitrary units; auxiliary gas flow, 30 arbitrary units. In addition, the ion trap was configured to operate in full scan (*m*/*z* 65–1200) and data-dependent MS/MS at 30% collision energy to obtain the corresponding fragment ions with an isolation amplitude of 3 *m*/*z*.

For the quantification of *S. aureus* carotenoids, there were no commercially available standards, so β-carotene was used as an external standard (Figure S2). Quantitative analysis was realized with a calibration curve over the range of 5–500 μg/mL, using the operating conditions described above. Limits of detection (LODs) and quantification (LOQs) corresponded to 1.5 and 5.0 μg/mL, respectively. Additionally, a molecular weight correction factor was applied to account for the difference in detector response, consistent with previous reports [48,49]. The linear regression curve of the standard was $y = 2512.4x - 4352.6$; $R^2 = 0.999$. Carotenoids concentration is displayed in μg/g dry cell weight.

4.10. Statistical Analysis

The different experiments were carried out with three biological replicates. Error bars represent the standard error of the mean (mean ± SD). Data obtained were subjected to analysis of variance (ANOVA) and the means were compared with Tukey's HSD using software Prism—GraphPad 8.0.2. Results with *p*-value ($p \leq 0.05$) were considered statistically significant.

5. Conclusions

S. aureus is a major bacterial human pathogen, and a leading cause of a variety of clinical conditions including bacteremia and infective endocarditis, as well as osteoarticular and skin and soft tissue infections. The emergence of methicillin-resistant strains has led to the search for alternative treatment strategies that include antimicrobial peptides and oxidative agents. However, these treatment strategies are sensitive to the composition and physical properties of *S. aureus* membranes, which can vary due to growth conditions. Here, we show that oxygen-restricting growth conditions decreases carotenoid content in *S. aureus* biofilms and planktonic cells. Changes in phospholipid composition are also observed, where cardiolipin is found to increase in proportion to other phospholipids in biofilms, and Lysyl-PG is found to increase under oxygen-restricting growth conditions in planktonic cells. These changes in composition lead to variations in the biophysical properties of the membranes. Under oxygen-restricting conditions, the reduction in carotenoids leads to a decrease in order at the membrane core. However, this decrease in membrane order is not tied to an increase in acyl chain order of the phospholipids, indicating that the presence of carotenoids may rigidify the membrane while maintaining disorder chains in the phospholipids. At low temperatures, oxygen-restricting conditions lead to a sharp increase in membrane order. This indicates that the presence of carotenoids maintains high levels of fluidity in *S. aureus* membranes at low temperatures. These results are relevant in understanding the physical state of the membrane under different environmental conditions, which vary in oxygen levels. This can lead to a better understanding of the susceptibility of *S. aureus* to different membrane active antimicrobial agents under different environments. In particular, certain tissues are characterized by having low oxygen levels, which can lead to reduced levels of carotenoids in *S. aureus* cells residing in these tissues.

Supplementary Materials: The following supporting information can be downloaded at: https://www.mdpi.com/article/10.3390/ijms241914906/s1.

Author Contributions: Conceptualization, L.Z.-C., G.-D.L. and C.L.; methodology, L.Z.-C., E.S. and G.-D.L.; validation, L.Z.-C., E.S. and G.-D.L.; formal analysis, L.Z.-C., E.S. and G.-D.L.; investigation, L.Z.-C., E.S., G.-D.L., C.C., M.M.-M. and C.L.; resources, C.C., M.M.-M. and C.L.; data curation, L.Z.-C., E.S., G.-D.L., C.C., M.M.-M. and C.L.; writing—original draft preparation, L.Z.-C. and C.L.; writing—review and editing, L.Z.-C., E.S., G.-D.L., C.C., M.M.-M. and C.L.; visualization, L.Z.-C., G.-D.L. and C.L.; supervision, M.M.-M. and C.L.; project administration, M.M.-M. and C.L.; funding acquisition, C.C., M.M.-M. and C.L. All authors have read and agreed to the published version of the manuscript.

Funding: The research was funded by MinCiencias through the Grant Program (Cod. 120480763040, RC 846-2018) and the support to No. 80740-532-2019 project. Also, financial support was received through the Faculty of Sciences of the Universidad de los Andes (INV-2019-84-1824).

Institutional Review Board Statement: Not applicable.

Informed Consent Statement: Not applicable.

Data Availability Statement: The data involved in this paper have been presented in articles and supporting materials in the form of diagrams or tables.

Acknowledgments: We would like to thank the Centro de Microscopía de la Universidad de los Andes (MicroCore) for their support in the preparation of samples, acquisition of images, and in providing internal funding for this research.

Conflicts of Interest: The authors declare no conflict of interest.

References

1. Tong, S.Y.C.; Davis, J.S.; Eichenberger, E.; Holland, T.L.; Vance, F. *Staphylococcus aureus* Infections: Epidemiology, Pathophysiology, Clinical Manifestations, and Management. *Clin. Microbiol. Rev.* **2015**, *28*, 603–661. [CrossRef] [PubMed]
2. Lowy, F. *Staphylococcus aureus* infections. *N. Engl. J. Med.* **1998**, *339*, 520–532. [CrossRef] [PubMed]
3. Kintarak, S.; Whawell, S.A.; Speight, M.; Packer, S.; Nair, S.P. Internalization of *Staphylococcus aureus* by Human Keratinocytes. *Infect. Immun.* **2004**, *72*, 5668–5675. [CrossRef]
4. Foster, T.J. Antibiotic resistance in *Staphylococcus aureus*. Current status and future prospects. *FEMS Microbiol. Rev.* **2017**, *41*, 430–449. [CrossRef] [PubMed]
5. Braungardt, H.; Singh, V.K. Impact of Deficiencies in Branched-Chain Fatty Acids and Staphyloxanthin in *Staphylococcus aureus*. *BioMed Res. Int.* **2019**, *2019*, 2603435. [CrossRef]
6. Perez-Lopez, M.I.; Mendez-Reina, R.; Trier, S.; Herrfurth, C.; Feussner, I.; Bernal, A.; Forero-Shelton, M.; Leidy, C. Variations in carotenoid content and acyl chain composition in exponential, stationary and biofilm states of *Staphylococcus aureus*, and their influence on membrane biophysical properties. *Biochim. Biophys. Acta Biomembr.* **2019**, *1861*, 978–987. [CrossRef]
7. Gruszecki, W.I.; Strzałka, K. Carotenoids as modulators of lipid membrane physical properties. *Biochim. Biophys. Acta Mol. Basis Dis.* **2005**, *1740*, 108–115. [CrossRef]
8. López, G.-D.; Suesca, E.; Álvarez-Rivera, G.; Rosato, A.E.; Ibáñez, E.; Cifuentes, A.; Leidy, C.; Carazzone, C. Carotenogenesis of *Staphylococcus aureus*: New insights and impact on membrane biophysical properties. *Biochim. Biophys. Acta Mol. Cell Biol. Lipids* **2021**, *1866*, 158941. [CrossRef]
9. Staubitz, P.; Neumann, H.; Schneider, T.; Wiedemann, I.; Peschel, A. MprF-mediated biosynthesis of lysylphosphatidylglycerol, an important determinant in staphylococcal defensin resistance. *FEMS Microbiol. Lett.* **2006**, *231*, 67–71. [CrossRef]
10. Manrique-Moreno, M.; Jemioła-Rzemińska, M.; Múnera-Jaramillo, J.; López, G.-D.; Suesca, E.; Leidy, C.; Strzałka, K. *Staphylococcus aureus* Carotenoids Modulate the Thermotropic Phase Behavior of Model Systems That Mimic Its Membrane Composition. *Membranes* **2022**, *12*, 945. [CrossRef]
11. Ocampo, J.; Afanador, N.; Vives, M.J.; Moreno, J.C.; Leidy, C. The antibacterial activity of phospholipase A2 type IIA is regulated by the cooperative lipid chain melting behavior in *Staphylococcus aureus*. *Biochim. Biophys. Acta Biomembr.* **2010**, *1798*, 1021–1028. [CrossRef] [PubMed]
12. Pérez-Gálvez, A.; Viera, I.; Roca, M. Carotenoids and Chlorophylls as Antioxidants. *Antioxidants* **2020**, *9*, 505. [CrossRef] [PubMed]
13. Kim, S.H.; Lee, C. Functional expression and extension of staphylococcal staphyloxanthin biosynthetic pathway in Escherichia coli. *J. Biol. Chem.* **2012**, *287*, 21575–21583. [CrossRef]
14. Hall, J.W.; Yang, J.; Guo, H.; Ji, Y. The *Staphylococcus aureus* AirSR Two-Component System Mediates Reactive Oxygen Species Resistance via Transcriptional Regulation of Staphyloxanthin Production. *Infect. Immun.* **2017**, *85*, e00838-16. [CrossRef]
15. Mishra, N.N.; Liu, G.Y.; Yeaman, M.R.; Nast, C.C.; Proctor, R.A.; McKinnell, J.; Bayer, A.S. Carotenoid-Related Alteration of Cell Membrane Fluidity Impacts *Staphylococcus aureus* Susceptibility to Host Defense Peptides. *Antimicrob. Agents Chemother.* **2011**, *55*, 526–531. [CrossRef]
16. Chaibenjawong, P.; Foster, S.J. Desiccation tolerance in *Staphylococcus aureus*. *Arch. Microbiol.* **2010**, *193*, 125–135. [CrossRef] [PubMed]
17. Crass, R.L.; Powell, K.L.; Huang, A.M. Daptomycin for the treatment of *Staphylococcus aureus* infections complicated by septic pulmonary emboli. *Diagn. Microbiol. Infect. Dis.* **2019**, *93*, 131–135. [CrossRef]
18. Liu, G.Y.; Essex, A.; Buchanan, J.T.; Datta, V.; Hoffman, H.M.; Bastian, J.F.; Fierer, J.; Nizet, V. *Staphylococcus aureus* golden pigment impairs neutrophil killing and promotes virulence through its antioxidant activity. *J. Exp. Med.* **2005**, *202*, 209–215. [CrossRef]
19. Wölk, C.; Youssef, H.; Guttenberg, T.; Marbach, H.; Vizcay-Barrena, G.; Shen, C.; Brezesinski, G.; Harvey, D.R. Phase Diagram for a Lysyl-Phosphatidylglycerol Analogue in Biomimetic Mixed Monolayers with Phosphatidylglycerol: Insights into the Tunable Properties of Bacterial Membranes. *ChemPhysChem* **2020**, *21*, 702–706. [CrossRef]
20. Zhang, Y.-M.; Rock, C.O. Membrane lipid homeostasis in bacteria. *Nat. Rev. Microbiol.* **2008**, *6*, 222–233. [CrossRef]
21. Hernández-Villa, L.; Manrique-Moreno, M.; Leidy, C.; Jemioła-Rzemińska, M.; Ortíz, C.; Strzałka, K. Biophysical evaluation of cardiolipin content as a regulator of the membrane lytic effect of antimicrobial peptides. *Biophys. Chem.* **2018**, *238*, 8–15. [CrossRef] [PubMed]
22. Renner, L.D.; Weibel, D.B. Cardiolipin microdomains localize to negatively curved regions of Escherichia coli membranes. *Proc. Natl. Acad. Sci. USA* **2011**, *108*, 62–69. [CrossRef] [PubMed]
23. Rocha-Roa, C.; Orjuela, J.D.; Leidy, C.; Cossío, P.; Santamaría, C.A. Cardiolipin prevents pore formation in phosphatidylglycerol bacterial membrane models. *FEBS Lett.* **2021**, *595*, 2701–2714. [CrossRef] [PubMed]
24. Fuchs, S.; Pané-Farré, J.; Kohler, C.; Hecker, M.; Engelmann, S. Anaerobic gene expression in *Staphylococcus aureus*. *J. Bacteriol.* **2007**, *189*, 4275–4289. [CrossRef]
25. Coleman, G.; Garbutt, I.; Demnitz, U. Ability of a *Staphylococcus aureus* isolate from a chronic osteomyelitic lesion to survive in the absence of air. *Eur. J. Clin. Microbiol.* **1983**, *2*, 595–597. [CrossRef]
26. Park, M.; Myers, R.; Marzella, L. Oxygen tensions and infections: Modulation of microbial growth, activity of antimicrobial agents, and immunologic responses. *Clin. Infect. Dis.* **1992**, *14*, 720–740. [CrossRef]

27. Ortiz-Prado, E.; Dunn, J.F.; Vasconez, J.; Castillo, D.; Viscor, G. Partial pressure of oxygen in the human body: A general review. *Am. J. Blood Res.* **2019**, *9*, 1–14.
28. Periasamy, S.; Joo, H.S.; Duong, A.C.; Bach, T.H.L.; Tan, V.Y.; Chatterjee, S.S.; Cheung, G.Y.; Otto, M. How Staphylococcus aureus biofilms develop their characteristic structure. *Proc. Natl. Acad. Sci. USA* **2012**, *109*, 1281–1286. [CrossRef]
29. Archer, N.K.; Mazaitis, M.J.; Costerton, J.G.; Leid, J.G.; Powers, M.E.; Shirtliff, M.E. *Staphylococcus aureus* biofilms. *Virulence* **2011**, *2*, 445–459. [CrossRef]
30. Heimburg, T. *Thermal Biophysics of Membranes*; Wiley: Hoboken, NJ, USA, 2007.
31. Chamberlain, N.R.; Mehrtens, B.G.; Xiong, Z.; Kapral, F.A.; Boardman, J.L.; Rearick, J.I. Correlation of carotenoid production, decreased membrane fluidity, and resistance to oleic acid killing in *Staphylococcus aureus* 18Z. *Infect. Immun.* **1991**, *59*, 4332–4337. [CrossRef]
32. Seel, W.; Baust, D.; Sons, D.; Albers, M.; Etzbach, L.; Fuss, J.; Lipski, A. Carotenoids are used as regulators for membrane fluidity by *Staphylococcus xylosus*. *Sci. Rep.* **2020**, *10*, 330. [CrossRef]
33. Lin-Vien, D.; Colthup, N.B.; Fateley, W.G.; Grasselli, J.G. *The Handbook of Infrared and Raman Characteristic Frequencies of Organic Molecules*; Elsevier Science & Technology: Amsterdam, The Netherlands, 1991.
34. Lewis, R.N.A.H.; McElhaney, R.N. Membrane lipid phase transitions and phase organization studied by Fourier transform infrared spectroscopy. *Biochim. Biophys. Acta Biomembr.* **2013**, *1828*, 2347–2358. [CrossRef] [PubMed]
35. Costerton, J.W.; Stewart, S.; Greenberg, E.P. Bacterial Biofilms: A Common Cause of Persistent Infections. *Science* **1999**, *284*, 1318–1322. [CrossRef] [PubMed]
36. Oldfield, E.; Feng, X. Resistance-resistant antibiotics. *Trends Pharmacol. Sci.* **2014**, *35*, 664–674. [CrossRef]
37. Clauditz, A.; Resch, A.; Wieland, K.P.; Peschel, A.; Götz, F. Staphyloxanthin Plays a Role in the Fitness of *Staphylococcus* aureus and Its Ability to Cope with Oxidative Stress. *ASM J.* **2006**, *74*, 4950–4953. [CrossRef] [PubMed]
38. Romantsov, T.; Guan, Z.; Wood, J.M. Cardiolipin and the osmotic stress responses of bacteria. *Biochim. Biophys. Acta* **2009**, *1788*, 2092–2100. [CrossRef] [PubMed]
39. Ohniwa, R.L.; Kitabayashi, K.; Morikawa, K. Alternative cardiolipin synthase Cls1 compensates for stalled Cls2 function in *Staphylococcus aureus* under conditions of acute acid stress. *FEMS Microbiol. Lett.* **2013**, *338*, 141–146. [CrossRef]
40. Zhang, T.H.; Muraih, J.K.; Nasim, T.; Hershowitz, J.; Victor, R.L.; Silverman, J.; Uwumarenogie, S.; Taylor, S.D.; Palmer, M. Cardiolipin prevents membrane translocation and permeabilization by Daptomycin. *J. Biol. Chem.* **2014**, *289*, 11584–11591. [CrossRef]
41. Ernst, C.M.; Peschel, A. Broad-spectrum antimicrobial peptide resistance by MprF-mediated aminoacylation and flipping of phospholipids. *Mol. Microbiol.* **2011**, *80*, 290–299. [CrossRef]
42. Kossakowska-Zwierucho, M.; Kaźmierkiewicz, R.; Bielawski, K.P.; Nakonieczna, J. Factors Determining *Staphylococcus aureus* Susceptibility to Photoantimicrobial Chemotherapy: RsbU Activity, Staphyloxanthin Level, and Membrane Fluidity. *Front. Microbiol.* **2016**, *7*, 1141. [CrossRef]
43. Grudzinski, W.; Nierzwicki, L.; Welc, R.; Reszczynska, E.; Luchowski, R.; Czub, J.; Gruszecki, W.I. Localization and Orientation of Xanthophylls in a Lipid Bilayer. *Sci. Rep.* **2017**, *7*, 9619. [CrossRef]
44. Lewis, R.N.A.H.; McElhaney, N. The physicochemical properties of cardiolipin bilayers and cardiolipin-containing lipid membranes. *Biomembranes* **2009**, *1788*, 2069–2079. [CrossRef] [PubMed]
45. Peng, Q.; Tang, X.; Dong, W.; Sun, N.; Yuan, W. A Review of Biofilm Formation of *Staphylococcus aureus* and Its Regulation Mechanism. *Antibiotics* **2022**, *12*, 12. [CrossRef] [PubMed]
46. Fleming, K.G. *Fluorescence Theory, de Encyclopedia of Spectroscopy and Spectrometry*; Elsevier: Amsterdam, The Netherlands, 2017; pp. 647–653.
47. Rodríguez–Beltrán, É.; López, G.-D.; Anzola, J.M.; Rodríguez–Castillo, J.G.; Carazzone, C.; Murcia, M.I. Heterogeneous fitness landscape cues, pknG high expression, and phthiocerol dimycocerosate low production of *Mycobacterium tuberculosis* ATCC25618 rpoB S450L in enriched broth. *Tuberculosis* **2022**, *132*, 102156. [CrossRef] [PubMed]
48. Castro-Puyana, M.; Pérez-Sánchez, A.; Valdésb, A.; Ibrahim, O.H.M.; Suárez-Álvarez, O.S. Pressurized liquid extraction of Neochloris oleoabundans for the recovery of bioactive carotenoids with anti-proliferative activity against human colon cancer cells. *Food Res. Int.* **2017**, *99*, 1048–1055. [CrossRef] [PubMed]
49. Dhooghe, L.; Mesia, K.; Kohtala, E.; Tona, L.; Pieters, L.; Vlietinck, A.; Apers, S. Development and validation of an HPLC-method for the determination of alkaloids in the stem bark extract of Nauclea pobeguinii. *Talanta* **2007**, *76*, 462–468. [CrossRef]

International Journal of *Molecular Sciences*

MDPI

Article

Scavenging of Cation Radicals of the Visual Cycle Retinoids by Lutein, Zeaxanthin, Taurine, and Melanin

Malgorzata Rozanowska [1,2,*], Ruth Edge [3], Edward J. Land [4], Suppiah Navaratnam [5], Tadeusz Sarna [6] and T. George Truscott [7]

1. Cardiff Institute of Tissue Engineering and Repair, Cardiff University, Cardiff CF10 3AX, UK
2. School of Optometry and Vision Sciences, Cardiff University, Cardiff CF24 4HQ, UK
3. Dalton Cumbrian Facility, The University of Manchester, Westlakes Science Park, Moor Row, Cumbria CA24 3HA, UK; ruth.edge@manchester.ac.uk
4. The Paterson Institute, The University of Manchester, Wilmslow Road, Manchester M20 4BX, UK; e.land@mighty-micro.co.uk
5. Biomedical Sciences Research Institute, University of Salford, Manchester M5 4WT, UK; navaratnam1000@gmail.com
6. Department of Biophysics, Faculty of Biochemistry, Biophysics and Biotechnology, Jagiellonian University, 30-387 Krakow, Poland; tadeusz.sarna@uj.edu.pl
7. School of Chemical and Physical Sciences, Lennard-Jones Building, Keele University, Staffordshire ST5 5BG, UK; t.g.truscott@keele.ac.uk

* Correspondence: rozanowskamb@cardiff.ac.uk; Tel.: +44-292087-5057

Abstract: In the retina, retinoids involved in vision are under constant threat of oxidation, and their oxidation products exhibit deleterious properties. Using pulse radiolysis, this study determined that the bimolecular rate constants of scavenging cation radicals of retinoids by taurine are smaller than 2×10^7 $M^{-1}s^{-1}$ whereas lutein scavenges cation radicals of all three retinoids with the bimolecular rate constants approach the diffusion-controlled limits, while zeaxanthin is only 1.4–1.6-fold less effective. Despite that lutein exhibits greater scavenging rate constants of retinoid cation radicals than other antioxidants, the greater concentrations of ascorbate in the retina suggest that ascorbate may be the main protectant of all visual cycle retinoids from oxidative degradation, while α-tocopherol may play a substantial role in the protection of retinaldehyde but is relatively inefficient in the protection of retinol or retinyl palmitate. While the protection of retinoids by lutein and zeaxanthin appears inefficient in the retinal periphery, it can be quite substantial in the macula. Although the determined rate constants of scavenging the cation radicals of retinol and retinaldehyde by dopa-melanin are relatively small, the high concentration of melanin in the RPE melanosomes suggests they can be scavenged if they are in proximity to melanin-containing pigment granules.

Keywords: retinal; vitamin A; free radicals; xanthophylls; carotenoids; vitamin E; vitamin C; retina; age-related macular degeneration; Stargardt's disease

1. Introduction

Upon absorption of light by the visual pigment in photoreceptive neurons of the vertebrate retina, its chromophore-11-*cis*-retinaldehyde is isomerized to all-*trans* configuration, which is followed by its hydrolysis from opsin and transient accumulation in photoreceptor outer segments (POS) until it is enzymatically reduced to all-*trans*-retinol [1,2]. To enable the regeneration of visual pigment, so it can absorb another photon of visible light to initiate a cascade of events leading to visual perception, the apoprotein needs to bind to another 11-*cis*-retinaldehyde, which needs to be delivered to photoreceptors mostly from the neighbouring retinal pigment epithelium (RPE). All-*trans*-retinol formed in POS is transported by chaperone proteins to the RPE [1–4]. In the RPE, all-*trans*-retinol, delivered either from photoreceptors or from blood, is the substrate for enzymes responsible for the synthesis of 11-*cis*-retinaldehyde. Firstly, it is esterified with fatty acids forming mainly all-*trans*-retinyl palmitate. All-*trans*-retinyl palmitate is used for retinoid storage or as a substrate of isomerohydrolase RPE65. All-*trans*-retinyl esters are stored in specialized lipid droplets called retisomes, and to enable them to become a substrate for RPE65 they need to be transferred to the endoplasmic reticulum (ER). It has been recently shown that patatin-like phospholipase domain containing 2 (PNPLA2) hydrolyses these esters to release all-*trans*-retinol so it can be transported to ER where it is esterified again by lecithin-retinol acyltransferase (LRAT) to become the substrate of RPE65 [5]. RPE65 converts all-*trans*-retinyl esters to 11-*cis*-retinol. Upon enzymatic oxidation of 11-*cis*-retinol, 11-*cis*-retinaldehyde is formed and is transported back to photoreceptive neurons by cellular and interphotoreceptor retinoid-binding proteins to regenerate the visual pigment and complete the major pathway of visual pigment chromophore synthesis in the visual cycle. Another pathway leading to regeneration of the visual pigment chromophore operates under exposure to light. Both RPE and Müller cells express retinal G-protein-coupled receptor (RGR) photoisomerase, which binds all-*trans*-retinaldehyde via Schiff-base linkage [6–8]. Upon absorption of light by RGR, all-*trans*-retinaldehyde is isomerized to 11-*cis* form, hydrolyzed from RGR, and transported by chaperone proteins to photoreceptors for regeneration of visual pigments.

In the retina, the trafficking of the visual cycle retinoids exposes them to the constant threat of oxidation due to the presence of potent photosensitizers exposed to light and under high oxygen tension and high concentration of polyunsaturated fatty acids susceptible to oxidation and propagation of lipid peroxidation via peroxyl radicals [9,10]. Lipid peroxidation can be facilitated in retinal diseases such as age-related macular degeneration (AMD), where RPE exhibits about a five-fold increased content of total iron and iron chelateable by desferrioxamine in comparison with RPE in age-matched normal retinas [11]. Increased levels of products of lipid peroxidation have been identified post-mortem in the human retinas affected by AMD, which is the predominant cause of vision loss in people above 50 years of age in developed countries [12–17]. It has been previously shown that retinoids, including retinol and retinaldehyde, can scavenge peroxyl radicals, thereby contributing to the antioxidant action of retinoids in lipid peroxidation where their antioxidant action can exceed that of vitamin E, a well-known chain-breaker in lipid peroxidation [18,19]. Scavenging of peroxyl radicals by retinoids can result in the formation of retinoid cation radicals. Moreover, retinoids can form cation radicals as a result of interaction with hydroxyl radicals or photoexcitation [20–22]. While retinaldehyde can be photoexcited by UV and blue light, the absorption spectra of retinol and retinyl palmitate extend very little into the visible range of light but show a relatively strong absorption at 320 nm. While most UV light is absorbed by the cornea and the lens of the adult human eye, the transmission window with a maximum at about 320 nm persists even in people above the age of 60 years [23].

Retinoid cation radicals can be damaging to amino acids and proteins [22]. For example, retinol cation radicals can oxidize several amino acids, such as cysteine, tryptophan, lysine, and arginine [22]. The formation of retinoid cation radicals can lead to their further degradation. It has been previously shown that a mixture of oxidation products of retinaldehyde retains the photosensitizing properties of retinaldehyde while becoming more (photo)toxic than the parent compound [24]. (Photo)toxicity of retinaldehyde and its degradation products is of particular relevance to blinding diseases caused by delayed clearance of all-*trans*-retinaldehyde, such as Stargardt's disease caused by dysfunction of enzymes responsible for all-*trans*-retinaldehyde removal: ATP-binding cassette transporter A4 (ABCA4) or retinol dehydrogenase 8 (RDH8) [25–27]. Stargardt's disease is usually diagnosed in childhood or early adulthood and progresses rapidly to severe loss of vision.

It has been previously shown that the cation radicals of retinol and retinaldehyde can be scavenged by vitamin E and vitamin C, presumably restoring the parent compound [18,28]. Vitamins C (ascorbate) and vitamin E, present mainly as α-tocopherol, are essential components of the vertebrate retina as well as of an antioxidant supplement which, together with zinc, lutein, and zeaxanthin, has been shown to decrease the risk of progression of a moderate form of AMD to its advanced form over a 10-year period from 49.2% to 47.3%, and the effect was statistically significant [29–33].

The potential formation of cation radicals of retinyl palmitate as a result of interaction with oxidizing radicals and the effect of vitamins C and E on retinyl palmitate cation radicals have not been investigated yet. In addition to vitamins E and C, the area where the trafficking of the visual cycle retinoids takes place contains other antioxidants. The POS and RPE of the human retina accumulate two carotenoids: lutein and zeaxanthin [34–37], which are thought to play important antioxidant functions in the retina [38,39]. The retinas of vertebrates contain high concentrations of taurine (2-amino-ethanesulfonic acid), which is essential for the viability of retinal neurons, including photoreceptors [40–42]. RPE contains eumelanin enclosed within melanosomes. which, especially in the young eye, appear mainly in the apical processes of the RPE extending into the layer of POS (reviewed in [43]). With ageing, RPE accumulates other melanin-containing granules—melanolysosomes and melanolipofuscin. Therefore, the aim of this study was to investigate the formation of retinyl palmitate cation radicals as a result of electron transfer from oxidizing radicals and the interactions of cation radicals of all three retinoids involved in the visual cycle with the major antioxidants present in the retina: ascorbate, lutein, zeaxanthin, taurine, and melanin.

The retinoid radical cations were generated by pulse radiolysis of nitrous oxide-saturated benzene or of aqueous buffered solution of potassium bromide, where retinoids were solubilized in Triton X-100 micelles. We observed retinyl palmitate cation radicals formation as a result of the interaction of retinyl palmitate with both dibromine anion radicals and benzene cation radicals. As a result of interaction with dibromine anion radicals, retinyl palmitate in micelles formed its cation radical with an absorption maximum at 590 nm, whereas in benzene, an electron transfer from retinyl palmitate to cation radicals of benzene resulted in the formation of retinyl palmitate cation radicals with an absorption maximum at 610 nm. Retinyl palmitate cation radicals were scavenged by α-tocopherol and by ascorbate so the bimolecular rate constants of scavenging could be determined. We observed no effect of taurine on the rate of cation radical decay for any of the retinoids. Synthetic dopa-melanin, used as a model of eumelanin present in the RPE, had no effect on the rate of decay of cation radicals of retinyl palmitate but scavenged cation radicals of retinol and retinaldehyde. Lutein and zeaxanthin scavenged cation radicals of all three retinoids with the bimolecular rate constants approaching the diffusion-controlled limits in benzene. The physiological relevance of the results is discussed in terms of concentrations of antioxidants in the retina. Despite that lutein and zeaxanthin tend to be more effective scavengers of retinoid cation radicals than ascorbate, the greater concentrations of ascorbate in the retina suggest that vitamin C may be the main protectant of visual cycle retinoids from oxidation. Lutein and zeaxanthin are likely to offer substantial contribution to retinoid cation radical scavenging oxidation only in the macula.

2. Results

2.1. Formation of Cation Radical of Retinyl Palmitate in Benzene

The pulse radiolysis of N_2O-saturated benzene in the presence of 1 mM of retinyl palmitate resulted in the formation of retinyl palmitate triplet state with an absorption maximum at about 410 nm and retinyl palmitate cation radicals exhibiting an absorption maximum at 610 nm and decaying with kinetics, which, after omitting the initial contribution to the decay from the triplet state, could be fitted with a single exponential decay (Figure 1A,B).

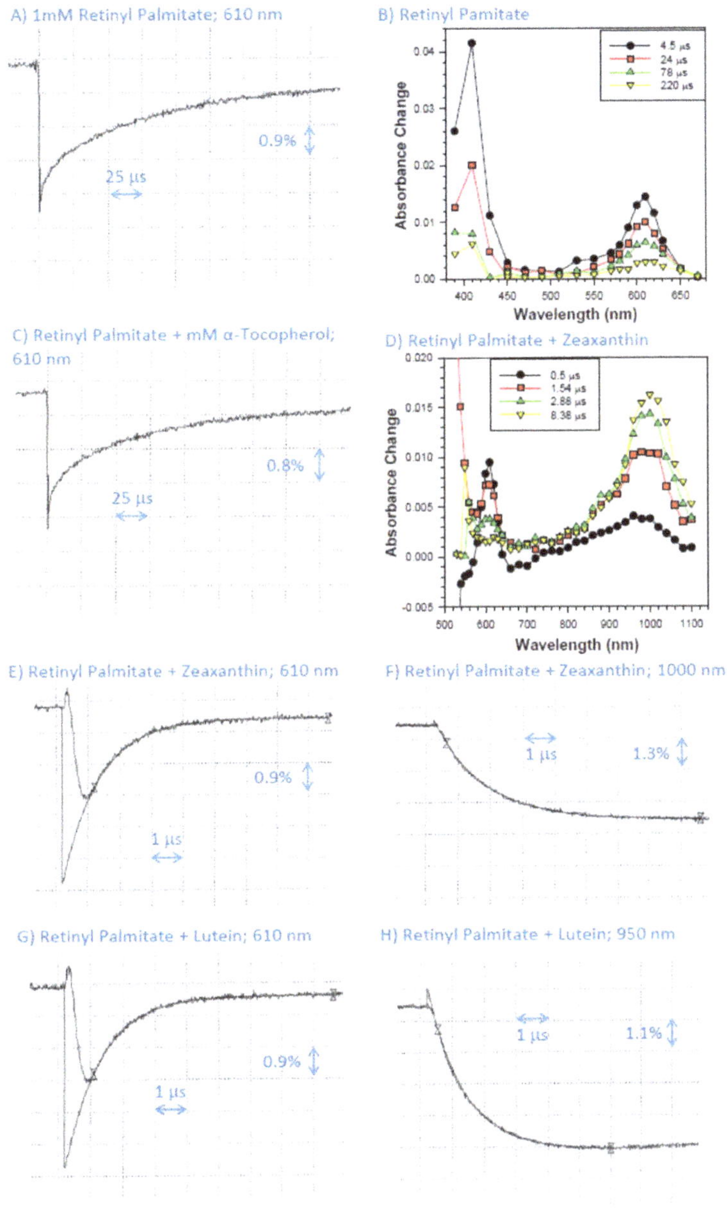

Figure 1. Representative kinetics of the formation and decay of the transient species monitored at 610 nm after pulse radiolysis of N_2O-saturated benzene with solubilized 1 mM of retinyl palmitate (**A**) and transient absorption spectra at indicated times after the pulse radiolysis of that solution (**B**). (**C**) Representative kinetics of the formation and decay of the transient species monitored at 610 nm after pulse radiolysis of N_2O-saturated benzene with solubilized 1 mM of retinyl palmitate in the presence of 0.1 mM α-tocopherol. (**D**) Transient absorption spectra at indicated times after the pulse radiolysis of N_2O-saturated benzene with solubilized 1 mM of retinyl palmitate and 0.1 mM zeaxanthin, and representative kinetics of the formation and decay of transient species monitored at

610 nm (**E**) and 1000 nm (**F**) after pulse radiolysis of that solution. (**G,H**) Representative kinetics of the formation and decay of transient species monitored at 610 nm (**G**) and 950 nm (**H**) after pulse radiolysis of N_2O-saturated benzene with solubilized 1 mM of retinyl palmitate in the presence of 0.1 mM lutein.

2.2. Interaction of Cation Radical of Retinyl Palmitate with α-Tocopherol, Lutein, and Zeaxanthin

In the presence of α-tocopherol, the decays of retinyl palmitate cation radicals were faster than in its absence, allowing for the determination of the bimolecular rate constant of scavenging as $(2.7 \pm 0.3) \times 10^7\ M^{-1}s^{-1}$ (Figure 1A,C; Table 1).

Table 1. Bimolecular rate constants of scavenging of retinoid cation radicals by lutein, zeaxanthin, vitamin E (α-tocopherol), vitamin C (ascorbic acid), taurine, and melanin (dopa-melanin). To determine the bimolecular rate constants of interactions between lipophilic antioxidants (lutein, zeaxanthin, and α-tocopherol) and retinoid cation radicals, both reactants were solubilized in benzene. To determine the bimolecular rate constants of interactions between retinoid cation radicals and hydrophilic antioxidants (ascorbate, taurine, dopa-melanin), the retinoids were incorporated into Triton X-100 micelles, whereas the hydrophilic antioxidants were solubilized directly in 10 mM of phosphate buffer pH 7.0. [a] from [21]; [b] from [28]; [c] from [18]; [d] from [22].

	Bimolecular Rates of Scavenging of Retinoid Radical Cations ($10^9\ M^{-1}s^{-1}$)		
	Retinyl Palmitate	**Retinaldehyde**	**Retinol**
Lutein	8.85 ± 0.25	11.5 ± 1.4	12.6 ± 0.4
Zeaxanthin	6.39 ± 0.02	6.5 ± 0.3	7.9 ± 0.3 5.76 ± 0.50 in methanol [a] 2.50 ± 0.20 in benzonitrile [a]
α-Tocopherol	0.027 ± 0.003	8.0 ± 0.3 [b]	0.080 ± 0.045 [a]
Ascorbate	0.58 ± 0.10	0.73 [c] 0.53 ± 0.08 in methanol [d]	0.12 [c]
Taurine	<0.02	<0.01	<0.002
Dopa-melanin	<0.02	0.016 ± 0.008	0.0051 ± 0.0001

In the presence of zeaxanthin, due to the high absorption coefficient of the zeaxanthin ground state, the changes in the transmittance changes could not be monitored below 520 nm wavelength (Figure 1D; Table 1). As expected, a formation of a zeaxanthin triplet state with an absorption maximum at about 510–520 nm could be seen, where the majority of the zeaxanthin triplet state appeared to be a result of energy transfer from the triplet state of retinyl palmitate (Figure 1D) [44,45]. The absorption of the zeaxanthin triplet state partly overlapped with the absorption of the retinyl palmitate cation radical. In the presence of zeaxanthin, the lifetime of retinyl palmitate cation radical was substantially shortened, and its decay was accompanied by the concomitant formation of zeaxanthin cation radical with an absorption maximum at about 1000 nm [46] (Figure 1D–F). The bimolecular rate constant of scavenging of retinyl palmitate cation radicals by zeaxanthin was $(8.80 \pm 0.05) \times 10^9\ M^{-1}s^{-1}$ when determined based on rates of decay at 610 nm and $(6.39 \pm 0.02) \times 10^9\ M^{-1}s^{-1}$ when determined based on rates of formation of zeaxanthin cation radical at 1000 nm (Figure 1D–F; Table 1). The apparent faster bimolecular rate constant of scavenging of retinyl palmitate cation radicals based on 610 nm data suggested that the zeaxanthin triplet state could affect the kinetics at 610 nm, and therefore we took the value obtained from the rates of formation of the zeaxanthin cation radical as a more accurate value (Table 1).

In the presence of lutein, the lifetime of retinyl palmitate cation radical was also substantially shortened, and its decay was accompanied by the concomitant formation of lutein cation radical with an absorption maximum at about 950 nm [46] (Figure 1D–F). The bimolecular rate constants of scavenging of retinyl palmitate cation radicals by lutein were

(8.97 ± 0.05) × 10^9 $M^{-1}s^{-1}$ and (8.85 ± 0.25) × 10^9 $M^{-1}s^{-1}$ when determined based on rates of decay at 610 nm and rates of formation of at 950 nm, respectively (Figure 1D–F; Table 1). While both values appear to be the same within the experimental uncertainty, for consistency we used the value based on rates of the formation of lutein cation radical (Table 1).

2.3. *Interaction of Retinyl Palmitate with Dibromine Radical Anions*

To create retinyl palmitate cation radical in Triton X-100 micelles, we used pulse radiolysis to create highly oxidizing dibromine radical, which we have shown previously abstracts electrons from retinol and retinaldehyde thereby creating cation radicals of these retinoids with absorption maxima at 590 nm. In case of retinyl palmitate, there was also a formation of a transient species with the absorption maximum at 590 nm (Figure 2A). In addition, there was another species formed with an absorption maximum at about 400 nm. Some of this species was formed within the electron pulse with a subsequent growth, which, judging by an isosbestic point at about 460 nm, was due to the decomposition of the radical cation of retinyl palmitate. Drawing analogy from data on carotenoids presented by Polyakov et al., it can be suggested that the transient at 400 nm is a neutral radical of retinyl palmitate formed as a result of proton loss from the cation radical [47].

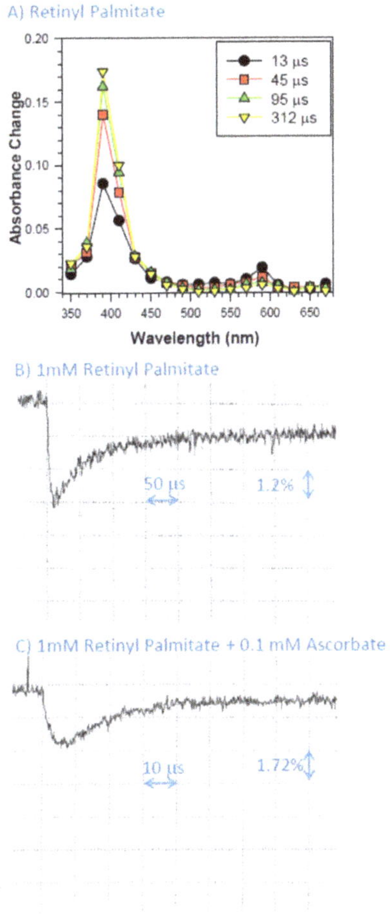

Figure 2. Transient absorption spectra after pulse radiolysis of aqueous solution saturated with N_2O and containing 10 mM phosphate, pH 7, 0.1 M KBr, and 1 mM retinyl palmitate incorporated in 2%

Triton X-100 micelles (**A**) and a representative kinetics of the formation and decay of retinyl palmitate cation radicals monitored at 590 nm after pulse radiolysis of that solution in the absence (**B**) and presence of 0.1 mM of ascorbate (**C**).

2.4. Interaction of Cation Radical of Retinyl Palmitate with Ascorbate, Taurine, and Dopa-Melanin

In the presence of ascorbate, the rates of decay of retinyl palmitate cation radical were increased, and the calculated bimolecular rate constant of scavenging of retinyl palmitate cation radicals was (5.8 ± 0.1) × 10^8 $M^{-1}s^{-1}$ (Figure 2; Table 1).

Taurine or dopa-melanin had no effect on the rate of decay of the retinyl palmitate cation radical, suggesting that if the scavenging occurs, the bimolecular rate constant is smaller than 2 × 10^7 $M^{-1}s^{-1}$ (Figure 3; Table 1).

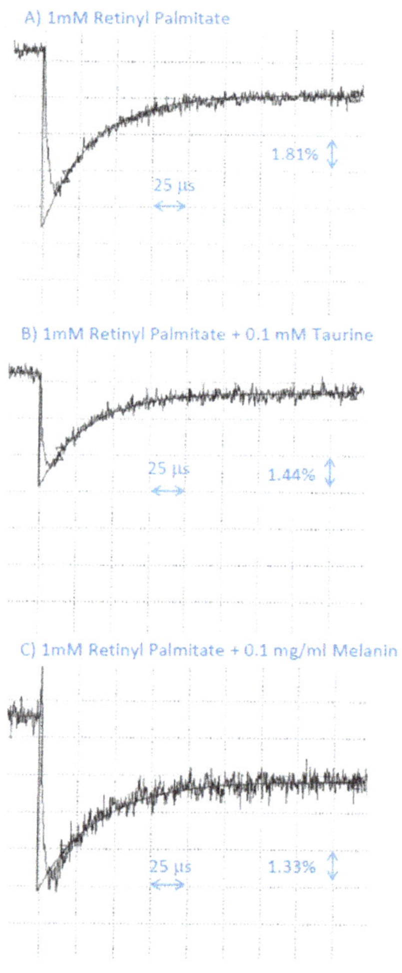

Figure 3. Representative kinetics of the formation and decay of retinyl palmitate cation radicals monitored at 590 nm after pulse radiolysis of aqueous solution saturated with N_2O and containing 10 mM phosphate, pH 7, 0.1 M KBr, and 1 mM retinyl palmitate incorporated in 2% Triton X-100 micelles, in the absence (**A**) and presence of 0.1 mM of taurine (**B**) or 0.1 mg/mL (equivalent to 0.67 mM monomers) dopa-melanin (**C**).

2.5. Interaction of Cation Radical of Retinaldehyde with Lutein, and Zeaxanthin

Similar to the previous findings, pulse radiolysis of N_2O-saturated benzene in the presence of 1 mM retinaldehyde resulted in the formation of its triplet state species and cation radical with absorption maximum at 610 nm [28,45]. While the triplet state decayed faster than the cation radical, it did contribute to the decay kinetics observed at 610 nm, and therefore we omitted that initial part from the fitting of the exponential decay to obtain a more accurate value of the rate of decays of the retinaldehyde cation radical (Figure 4A). In the presence of lutein, the lifetime of retinaldehyde cation radical was substantially shortened, and its decay was accompanied by the concomitant formation of lutein cation radical with an absorption maximum at about 950 nm (Figure 4A–D). The bimolecular rate constants of scavenging of retinaldehyde cation radicals by lutein were $(1.10 \pm 0.02) \times 10^{10}$ $M^{-1}s^{-1}$ and $(1.15 \pm 0.14) \times 10^{10}$ $M^{-1}s^{-1}$ when determined based on rates of decay at 610 nm and rates of formation of at 950 nm, respectively (Table 1). In the presence of zeaxanthin, the lifetime of retinaldehyde cation radical was also substantially shortened, and its decay was accompanied by the concomitant formation of zeaxanthin cation radical with an absorption maximum at about 1000 nm (Figure 4E,F). The bimolecular rate constants of scavenging of retinaldehyde cation radicals by zeaxanthin were $(7.63 \pm 0.55) \times 10^9$ $M^{-1}s^{-1}$ and $(6.48 \pm 0.29) \times 10^9$ $M^{-1}s^{-1}$ when determined based on rates of decay at 610 nm and rates of formation of at 950 nm, respectively.

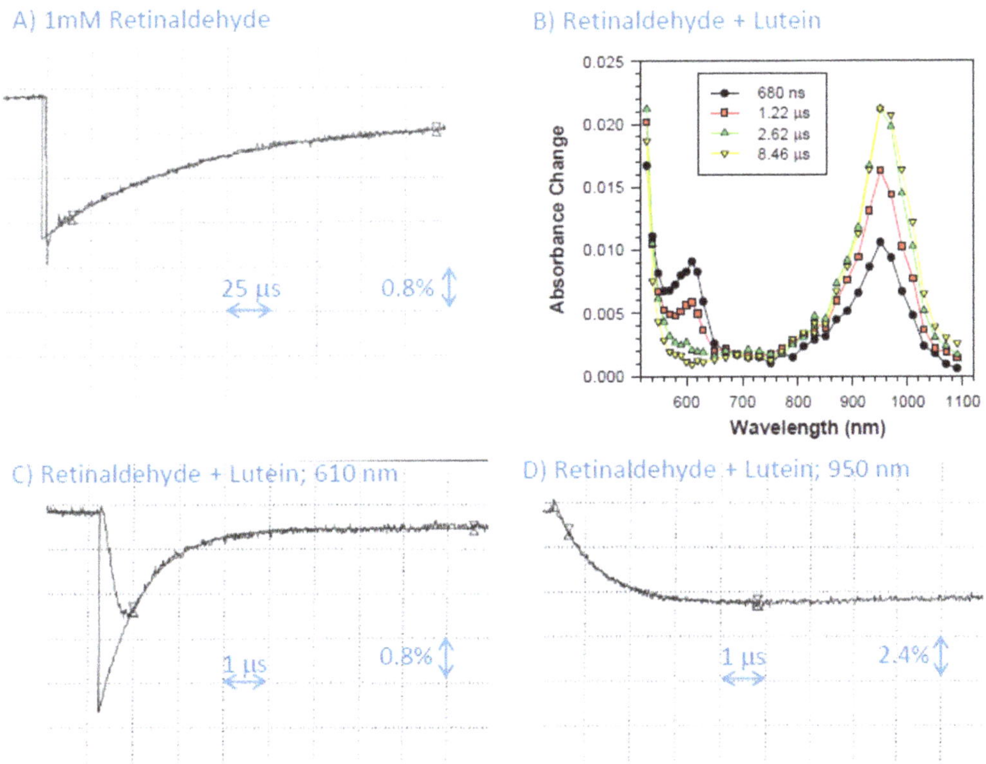

Figure 4. *Cont.*

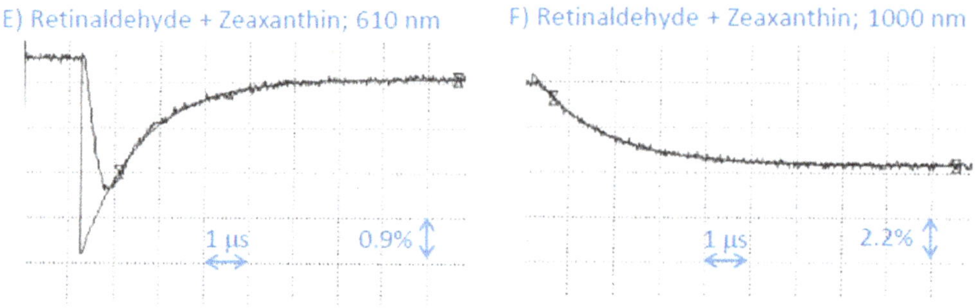

Figure 4. (**A**) Representative kinetics of the formation and decay of the transient species monitored at 610 nm after pulse radiolysis of N_2O-saturated benzene with solubilized 1 mM of retinaldehyde. (**B–D**): Transient absorption spectra at indicated times after the pulse radiolysis of N_2O-saturated benzene with solubilized 1 mM of retinaldehyde and 0.1 mM lutein (**B**) and representative kinetics of the of the formation and decay of the transient species after the pulse radiolysis of that solution monitored at 610 nm (**C**) and 950 nm (**D**) after pulse radiolysis of that solution. (**E**,**F**) Representative kinetics of the formation and decay of transient species monitored at 610 nm (**E**) and 1000 nm (**F**) after pulse radiolysis of N_2O-saturated benzene with solubilized 1 mM of retinaldehyde in the presence of 0.1 mM zeaxanthin.

2.6. *Interaction of Cation Radical of Retinaldehyde with Taurine, and Dopa-Melanin*

Taurine had no effect on the rate of decay of the retinaldehyde cation radical, suggesting that, if the scavenging occurs, the bimolecular rate constant is smaller than 1×10^7 $M^{-1}s^{-1}$ (Figure 5; Table 1). In the presence of dopa-melanin, the lifetime of retinaldehyde cation radical was shortened and the bimolecular rate constant of scavenging was $(1.6 \pm 0.8) \times 10^7$ $M^{-1}s^{-1}$ (Figure 5; Table 1).

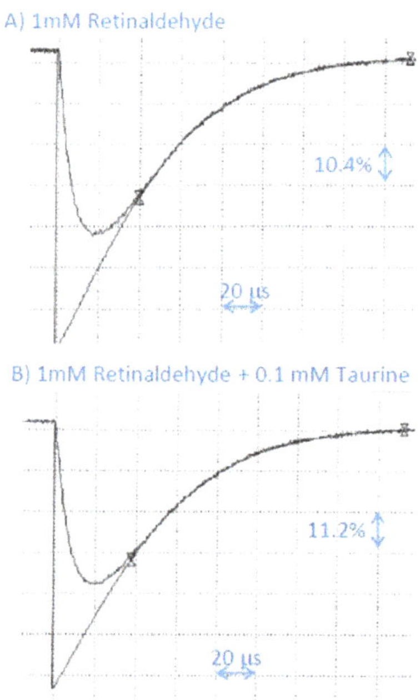

Figure 5. *Cont.*

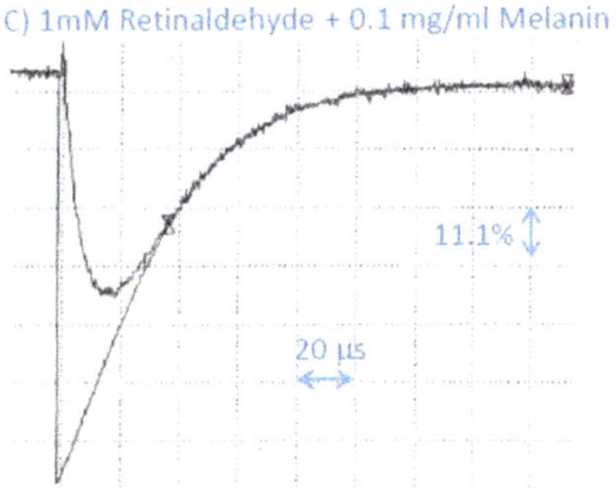

Figure 5. Representative kinetics of the formation and decay of retinaldehyde cation radicals monitored at 590 nm after pulse radiolysis of aqueous solution saturated with N_2O and containing 10 mM phosphate, pH 7, 0.1 M KBr, and 1 mM retinaldehyde incorporated in 2% Triton X-100 micelles, in the absence (**A**) and presence of 0.1 mM of taurine (**B**) or 0.1 mg/mL (equivalent to 0.67 mM monomers) dopa-melanin (**C**).

2.7. Interaction of Cation Radical of Retinol with Lutein, and Zeaxanthin

Pulse radiolysis of N_2O-saturated benzene in the presence of 1 mM retinol resulted in the formation of transient species similar to that observed before [28,45]. In the presence of lutein, the lifetime of retinol cation radical was substantially shortened, and its decay was accompanied by the concomitant formation of lutein cation radical with an absorption maximum at about 950 nm (Figure 6A–C). The bimolecular rate constants of scavenging of retinaldehyde cation radicals by lutein were $(1.01 \pm 0.05) \times 10^{10}$ $M^{-1}s^{-1}$ and $(1.26 \pm 0.04) \times 10^{10}$ $M^{-1}s^{-1}$ when determined based on rates of decay at 610 nm and rates of formation of at 950 nm, respectively (Table 1). In the presence of zeaxanthin, the lifetime of retinol cation radical was also substantially shortened, and its decay was accompanied by the concomitant formation of zeaxanthin cation radical with an absorption maximum at about 1000 nm (Figure 6D,E). The bimolecular rate constants of scavenging of retinaldehyde cation radicals by zeaxanthin were $(10.3 \pm 0.6) \times 10^{10}$ $M^{-1}s^{-1}$ and $(7.9 \pm 0.3) \times 10^{9}$ $M^{-1}s^{-1}$ when determined based on rates of decay at 610 nm and rates of formation of at 950 nm, respectively. These scavenging rates are somewhat greater than the rate of scavenging of retinol cation radicals by zeaxanthin in methanol of $(5.8 \pm 0.8) \times 10^{9}$ $M^{-1}s^{-1}$ [21].

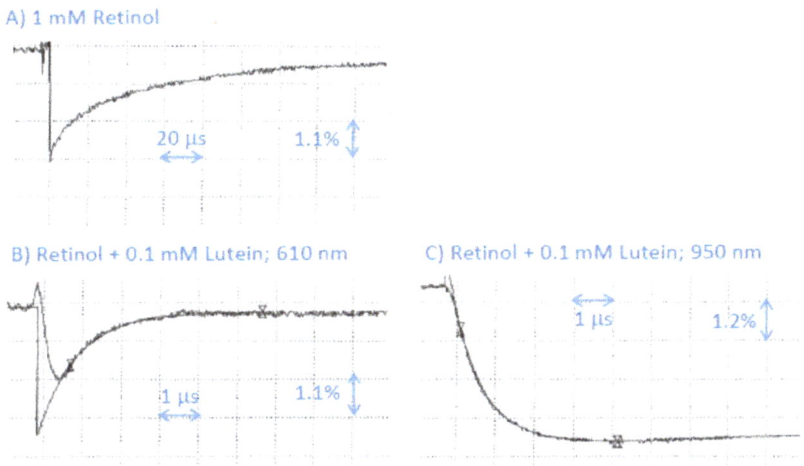

Figure 6. *Cont.*

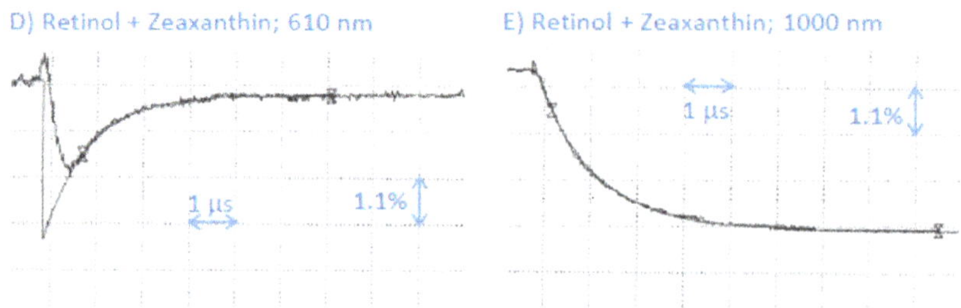

Figure 6. (**A**) Representative kinetics of the formation and decay of the transient species monitored at 610 nm after pulse radiolysis of N_2O-saturated benzene with solubilized 1 mM of retinol. (**B**,**C**) Representative kinetics of the formation and decay of transient species monitored at 610 nm (**B**) and 950 nm (**C**) after pulse radiolysis of N_2O-saturated benzene with solubilized 1 mM of retinol in the presence of 0.1 mM lutein. (**D**,**E**) Representative kinetics of the formation and decay of transient species monitored at 610 nm (**D**) and 1000 nm (**E**) after pulse radiolysis of N_2O-saturated benzene with solubilized 1 mM of retinaldehyde in the presence of 0.1 mM zeaxanthin.

2.8. Interaction of Cation Radical of Retinol with Taurine, and Dopa-Melanin

Like for other retinoids, taurine had no effect on the rate of decay of the retinol cation radical, suggesting that, if the scavenging occurs, the bimolecular rate constant is smaller than 2×10^6 $M^{-1}s^{-1}$ (Figure 7; Table 1). In the presence of dopa-melanin, the lifetime of retinol cation radical was shortened and the bimolecular rate constant of scavenging was $(5.1 \pm 0.1) \times 10^6$ $M^{-1}s^{-1}$ (Figure 5; Table 1).

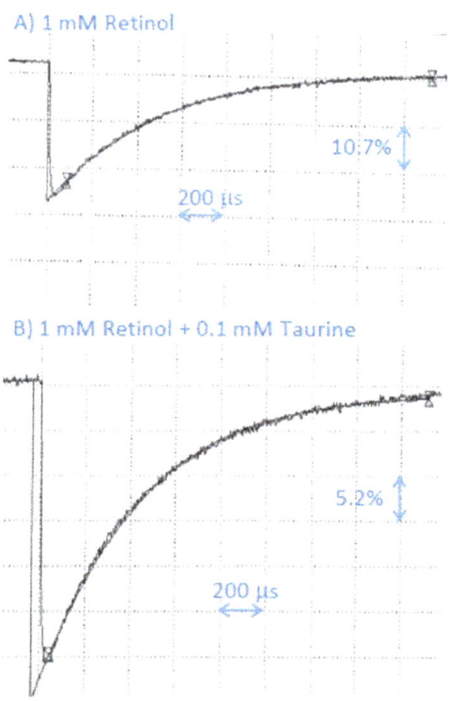

Figure 7. *Cont.*

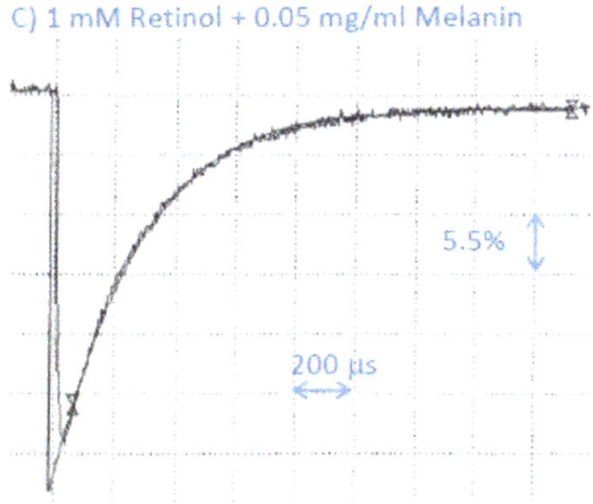

Figure 7. Representative kinetics of the formation and decay of retinol cation radicals monitored at 590 nm after pulse radiolysis of aqueous solution saturated with N_2O and containing 10 mM phosphate, pH 7, 0.1 M KBr, and 1 mM retinol incorporated in 2% Triton X-100 micelles, in the absence (**A**) and presence of 0.1 mM of taurine (**B**) or 0.1 mg/mL (equivalent to 0.67 mM monomers) dopa-melanin (**C**).

3. Discussion

3.1. Scavenging of Retinyl Palmitate Cation Radicals by Retinal Antioxidants

We have determined that pulse radiolysis of N_2O-saturated benzene in the presence of retinyl palmitate leads to the formation of retinyl palmitate cation radicals with absorption maxima at 610 nm, whereas pulse radiolysis of buffered aqueous solutions of KBr in the presence of retinyl palmitate solubilized in Triton-X micelles leads to the formation of retinyl palmitate cation radicals with absorption maxima at 590 nm. Under the experimental conditions used, cation radicals of retinyl palmitate are not scavenged by melanin or taurine, which allowed us to determine that if they do scavenge retinyl palmitate cation radicals, the upper limits of the bimolecular rate constants are smaller than 2×10^7 $M^{-1}s^{-1}$. The cation radicals of retinyl palmitate are scavenged by lutein, zeaxanthin, α-tocopherol, and ascorbate with bimolecular rate constants of $(8.85 \pm 0.25) \times 10^9$ $M^{-1}s^{-1}$, $(6.39 \pm 0.02) \times 10^9$ $M^{-1}s^{-1}$, $(2.7 \pm 0.3) \times 10^7$ $M^{-1}s^{-1}$ and $(5.8 \pm 1.0) \times 10^8$ $M^{-1}s^{-1}$, respectively. Interestingly, α-tocopherol appears to be less efficient than both carotenoids in scavenging retinyl palmitate cation radicals. To evaluate the relative contributions of ascorbate, α-tocopherol, lutein, and zeaxanthin to scavenging of retinyl palmitate cation radicals in the retina, the concentrations of these antioxidants need to be considered in the RPE where retinyl palmitate accumulates.

Sommerburg et al. determined the concentration of lutein and zeaxanthin in macular and peripheral RPE from 11 cadavers. Lutein and zeaxanthin concentrations from the macular area of 6 mm in diameter are (0.27 ± 0.07) ng/mm^2 and (0.18 ± 0.05) ng/mm^2, respectively, whereas outside that area they are (7.05 ± 0.9) ng/tissue and (1.97 ± 0.33) ng/tissue, respectively [34]. Taking 13.63 cm^2 as the total retinal surface [48] and 8 µm as the height of the RPE layer allows us to estimate the molar concentrations of these carotenoids: 33.8 and 22.5 µM for the average concentration of lutein and zeaxanthin in the macular area (6 mm in diameter) and 1.14 and 0.32 µM in the periphery.

The measurements of Rapp et al. [35] and Sommerburg et al. [34] were done before supplements of lutein and zeaxanthin became commercially available. Rapp et al. demonstrated that there is a linear positive correlation between the content of lutein and zeaxanthin in POS and the content of these carotenoids in the central macula of 3 mm in diameter, which is the area in the retina where lutein and zeaxanthin accumulate in the greatest concentration, mainly in the inner retinal layers [35,38,49]. It has been shown that the content of these carotenoids in the retina can be increased on average 2.9-fold by increased dietary intake and/or supplementation [50]. Therefore, it can be considered that the average concentrations of lutein and zeaxanthin in the RPE can also be increased 2.9-fold, achieving 98 and 65 μM for the average concentration of lutein and zeaxanthin in the macular area and 3.3 and 0.9 μM in the periphery. The highest concentrations of lutein and zeaxanthin reported in persons with high intake of these carotenoids were 5.1- and 5.9-fold greater for lutein and zeaxanthin, respectively, than the average values for people with normal intake. This gives us the estimate of the maximal concentrations of these carotenoids in the RPE: 172 and 133 μM for the maximal concentration of lutein and zeaxanthin in the macular area and 5.8 and 1.9 μM in the periphery.

The concentrations of lutein and zeaxanthin in different retinal areas allow evaluation of the effectiveness of their potential scavenging of retinoid cation radicals (Table 2). Table 2 shows the products of multiplications of antioxidant concentrations in different retinal areas/tissues and the bimolecular rate constant of retinyl palmitate cation radical scavenging. This allows comparison of the contribution of different antioxidants to such scavenging. In most cases, these data are based on the average concentration of antioxidants in the entire RPE cells. It can be seen that, without supplementation, lutein may be 2.1-fold more effective than zeaxanthin in scavenging retinyl palmitate cation radicals in the macular RPE and 5-fold more effective than zeaxanthin in the periphery. However, the overall effectiveness of both carotenoids strongly decreases in the periphery: 30 times for lutein and 71 times for zeaxanthin.

Table 2. Effectiveness of scavenging of retinyl palmitate cation radicals in RPE, which is the main storage site of retinyl esters in the retina. The multiplication products are based on average concentrations of lutein and zeaxanthin of the macular and peripheral RPE isolated from retinas of people with normal lutein/zeaxanthin intake, increased intake, and the highest reported and on average and maximal concentrations of vitamin E in the RPE. It is not clear what the concentrations of ascorbate are in the RPE; therefore, the calculations are done for 1 mM detected in brain glial cells, 2 mM in the vitreous body, and 10 mM in the brain neurons. See the main body of text for more details on concentrations of antioxidants in the retina.

	Effectiveness of Scavenging of Retinyl Palmitate Radical Cations in the RPE Scavenging Rate Constant × Antioxidant Concentration (10^3 s^{-1})		
	Lutein/Zeaxanthin		
	Normal Intake	Increased Intake	Maximum
[Macular RPE lutein] (μM)	33.8	97.9	172
[Macular RPE lutein]k_Q	299	866	1520
[Macular RPE zeaxanthin] (μM)	22.5	65.3	133
[Macular RPE zeaxanthin]k_Q	144	417	848
[Peripheral RPE lutein] (μM)	1.14	3.30	5.8
[Peripheral RPE lutein]k_Q	10.1	29.2	51.3
[Peripheral RPE zeaxanthin] (μM)	0.32	0.92	1.87
[Peripheral RPE zeaxanthin]k_Q	2.0	5.9	12.0
	Vitamin E (α-tocopherol)		
	Average		Maximum
[RPE Vitamin E] (μM)	115		230
[RPE Vitamin E]k_Q	3.11		6.21
	Vitamin C (ascorbate)		
[Ascorbate] (mM)	1 mM	2 mM	10 mM
[Ascorbate]k_Q	580	1160	5800

It is likely that antioxidants such as carotenoids have increased concentrations in certain subcellular compartments [36]. It has been shown that lutein can act as a substrate of the same enzyme, RPE65, which converts all-trans-retinyl palmitate into 11-*cis*-retinol [51]. In the case of lutein, RPE65 converts it into meso-zeaxanthin, which then accumulates in the neural retina, particularly in the macula. These data are based on studies on developing chicken eyes and cultured cells, expressing either human or chicken RPE65. The accumulation of meso-zeaxanthin in the chicken retina coincides with an increased expression of RPE65 and can be inhibited by an inhibitor of RPE65, which is also effective in inhibiting the visual pigment chromophore synthesis. These results, together with the well-documented accumulation of meso-zeaxanthin in the macula of the human retina, suggest that lutein and retinyl palmitate can colocalize in proximity to RPE65, which is present in the endoplasmic reticulum. Arguably, the retinoid cation radical may be involved in the combined reaction of isomerization and hydrolysis of all-trans-retinyl palmitate into 11-cis-retinol [52–54]. This raises a possibility, worth investigating, that lutein may act as an inhibitor of RPE65 activity towards retinyl palmitate not only as a competitive substrate but also as a scavenger of retinoid radical cation if it is involved in that process.

Another place of possible co-localization of lutein/zeaxanthin with retinyl palmitate is specialized lipid droplets, retinosomes, which are the major storage site of retinyl esters in the retina [1].

Friedrichson et al. have determined that the concentrations of vitamin E in the RPE are similar in the macula and periphery of the human retina [29]. Based on a similar approach as for carotenoids above, the calculated average concentration of vitamin E is 115 μM. This concentration exceeds almost four-fold the concentration of this vitamin in blood plasma (25–35 μM) [55,56]. Based on the highest content of vitamin E determined in that study being about twice greater than the average, the highest concentration of vitamin E was determined as 230 μM. As can be seen in Table 2, vitamin E may be a slightly more effective scavenger of retinyl palmitate cation radicals than zeaxanthin in the peripheral retina but not in the macula, where zeaxanthin is at least an order of magnitude more effective. The maximum concentration of vitamin E is not enough to provide similar scavenging as the average concentration of lutein in the periphery. In the macula, lutein has the potential to provide scavenging close to two orders of magnitude greater than α-tocopherol despite the average concentration of lutein is lower than that of vitamin E.

The concentration of ascorbate was determined in the vitreous as 2 mM [30], and, due to the lack of any barriers, it can be assumed to be similar in the interphotoreceptor matrix. This concentration hugely exceeds the typical concentrations of ascorbate in the blood plasma of 0.05 to 0.08 mM [56,57]. While, to our knowledge, there are no measurements of ascorbate in the human retinal neurons or RPE, the high expression of sodium-dependent vitamin C transporters in the basal plasma membrane of human RPE and, to a smaller extent, in photoreceptors and other cells of the neural retina suggests that RPE and retinal neurons can accumulate similar concentrations of ascorbate as glial cells and neurons in the brain: about 1 and 10 mM [30,58,59], respectively. Vitamin C, even at the lowest of these concentrations, appears to be able to provide the greatest contribution to scavenging of retinyl palmitate cation radicals in the peripheral RPE where its efficiency can be greater than that of maximal concentrations of lutein or vitamin E. In the macula, it may provide better scavenging than average concentrations of lutein and zeaxanthin combined.

3.2. *Scavenging of Retinol Cation Radicals by Retinal Antioxidants*

The cation radicals of retinol are scavenged by lutein and zeaxanthin more effectively than by α-tocopherol or ascorbate (Table 1). To evaluate their potential contributions to scavenging retinol cation radicals, we considered their concentrations in the retina (Table 3). It has been shown that about 15–25% of retinal carotenoids are present in POS [34,35]. Sommerburg et al. determined the content of lutein and zeaxanthin in POS isolated from the entire human retina so their molar concentrations can be estimated as 1.01 μM for lutein and 0.47 μM for zeaxanthin [34,37]. It is reasonable to expect that in the macula, where the concentrations of these carotenoids in the RPE are several-fold greater than in the periphery and their concentrations in the inner retina of the central macula reach millimolar values, the concentrations of lutein and zeaxanthin in POS will be also greater than in the periphery. Therefore, we have estimated these concentrations based on 2.1% contribution of the macular area of 6 mm in diameter to the total retinal surface of 13.63 cm^2 and the ratio of respected carotenoid concentrations in the RPE of the macula and periphery. This approach gives 18.8 and 13.7 μM concentrations of lutein and zeaxanthin, respectively, in the macula and 0.63 and 0.18 μM in the periphery. It is likely that the concentrations are underestimated, possibly several-fold, due to losses of POS occurring in the isolation procedure. Table 3 shows that the potential effectiveness of retinol radical scavenging by lutein and zeaxanthin is the greatest in the RPE and POS in the macular area, and it drops at least an order of magnitude in the periphery.

Table 3. Effectiveness of scavenging of retinol cation radicals by antioxidants in pathways of retinol trafficking in the retina. The multiplication products are based on average concentrations of lutein and zeaxanthin of the macular and peripheral RPE and POS from retinas of people with normal lutein/zeaxanthin intake, increased intake, and the highest reported on average and maximal concentrations of vitamin E in the RPE and neural retinas dissected from macula and periphery. It is not clear what the concentrations of ascorbate are in the RPE; therefore, the calculations are done based on ascorbate concentrations of 1 mM in brain glial cells, 2 mM concentration of ascorbate in the vitreous body, and 10 mM concentration of ascorbate in the brain neurons. See the main body of text for more details on concentrations of antioxidants in the retina.

	Effectiveness of Scavenging of Retinol Radical Cations in the RPE and Neural Retina Scavenging Rate Constant × Antioxidant Concentration (10^3 s^{-1})		
	Lutein/Zeaxanthin		
	Normal Intake	Increased Intake	Maximum
[Macular RPE lutein] (μM)	33.8	97.9	172
[Macular RPE lutein]k_Q	425	1233	2169
[Macular RPE zeaxanthin] (μM)	22.5	65.3	133
[Macular RPE zeaxanthin]k_Q	178	515	1049
[Peripheral RPE lutein] (μM)	1.14	3.30	5.8
[Peripheral RPE lutein]k_Q	14	42	73
[Peripheral RPE zeaxanthin] (μM)	0.32	0.92	1.87
[Peripheral RPE zeaxanthin]k_Q	2.5	7.3	14.8
[Macular POS lutein] (μM)	18.8	54.5	95.9
[Macular POS lutein]k_Q	237	687	1209
[Macular POS zeaxanthin] (μM)	13.7	39.7	80.7
[Macular POS zeaxanthin]k_Q	108	314	638
[Peripheral POS lutein] (μM)	0.63	1.84	3.23
[Peripheral POS lutein]k_Q	8.0	23.1	40.7
[Peripheral POS zeaxanthin] (μM)	0.19	0.56	1.14
[Peripheral POS zeaxanthin]k_Q	1.5	4.4	9.0
	Vitamin E (α-tocopherol)		
	Average		Maximum
[RPE Vitamin E] (μM)	115		230
[RPE Vitamin E]k_Q	9.2		18.4
[neural vitamin E; macula] (μM)	46		78
[neural vitamin E; macula]k_Q	3.7		6.3
[neural vitamin E; periphery] (μM)	77		124
[neural vitamin E; periphery]k_Q	6.1		9.9

Table 3. *Cont.*

	Effectiveness of Scavenging of Retinol Radical Cations in the RPE and Neural Retina Scavenging Rate Constant × Antioxidant Concentration (10^3 s^{-1})		
		Vitamin C (ascorbate)	
[Ascorbate] (mM)	1 mM	2 mM	10 mM
[Ascorbate]k_Q	120	240	1200
		Melanin in RPE melanosomes	
[Young RPE melanin] (mM)		398	
[Young RPE melanin]k_Q		2030	
[Old RPE melanin] (mM)		310	
[Old RPE melanin]k_Q		1581	

It has been determined that the interphotoreceptor retinoid-binding protein (IRBP) binds lutein and zeaxanthin with similar affinity as retinoids [60]; therefore, it is likely that lutein and zeaxanthin are present in this area on their way to photoreceptors and the inner retina. To our knowledge, there are no reports on concentrations of lutein/zeaxanthin in the interphotoreceptor matrix and on how their binding to IRBP affects their interactions with free radicals; therefore, we cannot evaluate their potential contribution to scavenging of retinoid cation radicals in this area. We also have no such data for another lipophilic antioxidant, vitamin E, which is expected to be mostly solubilized in lipid membranes.

Based on data of Friedrichson et al., it has been determined that molar concentrations of vitamin E in the neural retina of the macular area and periphery are 46 and 77 μM, respectively [29,37]. Comparison of calculated effectiveness of retinol cation radicals scavenging clearly shows that, in comparison with lutein and zeaxanthin, α-tocopherol can provide only a minor contribution to the scavenging of the retinol cation radical in the macula but is similar or exceeds the contribution of these carotenoids in the periphery despite that the rate constant for scavenging of retinol cation radicals is 158 times greater for lutein than for α-tocopherol (Table 3).

Assuming 10 mM concentrations of ascorbate in photoreceptors, 2 mM in the interphotoreceptor matrix, and 1 mM in RPE, it can be estimated that, outside the macula, ascorbate can provide a greater contribution to scavenging of retinol cation radicals than any other antioxidant (Table 2). In the macula, ascorbate may provide a contribution to scavenging similar to that of lutein and zeaxanthin, both in POS and RPE. It can be expected that ascorbate can exert its antioxidant action at the interfaces of aqueous solution and lipid membranes as well as in the aqueous milieu. While it can be expected that, in the normal retina, the majority of retinol in the interphotoreceptor matrix is bound to IRBP [3], some may be solubilized directly in its aqueous part. It has been determined that the solubility of retinol in buffered water varies from 51 to 79 nM, depending on the evaluation method, and that ascorbate and α-tocopherol (in ethanolic/water solutions) can protect it from oxidative degradation in such environment [61]. Our results provide a potential mechanism that can be responsible for this protective effect and suggest that ascorbate could serve as the main protectant of retinol in the interphotoreceptor matrix.

Although the determined rate constant of the interaction of the cation radicals of retinol with synthetic dopa-melanin is relatively low, $(5.1 \pm 0.1) \times 10^6$ $M^{-1}s^{-1}$, the obtained data do not rule out a possible protective role of melanin against oxidation by this radical in the RPE. In the retina, melanin is present in melanosomes and in melanolipofuscin granules (reviewed in [43]). Melanosomes, particularly in the young retina, are abundant in the apical processes of the RPE surrounding POS. Particularly these melanosomes are in the trafficking pathway of retinol diffusing, either as free retinol or bound by cellular retinol-binding protein (CRBP), toward the cell body where it becomes esterified. With ageing, melanosomes redistribute throughout the entire RPE cell while melanolysosomes and melanolipofuscin granules start to accumulate mainly in the cell body. It has been determined that melanin content in human RPE melanosomes isolated from 20–30-year-old cadavers is (0.18 ± 0.01) pg/melanosomes and decreases to (0.14 ± 0.2) pg/melanosomes for melanosomes isolated from 60–90-year-old cadavers [62]. Taking 2 μm as the long axis and 0.6 μm as the short axis of melanosome and calculating its volume as the volume of an ellipsoid and 150 g/mol as the molecular weight of melanin monomer, the concentration of melanin within melanosome can be estimated as 398 mM in young RPE and 310 mM in old RPE. Due to this high local concentration of melanin in RPE melanosomes and the strong affinity of melanin for positively charged molecules, it can be expected that retinol cation radicals formed in the proximity of RPE melanosomes will be effectively scavenged by melanin. It can be also expected that melanosomal concentrations of amino acids susceptible to oxidation by retinol cation radicals, namely, cysteine, tryptophan, lysine, and arginine, are orders of magnitude smaller than that of melanin. These amino acids scavenge retinol cation radicals with similar bimolecular rate constants as melanin, that is 2.6×10^5, 7.1×10^5, 1.4×10^7, 1.5×10^7, and 5.0×10^7 $M^{-1}s^{-1}$ for cysteine, tryptophan, lysine, and arginine, respectively [22]. Therefore, it can be expected that, due to the high concentration of melanin within melanosomes, they can be effectively protected from oxidation by retinol cation radicals.

3.3. Scavenging of Retinaldehyde Cation Radicals by Retinal Antioxidants

The cation radicals of retinaldehyde are scavenged by lutein and zeaxanthin with similar bimolecular rate constants as that of α-tocopherol and about 16- and 9-fold greater, respectively, than by ascorbate (Table 1). In the case of scavenging of retinaldehyde cation radicals in the retina, α-tocopherol and ascorbate may provide a major contribution both in the RPE and neural retina (Table 4). Lutein, followed by zeaxanthin, may provide substantial protection in the RPE and POS in the macular area but only minor protection in the periphery.

Table 4. Effectiveness of scavenging of retinaldehyde cation radicals in pathways of retinal trafficking in the retina. See Table 3 caption for further details.

	Effectiveness of Scavenging of Retinaldehyde Radical Cations in the RPE and Neural Retina Scavenging Rate Constant × Antioxidant Concentration (10^3 s^{-1})		
	Lutein/Zeaxanthin		
	Normal Intake	Increased Intake	Maximum
[Macular RPE lutein] (μM)	33.8	97.9	172
[Macular RPE lutein]k_Q	388	1126	1979
[Macular RPE zeaxanthin] (μM)	22.5	65.3	133
[Macular RPE zeaxanthin]k_Q	146	424	863
[Peripheral RPE lutein] (μM)	1.14	3.30	5.8
[Peripheral RPE lutein]k_Q	13	38	67
[Peripheral RPE zeaxanthin] (μM)	0.32	0.92	1.87
[Peripheral RPE zeaxanthin]k_Q	2.1	6.0	12.2
[Macular POS lutein] (μM)	18.8	54.5	95.9
[Macular POS lutein]k_Q	217	627	1103
[Macular POS zeaxanthin] (μM)	13.7	39.7	80.7
[Macular POS zeaxanthin]k_Q	89	258	525
[Peripheral POS lutein] (μM)	0.63	1.84	3.23
[Peripheral POS lutein]k_Q	7.3	21.1	37.1
[Peripheral POS zeaxanthin] (μM)	0.19	0.56	1.14
[Peripheral POS zeaxanthin]k_Q	1.3	3.6	7.4

Table 4. *Cont.*

	Effectiveness of Scavenging of Retinaldehyde Radical Cations in the RPE and Neural Retina Scavenging Rate Constant × Antioxidant Concentration (10^3 s^{-1})		
	Vitamin E (α-tocopherol)		
	Average		Maximum
[RPE Vitamin E] (μM)	115		230
[RPE Vitamin E]k_Q	920		1840
[neural vitamin E; macula] (μM)	46		78
[neural vitamin E; macula]k_Q	367		627
[neural vitamin E; periphery] (μM)	77		124
[neural vitamin E; periphery]k_Q	613		991
	Vitamin C (ascorbate)		
[Ascorbate] (mM)	1 mM	2 mM	10 mM
[Ascorbate]k_Q	730	1460	7300
	Melanin in RPE melanosomes		
[Young RPE melanin] (mM)		398	
[Young RPE melanin]k_Q		6368	
[Old RPE melanin] (mM)		310	
[Old RPE melanin]k_Q		4960	

The rate constant of scavenging of retinaldehyde cation radicals by dopa-melanin is about 3.1-fold greater than that for retinol cation radicals. Similar considerations as in the case of potential interactions of retinol cation radicals with melanin-containing pigment granules discussed above can be applied to the potential interactions of melanin with retinaldehyde cation radicals. After its release from the visual pigments, all-trans-retinaldehyde may escape its enzymatic reduction to all-trans-retinol in POS and diffuse to the apical processes of the RPE filled with melanosomes. Moreover, the tips of POS are phagocytosed daily by the RPE and enter its phagolysosomal system, where retinaldehydes are released as a result of proteolysis of visual pigments. In Stargardt's disease caused by mutations resulting in the loss of function in ATP-binding cassette transporter subfamily A member 4 (ABCA4) [27,63], the removal of retinaldehydes from phagocytosed tips of POS can be further compromised, thereby increasing the risk of its oxidation. Such retinaldehydes may appear in proximity to melanolysosomes and melanolipofuscin. Also, the newly synthesized 11-cis-retinaldehyde may be in sufficient proximity to melanolysosomes and melanolipofuscin to facilitate its interaction with melanin once it becomes oxidized to cation radical.

Another enzyme responsible for the clearance of all-trans-retinaldehydes from POS is retinol dehydrogenase 8 (RDH8), which is the main enzyme there responsible for the reduction of all-trans-retinaldehyde to all-trans-retinol [1,64]. RDH8 dysfunction also causes Stargardt's disease [26]. It can be expected that the delayed clearance of all-trans-retinaldehyde due to the loss of function of ABCA4 or RDH8 can affect mostly POS and subsequently the entire photoreceptor cell and RPE [64,65].

Retinaldehyde can accumulate in POS in substantial concentrations also in the healthy retina as a result of exposure of the dark-adapted retina to bright light that causes photoactivation of most visual pigments followed by the release of all-trans-retinaldehyde at concentrations up to 3.8 mM [1,64,65]. With ageing of the retina, there is an increased frequency of POS losing their ordered appearance and becoming convoluted and mismanaged [66]. This raises a possibility that retinoid trafficking and their enzymatic transformations can be affected and clearance of retinaldehydes delayed. Retinaldehydes are potent photosensitizers that upon absorption of UV or blue light photosensitize the generation of singlet oxygen, free radicals, and lipid peroxidation. POS contain a high concentration of polyunsaturated fatty acids, including docosahexaenoate with 6 unsaturated double bonds, making it extremely susceptible to oxidation by singlet oxygen or free radicals and subsequent propagation of lipid peroxidation, especially in the aged retina and retina affected by AMD, which exhibits increased levels of iron [9–11,67]. This oxidative environment, to which retinaldehydes contribute, can lead to the formation of retinaldehyde cation radicals. Moreover, further photooxidative degradation of retinaldehydes leads to the formation of products with increased cytotoxicity and phototoxicity in comparison to retinaldehyde [24]. Therefore, the adequate antioxidant protection of the retina is of utmost importance. Lutein, zeaxanthin, and vitamins E and C can provide antioxidant protection by quenching excited triplet states of retinaldehydes and singlet oxygen and by inhibiting lipid peroxidation [36,68,69]. The results of this study suggest that scavenging retinaldehyde cation radical may be another way these antioxidants can offer protection.

3.4. Summary and Wider Physiological Relevance

In summary, we have shown the formation of retinyl palmitate cation radicals both in benzene and in Triton X-100 micelles solubilized in phosphate buffer. Vitamin C scavenged retinyl palmitate radical cation with a bimolecular rate constant of $(5.8 \pm 0.1) \times 10^8$ $M^{-1}s^{-1}$. This value is 4.8 times greater than the bimolecular rate constant of scavenging by ascorbate of cation radicals of retinol but 21% smaller than that for scavenging of cation radicals of retinaldehyde determined previously [18]. Vitamin E was 3- and 296-fold less effective in scavenging cation radicals of retinyl palmitate than in scavenging cation radicals of retinol or retinaldehyde, respectively.

We observed no effect of taurine, up to the highest concentration studied of 0.1 mM, on the rates of cation radical decays for any of the retinoids studied. Dopa-melanin had no effect on the decays of cation radicals of retinyl palmitate but did scavenge cation radicals of retinol and retinaldehyde with the bimolecular rate constants $(5.1 \pm 0.1) \times 10^6$ $M^{-1}s^{-1}$ and $(1.6 \pm 0.8) \times 10^7$ $M^{-1}s^{-1}$.

Lutein and zeaxanthin scavenged cation radicals of all three retinoids with the bimolecular rate constants approaching the diffusion-controlled limits in benzene, ranging from 6.4×10^9 $M^{-1}s^{-1}$ for retinyl palmitate cation radical scavenged by zeaxanthin to 12.6×10^9 $M^{-1}s^{-1}$ for retinol cation radical scavenged by lutein. Despite lutein exhibiting greater scavenging rate constants of retinoid cation radicals than other antioxidants, the greater concentrations of ascorbate in the retina suggest that vitamin C may be the main protectant of the visual cycle retinoids from oxidative degradation. Vitamin E may play a substantial role in the protection of retinaldehyde but may be relatively inefficient in the protection of retinol or retinyl palmitate. While the protection of retinoids by lutein and zeaxanthin appears inefficient in the retinal periphery, it can be quite substantial in the macula. It needs to be stressed that the calculations are based on carotenoid concentrations averaged from the entire macula. Lutein and zeaxanthin are concentrated in the centre of the macula of about 3 mm in diameter. Therefore, there is a possibility that the contributions of lutein and zeaxanthin in that area may be much greater than those estimated in Tables 1–3. Moreover, Bhosale et al. reported a great inter-individual variability in the contents of these carotenoids, which, in the macular area of 4 mm in diameter, range from about 1 ng up to about 115 ng [50]. This indicates that the inter-individual ability in scavenging of retinoid cation radicals by lutein and zeaxanthin can vary by a factor of over 100.

Although the determined rate constants of the interaction of the cation radicals of retinol and retinaldehyde with synthetic dopa-melanin are relatively low, the obtained data do not rule out a possible protective role of melanin in the RPE. This is because the local concentration of melanin in RPE melanosomes is very high, which would facilitate effective scavenging of the cation radicals by melanin if the cation radicals are formed in the proximity of RPE melanosomes.

While scavenging of retinoid cation radicals by lutein and zeaxanthin is a good thing because it prevents further oxidative degradation of retinoids and protects amino acids that otherwise could be oxidized by retinoid cation radicals, it is worth considering that the products of these reactions: cation radicals of lutein and zeaxanthin retain the ability to oxidize certain amino acids, namely tyrosine and cysteine [70]. However, the radical cations of these carotenoids can be reduced, thereby recycled to the parent molecules, by α-tocopherol, ascorbate, and melanin [70,71]. Therefore, it can be expected that these antioxidants working in combination with lutein and zeaxanthin can offer a synergistic protection. It has been previously shown that zeaxanthin in combination with α-tocopherol and/or ascorbate can often offer synergistic protection of unsaturated lipids or cultured cells exposed to light in the presence of photosensitizers [36,37,69,72–74]. The synergistic effect in protection against photooxidative damage, where both singlet oxygen and free radicals were generated, was ascribed to the ability of α-tocopherol and/or ascorbate to protect zeaxanthin from its oxidative degradation, thereby functioning for longer as an unsurpassed singlet oxygen quencher. This effect was particularly striking in a study on cultured ARPE-19 cells exposed to visible light and retinaldehyde-containing liposomes as a model of POS, where single antioxidants offered very small protective effects but their combinations provided a substantial synergistic effect [37]. It can be argued that scavenging of cation radicals of retinaldehyde could contribute to that protective effect.

The main limitation of this study is the extrapolation of the results obtained in a homogenous solution and Triton-X micelles to the retina and using for calculations the antioxidant concentrations that were averaged over the entire RPE cells or, in the case of vitamin E, over the entire neural retina. Benzene was meant to mimic the non-polar environment of the lipid membrane. However, the real retinal lipid membranes are complex structures with polar groups of retinol, retinaldehyde, lutein, and zeaxanthin likely to face the lipid-aqueous interface. Gao et al. reported that the polarity of the environment can exert substantial effects on the oxidation potentials of retinol and β-carotene and two other carotenoids [75]. Therefore, it would be of great interest to determine the dependence of the oxidation potential on the solvent polarity also for retinyl palmitate, retinaldehyde, lutein, and zeaxanthin. The local concentrations of carotenoids and vitamin E are likely to vary in different intracellular compartments. Alpha-tocopherol can be expected to be present in lipid membranes, but it is not so clear whether this is the case for lutein and zeaxanthin, which can bind to proteins such as pi isoform of glutathione S-transferase (GSTP1) or steroidogenic acute regulatory domain 3 (StARD3) [39]. Binding of carotenoid to protein may greatly influence their diffusion and antioxidant properties. While substantial progress has been made in understanding the selective uptake and accumulation of lutein and zeaxanthin in the retina [76–80], it is still unclear if free lutein and/or zeaxanthin are present in the POS or RPE or if they are bound to proteins.

The results of this study are of particular importance for diseases associated with delayed clearance of retinaldehydes and/or increased oxidative stress in the retina. It can be expected that the delayed clearance of retinaldehydes occurs in Stargardt's disease caused by mutations in ABCA4 or RDH8 [25–27]. Upon absorption of ultraviolet or blue light, 11-*cis*- and all-*trans*-retinaldehydes can form an excited triplet state, which in the presence of oxygen can transfer the excess of energy to oxygen, generating an excited form of oxygen, singlet oxygen, or can transfer an electron generating superoxide anion radical [9,10,65]. Oxidative degradation of all-*trans*-retinaldehyde leads to a mixture of products retaining the ability to generate singlet oxygen and exhibiting increased (photo)toxic properties to cultured RPE cells [24]. These reactive oxygen species can induce oxidative damage to lipids, nucleic acids, and proteins. The latter includes ABCA4, which, as a result of exposure to visible light in the presence of all-*trans*-retinaldehyde, can be inactivated as an enzyme and form aggregates [81]. While it is not clear whether the damage to ABCA4 is due to reactive oxygen species produced by photoexcited retinaldehyde in the presence of oxygen or its oxidation products [9,10,24,65], it cannot be excluded that retinaldehyde oxidation products could be at least partly involved.

Mouse models of Stargardt's disease: single gene knockouts: *abca4-/-* and particularly *rdh8-/-*, or double knockout *abca4(-/-)rdh8(-/-)* mice exhibit delayed clearance of all-*trans*-retinaldehyde after exposure to light and increased susceptibility to retinal degeneration manifesting itself as loss of photoreceptive neurons, which is accelerated when animals are raised in cyclic light as opposed to when rearing in the dark [2,64,82–88]. There is a growing body of evidence demonstrating that these single-knockout and double-knockout mice exhibit high susceptibility to retinal injury by acute exposure to light [64,89–111]. Also, the loss of function of retinol dehydrogenase 12 (RDH12), which is expressed in photoreceptor inner segments where it reduces retinaldehydes to retinol, leads to the increased susceptibility of the retina to light-induced injury [112]. Triple knockout mice *abca4(-/-)rdh8(-/-)rdh12(-/-)* raised under cyclic light develop retinal degeneration at the age of 6 weeks, whereas at least 3 months are needed to demonstrate a similar level of retinal degeneration for double knockout mice *abca4(-/-)rdh8(-/-)* [113]. Mutations of *RDH12* can cause early onset severe cone-rod dystrophy, Leber congenital amaurosis, retinitis pigmentosa, and pseudocoloboma-like maculopathy [114–118].

As mentioned in the Introduction, the AMD retina is at increased risk of oxidative damage due to an increased content of iron, and several products of lipid and protein oxidation have been detected there [11–13]. It can be expected that, in in the aged retina, and even more so in AMD, the retinoid trafficking and rapid reduction of all-*trans*-retinaldehyde to all-*trans*-retinol can be impaired due to the disordered structure of photoreceptor outer segments [66]. This is consistent with delayed dark adaptation observed in elderly patients and even more so in AMD. One of the potential factors responsible for the delayed dark adaptation could be an inadequate supply of 11-*cis*-retinaldehyde to visual pigments. This hypothesis of decreased availability of 11-*cis*-retinaldehyde was investigated by Hanneken et al. who found no difference in the ability of normal and AMD eyes to recover all rhodopsin during dark-adaptation post-mortem, suggesting that availability of 11-*cis*-retinaldehyde is not a limiting factor [119]. However, they allowed several hours for the dark adaptation to occur, so delayed trafficking of 11-*cis*-retinaldehydes to the convoluted POS in AMD retina cannot be excluded. Also, such convoluted POS may impair the trafficking of enzymatic co-factors, namely ATP and NADPH, needed by ABCA4 and RDH8 for the flipping of all-*trans*-retinaldehyde to the outer leaflet of the disk lipid membrane and its reduction to all-*trans*-retinol. Then all-trans-retinol needs to diffuse outside the convoluted POS to bind to a chaperone protein, interphotoreceptor retinoid binding protein (IRBP), for its transport to the RPE. The delayed clearance of retinoids from POS in the environment exposed to light and rich in polyunsaturated fatty acids and iron may allow for their oxidation [11,67]; therefore, adequate antioxidant protection is essential to prevent the downstream deleterious effects.

The results of this study demonstrating the effectiveness of lutein, zeaxanthin, α-tocopherol, and ascorbate in scavenging retinoid cation radicals are consistent with the results of a large multicentre randomized placebo-controlled clinical trial Age-Related Eye Disease Study 2 (AREDS2) showing that a daily supplementation with zinc, vitamins C and E, lutein, and zeaxanthin slows down the progression of AMD, particularly in persons with low dietary intake of these antioxidants [31–33]. AREDS2 also demonstrated that including lutein and zeaxanthin in the antioxidant mixture brings about a greater benefit than having no carotenoids or β-carotene. It is thought that lutein, zeaxanthin, α-tocopherol, and ascorbate can exert their beneficial effect by scavenging lipid-derived free radicals and superoxide and, in the case of lutein and zeaxanthin, by quenching of singlet oxygen and excited states of photosensitizers as well as by absorption of blue light in the innermost layers of the retina in the centre of the macula, thereby preventing blue light reaching POS and RPE in this area [38,39,68,69,120]. The results of this study demonstrate that the scavenging of retinoid cation radicals can be added to the repertoire of antioxidant actions of AREDS2 antioxidants.

4. Materials and Methods

4.1. Materials

Lutein and zeaxanthin were from DSM Nutritional Products AG (Basel, Switzerland). Retinoids (all-*trans*-retinaldehyde, all-*trans*-retinol, all-*trans*-retinyl palmitate), α-tocopherol, ascorbic acid, taurine, dihydroxyphenylalanine (dopa), Triton X-100, potassium bromide, disodium hydrogen phosphate, potassium dihydrogen phosphate, and thiocyanate were purchased from Sigma Aldrich (St. Louis, MO, USA). Benzene was purchased from VWR BDH Chemicals (Reading, UK). Cylinders with gases: nitrous oxide, argon, and nitrogen were purchased from BOC Ltd. (Woking, UK).

4.2. Synthesis of Dopa-Melanin

Dopa-melanin was used as a model of eumelanin present in the RPE [43,76]. It was synthesized by autooxidation of dopa as described previously [71,121]. For calculations of bimolecular rate constants of scavenging, the molecular weight of dopa-melanin monomer of 150 g/mol was used.

4.3. Preparation of Triton X-100 Micelles with Retinoids

All handling of retinoids was done under dim light. The retinoids and Triton X-100 were solubilized in argon-saturated benzene in a round flask, and the solvent was removed on rotary evaporator (Buchi, Flawil, Switzerland) while a thin film of retinoids and Triton X-100 was formed on the inner surface of the flask. To saturate the benzene with argon, a round flask was attached to the rotary evaporated, the pressure was lowered to 40 mbar, and then argon was allowed to the system to reach the atmospheric pressure. The remnants of benzene were removed from the flask under vacuum. Before removing the flask from the rotary evaporator, argon was allowed to fill the flask until the atmospheric pressure was reached. Then the flask was detached from the rotary evaporator and closed with a stopper. Just before the experiments, the solution of 0.1 M KBr in 10 mM phosphate buffer was slowly added to allow for formation of Triton-X micelles with incorporated retinoids.

4.4. Generation of Retinoid Cation Radicals and Monitoring Their Interaction with Antioxidants

The retinoid radical cations were generated by pulse radiolysis of nitrous oxide-saturated benzene with 1 mM retinoids or of aqueous buffered solution of potassium bromide where 1 mM retinoids were solubilized in 2% Triton X-100 micelles as described previously [18,28]. To saturate samples with nitrous oxide, a solution was placed in 25, 50, or 100 mL flasks with caps allowing for the flow of connected gas to the bottom of the flask and gently bubbling for 30 min to replace the air. The taps on the cap were closed and the flask was transferred to the pulse radiolysis room where the taps were connected to the optical cell and nitrogen gas. For radiolysis of the solution, a 20 to 500 ns pulse of electrons from a 9–12 MeV linear particle accelerator (Metropolitan-Vickers, Manchester, UK) or a 12-MeV RDL, 3-GHz electron linear accelerator with a pulse duration from 0.22 to 2 μs was used. As targeted a quartz flow-through cell with an optical path-length of 2.5 cm filled with a solution in benzene or water was used. The solution was supplied to the cell under pressure of nitrogen from a stock flask. Thiocyanate was used as a dosimeter. Changes in the transmittance of the solution at a selected wavelength were monitored at 90° by transient absorption spectroscopy.

In the case of benzene, the pulse of electrons caused the ionization of the solvent and the formation of benzene cation radicals, free electrons, and excited states. The electrons were scavenged by nitrous oxide, resulting in the formation of dinitrogen and anion radical of atomic oxygen. The oxygen anion radical was scavenged by benzene forming a radical anion adduct. The radical anion adduct was scavenged also by nitrous oxide, and the product reacted with benzene forming stable products. The cation radicals of benzene abstracted electrons from retinoids forming retinoid cation radicals [18,28].

In the case of the aqueous solution of 0.1 M potassium bromide and 10 mM phosphate buffer, pH7.0, radiolysis of water by a pulse of electrons generated hydroxyl radicals and solvated electrons. Scavenging of solvated electrons by nitrous oxide in the presence of water generated more hydroxyl radicals. Hydroxyl radicals were scavenged by bromide anions generating bromine radicals. The interaction of bromide anions with bromine radicals resulted in the formation of a highly oxidizing dibromine radical anion. The interaction of the dibromine radical anion with retinoids solubilized in Triton X-100 micelles led to the formation of retinoid cation radicals [18].

To study the interaction between retinoid cation radicals with antioxidants, the selected antioxidant was present in the solution at concentrations up to 0.1 mM, and their effect on the kinetics of retinoid cation radicals was determined. In addition, for experiments involving lutein or zeaxanthin, we also monitored the formation of the carotenoid cation radical.

5. Conclusions

In conclusion, we have determined the bimolecular rate constant of scavenging cation radicals of retinoids involved in vision by lutein, zeaxanthin, dopa-melanin, and, in case of retinyl palmitate cation radicals, also by ascorbate and α-tocopherol. No increase in the rate of decay of retinoid cation radicals was observed in the presence of taurine, so it has been evaluated that the bimolecular rate constants of scavenging cation radicals of retinoids by taurine must be smaller than $2 \times 10^7\ M^{-1}s^{-1}$. Lutein scavenges cation radicals of all three retinoids with the bimolecular rate constants approaching the diffusion-controlled limits, while zeaxanthin is only 1.4–1.6-fold less effective. Despite lutein exhibiting greater scavenging rate constants of retinoid cation radicals than other antioxidants, the greater concentrations of ascorbate in the retina suggest that vitamin C may be the main protectant of the visual cycle retinoids from oxidative degradation. Vitamin E may play a substantial role in the protection of retinaldehyde but is relatively inefficient in the protection of retinol or retinyl palmitate. While the protection of retinoids by lutein and zeaxanthin appears inefficient in the retinal periphery, it can be quite substantial in the macula. Although the determined rate constants of the interaction of the cation radicals of retinol and retinaldehyde with synthetic dopa-melanin are relatively low, the high concentration of melanin in the RPE melanosomes suggest that melanin can be effective in scavenging the retinoid cation radicals in proximity to melanin-containing pigment granules.

Author Contributions: Conceptualization, M.R., T.S. and T.G.T.; Methodology, R.E., E.J.L., S.N. and T.G.T.; Performing experiments, M.R., R.E., E.J.L. and S.N.; Data Analysis, M.R., R.E., E.J.L., S.N., T.S. and T.G.T.; Writing—Original Draft Preparation, M.R.; Writing—Review & Editing, M.R., R.E., S.N., T.S. and T.G.T.; Funding Acquisition, M.R., R.E., S.N., T.S. and T.G.T. All authors have read and agreed to the published version of the manuscript.

Funding: This research was funded by the Wellcome Trust Travelling Fellowship to M.R. and the European Union Vth Framework Programme (Contract ICAI-CT-2000-70012). The pulse radiolysis experiments were performed at the Christie Hospital, Manchester or at the Free Radical Research Facility at Daresbury Synchrotron Laboratory and supported by the European Commission Human Potential Programme Access to Infrastructure (Contract HPRI-CT-2002-00183). This research was funded in part by the Wellcome Trust. For the purpose of Open Access, the author has applied a CC BY public copyright license to any Author Accepted Manuscript version arising from this submission.

Institutional Review Board Statement: Not applicable.

Informed Consent Statement: Not applicable.

Data Availability Statement: All data are presented in the manuscript.

Acknowledgments: R.E. would like to acknowledge the support of the University of Manchester's Dalton Cumbrian Facility (DCF), a partner in the National Nuclear User Facility, the EPSRC UK National Ion Beam Centre, and the Henry Royce Institute. We thank DSM Nutritional Products AG (Basel, Switzerland) for the generous gift of lutein and zeaxanthin.

Conflicts of Interest: The authors declare no conflicts. The funders had no role in the design of the study; in the collection, analyses, or interpretation of data; in the writing of the manuscript, or in the decision to publish the results.

References

1. Kiser, P.D.; Golczak, M.; Palczewski, K. Chemistry of the retinoid (visual) cycle. *Chem. Rev.* **2014**, *114*, 194–232. [CrossRef] [PubMed]
2. Kiser, P.D.; Palczewski, K. Pathways and disease-causing alterations in visual chromophore production for vertebrate vision. *J. Biol. Chem.* **2021**, *296*, 100072. [CrossRef] [PubMed]
3. Chen, C.; Adler, L.t.; Goletz, P.; Gonzalez-Fernandez, F.; Thompson, D.A.; Koutalos, Y. Interphotoreceptor retinoid-binding protein removes all-trans-retinol and retinal from rod outer segments, preventing lipofuscin precursor formation. *J. Biol. Chem.* **2017**, *292*, 19356–19365. [CrossRef] [PubMed]
4. Crouch, R.K.; Hazard, E.S.; Lind, T.; Wiggert, B.; Chader, G.; Corson, D.W. Interphotoreceptor retinoid-binding protein and alpha-tocopherol preserve the isomeric and oxidation state of retinol. *Photochem. Photobiol.* **1992**, *56*, 251–255. [CrossRef] [PubMed]
5. Hara, M.; Wu, W.; Malechka, V.V.; Takahashi, Y.; Ma, J.X.; Moiseyev, G. PNPLA2 mobilizes retinyl esters from retinosomes and promotes the generation of 11-cis-retinal in the visual cycle. *Cell Rep.* **2023**, *42*, 112091. [CrossRef] [PubMed]
6. Tworak, A.; Kolesnikov, A.V.; Hong, J.D.; Choi, E.H.; Luu, J.C.; Palczewska, G.; Dong, Z.; Lewandowski, D.; Brooks, M.J.; Campello, L.; et al. Rapid RGR-dependent visual pigment recycling is mediated by the RPE and specialized Muller glia. *Cell Rep.* **2023**, *42*, 112982. [CrossRef] [PubMed]
7. Zhang, J.; Choi, E.H.; Tworak, A.; Salom, D.; Leinonen, H.; Sander, C.L.; Hoang, T.V.; Handa, J.T.; Blackshaw, S.; Palczewska, G.; et al. Photic generation of 11-cis-retinal in bovine retinal pigment epithelium. *J. Biol. Chem.* **2019**, *294*, 19137–19154. [CrossRef]
8. Morshedian, A.; Kaylor, J.J.; Ng, S.Y.; Tsan, A.; Frederiksen, R.; Xu, T.; Yuan, L.; Sampath, A.P.; Radu, R.A.; Fain, G.L.; et al. Light-Driven Regeneration of Cone Visual Pigments through a Mechanism Involving RGR Opsin in Muller Glial Cells. *Neuron* **2019**, *102*, 1172–1183.e5. [CrossRef]
9. Rozanowska, M.; Rozanowski, B.; Boulton, M. Photobiology of the retina: Light damage to the retina. In *Photobiological Sciences Online*; Smith, K.C., Ed.; American Society for Photobiology: Herndon, VA, USA, 2009; Available online: http://www.photobiology.info (accessed on 22 January 2021).
10. Różanowska, M.; Różanowski, B. Visual Transduction and Age-Related Changes in Lipofuscin. In *Ophthalmology Research: The Visual Transduction Cascade*; Tombran-Tink, J., Barnstable, C.J., Eds.; The Humana Press Inc.: Totowa, NJ, USA, 2008; pp. 405–446.
11. Hahn, P.; Milam, A.H.; Dunaief, J.L. Maculas affected by age-related macular degeneration contain increased chelatable iron in the retinal pigment epithelium and Bruch's membrane. *Arch. Ophthalmol.* **2003**, *121*, 1099–1105. [CrossRef]
12. Gu, X.; Meer, S.G.; Miyagi, M.; Rayborn, M.E.; Hollyfield, J.G.; Crabb, J.W.; Salomon, R.G. Carboxyethylpyrrole protein adducts and autoantibodies, biomarkers for age-related macular degeneration. *J. Biol. Chem.* **2003**, *278*, 42027–42035. [CrossRef]
13. Crabb, J.W.; Miyagi, M.; Gu, X.; Shadrach, K.; West, K.A.; Sakaguchi, H.; Kamei, M.; Hasan, A.; Yan, L.; Rayborn, M.E.; et al. Drusen proteome analysis: An approach to the etiology of age-related macular degeneration. *Proc. Natl. Acad. Sci. USA* **2002**, *99*, 14682–14687. [CrossRef] [PubMed]
14. Wong, W.L.; Su, X.; Li, X.; Cheung, C.M.; Klein, R.; Cheng, C.Y.; Wong, T.Y. Global prevalence of age-related macular degeneration and disease burden projection for 2020 and 2040: A systematic review and meta-analysis. *Lancet Glob. Health* **2014**, *2*, e106–e116. [CrossRef] [PubMed]
15. Pennington, K.L.; DeAngelis, M.M. Epidemiology of age-related macular degeneration (AMD): Associations with cardiovascular disease phenotypes and lipid factors. *Eye Vis.* **2016**, *3*, 34. [CrossRef] [PubMed]
16. Rein, D.B.; Wittenborn, J.S.; Burke-Conte, Z.; Gulia, R.; Robalik, T.; Ehrlich, J.R.; Lundeen, E.A.; Flaxman, A.D. Prevalence of Age-Related Macular Degeneration in the US in 2019. *JAMA Ophthalmol.* **2022**, *140*, 1202–1208. [CrossRef] [PubMed]
17. Creuzot-Garcher, C.P.; Srour, M.; Baudin, F.; Daien, V.; Dot, C.; Nghiem-Buffet, S.; Girmens, J.F.; Coulombel, N.; Ponthieux, A.; Delcourt, C. Incidence and Prevalence of Neovascular Age-Related Macular Degeneration in France between 2008 and 2018: The LANDSCAPE Study. *Ophthalmol. Sci.* **2022**, *2*, 100114. [CrossRef] [PubMed]
18. Rozanowska, M.; Cantrell, A.; Edge, R.; Land, E.J.; Sarna, T.; Truscott, T.G. Pulse radiolysis study of the interaction of retinoids with peroxyl radicals. *Free Radic. Biol. Med.* **2005**, *39*, 1399–1405. [CrossRef] [PubMed]
19. Jakaria, M.; Belaidi, A.A.; Bush, A.I.; Ayton, S. Vitamin A metabolites inhibit ferroptosis. *Biomed. Pharmacother.* **2023**, *164*, 114930. [CrossRef] [PubMed]
20. Lo, K.K.; Land, E.J.; Truscott, T.G. Primary intermediates in the pulsed irradiation of retinoids. *Photochem. Photobiol.* **1982**, *36*, 139–145. [CrossRef]
21. El-Agamey, A.; Fukuzumi, S. Laser flash photolysis study on the retinol radical cation in polar solvents. *Org. Biomol. Chem.* **2011**, *9*, 1046–6437. [CrossRef]
22. El-Agamey, A.; Melo, T.B.; Sliwka, H.R. Exploring the reactivity of retinol radical cation toward organic and biological molecules: A laser flash photolysis study. *J. Photochem. Photobiol. B* **2017**, *170*, 33–39. [CrossRef]
23. Artigas, J.M.; Felipe, A.; Navea, A.; Fandino, A.; Artigas, C. Spectral transmission of the human crystalline lens in adult and elderly persons: Color and total transmission of visible light. *Investig. Ophthalmol. Vis. Sci.* **2012**, *53*, 4076–4084. [CrossRef] [PubMed]
24. Rozanowska, M.; Handzel, K.; Boulton, M.E.; Rozanowski, B. Cytotoxicity of all-trans-retinal increases upon photodegradation. *Photochem. Photobiol.* **2012**, *88*, 1362–1372. [CrossRef] [PubMed]
25. Cremers, F.P.M.; Lee, W.; Collin, R.W.J.; Allikmets, R. Clinical spectrum, genetic complexity and therapeutic approaches for retinal disease caused by ABCA4 mutations. *Prog. Retin. Eye Res.* **2020**, *79*, 100861. [CrossRef] [PubMed]

26. Zampatti, S.; Peconi, C.; Calvino, G.; Ferese, R.; Gambardella, S.; Cascella, R.; Sebastiani, J.; Falsini, B.; Cusumano, A.; Giardina, E. A Splicing Variant in RDH8 Is Associated with Autosomal Recessive Stargardt Macular Dystrophy. *Genes* **2023**, *14*, 1659. [CrossRef] [PubMed]
27. Molday, R.S.; Garces, F.A.; Scortecci, J.F.; Molday, L.L. Structure and function of ABCA4 and its role in the visual cycle and Stargardt macular degeneration. *Prog. Retin. Eye Res.* **2022**, *89*, 101036. [CrossRef] [PubMed]
28. Rozanowska, M.; Edge, R.; Land, E.J.; Navaratnam, S.; Sarna, T.; Truscott, T.G. Scavenging of Retinoid Cation Radicals by Urate, Trolox, and α, β, γ, and δ-Tocopherols. *Int. J. Mol. Sci.* **2019**, *20*, 2799. [CrossRef] [PubMed]
29. Friedrichson, T.; Kalbach, H.L.; Buck, P.; van Kuijk, F.J. Vitamin E in macular and peripheral tissues of the human eye. *Curr. Eye Res.* **1995**, *14*, 693–701. [CrossRef]
30. Ma, N.; Siegfried, C.; Kubota, M.; Huang, J.; Liu, Y.; Liu, M.; Dana, B.; Huang, A.; Beebe, D.; Yan, H.; et al. Expression Profiling of Ascorbic Acid-Related Transporters in Human and Mouse Eyes. *Investig. Ophthalmol. Vis. Sci.* **2016**, *57*, 3440–3450. [CrossRef]
31. Age-Related Eye Disease Study 2 Research, G.; Chew, E.Y.; Clemons, T.E.; Sangiovanni, J.P.; Danis, R.P.; Ferris, F.L., 3rd; Elman, M.J.; Antoszyk, A.N.; Ruby, A.J.; Orth, D.; et al. Secondary analyses of the effects of lutein/zeaxanthin on age-related macular degeneration progression: AREDS2 report No. 3. *JAMA Ophthalmol.* **2014**, *132*, 142–149. [CrossRef]
32. Evans, J.R.; Lawrenson, J.G. Antioxidant vitamin and mineral supplements for slowing the progression of age-related macular degeneration. *Cochrane Database Syst. Rev.* **2017**, *7*, CD000254. [CrossRef]
33. Chew, E.Y.; Clemons, T.E.; Agron, E.; Domalpally, A.; Keenan, T.D.L.; Vitale, S.; Weber, C.; Smith, D.C.; Christen, W.; Group, A.R. Long-term Outcomes of Adding Lutein/Zeaxanthin and omega-3 Fatty Acids to the AREDS Supplements on Age-Related Macular Degeneration Progression: AREDS2 Report 28. *JAMA Ophthalmol.* **2022**, *140*, 692–698. [CrossRef] [PubMed]
34. Sommerburg, O.G.; Siems, W.G.; Hurst, J.S.; Lewis, J.W.; Kliger, D.S.; van Kuijk, F.J. Lutein and zeaxanthin are associated with photoreceptors in the human retina. *Curr. Eye Res.* **1999**, *19*, 491–495. [CrossRef] [PubMed]
35. Rapp, L.M.; Maple, S.S.; Choi, J.H. Lutein and zeaxanthin concentrations in rod outer segment membranes from perifoveal and peripheral human retina. *Investig. Ophthalmol. Vis. Sci.* **2000**, *41*, 1200–1209.
36. Rozanowska, M.B.; Czuba-Pelech, B.; Landrum, J.T.; Rozanowski, B. Comparison of Antioxidant Properties of Dehydrolutein with Lutein and Zeaxanthin, and their Effects on Cultured Retinal Pigment Epithelial Cells. *Antioxidants* **2021**, *10*, 753. [CrossRef] [PubMed]
37. Rozanowska, M.B.; Czuba-Pelech, B.; Rozanowski, B. Is There an Optimal Combination of AREDS2 Antioxidants Zeaxanthin, Vitamin E and Vitamin C on Light-Induced Toxicity of Vitamin A Aldehyde to the Retina? *Antioxidants* **2022**, *11*, 1132. [CrossRef] [PubMed]
38. Arunkumar, R.; Gorusupudi, A.; Bernstein, P.S. The macular carotenoids: A biochemical overview. *Biochim. Biophys. Acta Mol. Cell Biol. Lipids* **2020**, *1865*, 158617. [CrossRef]
39. Bernstein, P.S.; Li, B.; Vachali, P.P.; Gorusupudi, A.; Shyam, R.; Henriksen, B.S.; Nolan, J.M. Lutein, zeaxanthin, and meso-zeaxanthin: The basic and clinical science underlying carotenoid-based nutritional interventions against ocular disease. *Prog. Retin. Eye Res.* **2016**, *50*, 34–66. [CrossRef]
40. Froger, N.; Moutsimilli, L.; Cadetti, L.; Jammoul, F.; Wang, Q.P.; Fan, Y.; Gaucher, D.; Rosolen, S.G.; Neveux, N.; Cynober, L.; et al. Taurine: The comeback of a neutraceutical in the prevention of retinal degenerations. *Prog. Retin. Eye Res.* **2014**, *41*, 44–63. [CrossRef]
41. Garcia-Ayuso, D.; Di Pierdomenico, J.; Martinez-Vacas, A.; Vidal-Sanz, M.; Picaud, S.; Villegas-Perez, M.P. Taurine: A promising nutraceutic in the prevention of retinal degeneration. *Neural Regen. Res.* **2024**, *19*, 606–610. [CrossRef]
42. Di Pierdomenico, J.; Martinez-Vacas, A.; Picaud, S.; Villegas-Perez, M.P.; Garcia-Ayuso, D. Taurine: An essential amino sulfonic acid for retinal health. *Neural Regen. Res.* **2023**, *18*, 807–808. [CrossRef]
43. Różanowska, M. Properties and Functions of Ocular Melanins and Melanosomes. In *Melanins and Melanosomes: Biosynthesis, Biogenesis, Physiological, and Pathological Functions*; Riley, P.A., Borovansky, J., Eds.; John Wiley and Sons Inc.: Hoboken, NJ, USA, 2011; pp. 187–224.
44. Rózanowska, M.B.; Pawlak, A.; Rózanowski, B. Products of docosahexaenoate oxidation as contributors to photosensitising properties of retinal lipofuscin. *Int. J. Mol. Sci.* **2021**, *22*, 3525. [CrossRef] [PubMed]
45. Burke, M.; Land, E.J.; McGarvey, D.J.; Truscott, T.G. Carotenoid triplet state lifetimes. *J. Photochem. Photobiol. B* **2000**, *59*, 132–138. [CrossRef] [PubMed]
46. Edge, R.; Land, E.J.; McGarvey, D.; Mulroy, L.; Truscott, T.G. Relative one-electron reduction potentials of carotenoid radical cations and the interactions of carotenoids with the vitamin E radical cation. *J. Am. Chem. Soc.* **1998**, *120*, 4087–4090. [CrossRef]
47. Polyakov, N.E.; Focsan, A.L.; Gao, Y.L.; Kispert, L.D. The Endless World of Carotenoids-Structural, Chemical and Biological Aspects of Some Rare Carotenoids. *Int. J. Mol. Sci.* **2023**, *24*, 9885. [CrossRef] [PubMed]
48. Nagra, M.; Gilmartin, B.; Thai, N.J.; Logan, N.S. Determination of retinal surface area. *J. Anat.* **2017**, *231*, 319–324. [CrossRef]
49. Li, B.; George, E.W.; Rognon, G.T.; Gorusupudi, A.; Ranganathan, A.; Chang, F.Y.; Shi, L.; Frederick, J.M.; Bernstein, P.S. Imaging lutein and zeaxanthin in the human retina with confocal resonance Raman microscopy. *Proc. Natl. Acad. Sci. USA* **2020**, *117*, 12352–12358. [CrossRef]
50. Bhosale, P.; Zhao, D.Y.; Bernstein, P.S. HPLC measurement of ocular carotenoid levels in human donor eyes in the lutein supplementation era. *Investig. Ophthalmol. Vis. Sci.* **2007**, *48*, 543–549. [CrossRef]

51. Shyam, R.; Gorusupudi, A.; Nelson, K.; Horvath, M.P.; Bernstein, P.S. RPE65 has an additional function as the lutein to *meso*-zeaxanthin isomerase in the vertebrate eye. *Proc. Natl. Acad. Sci. USA* **2017**, *114*, 10882–10887. [CrossRef]
52. Poliakov, E.; Parikh, T.; Ayele, M.; Kuo, S.; Chander, P.; Gentleman, S.; Redmond, T.M. Aromatic lipophilic spin traps effectively inhibit RPE65 isomerohydrolase activity. *Biochemistry* **2011**, *50*, 6739–6741. [CrossRef]
53. Kiser, P.D.; Zhang, J.; Badiee, M.; Li, Q.; Shi, W.; Sui, X.; Golczak, M.; Tochtrop, G.P.; Palczewski, K. Catalytic mechanism of a retinoid isomerase essential for vertebrate vision. *Nat. Chem. Biol.* **2015**, *11*, 409–415. [CrossRef]
54. Kiser, P.D. Retinal pigment epithelium 65 kDa protein (RPE65): An update. *Prog. Retin. Eye Res.* **2022**, *88*, 101013. [CrossRef] [PubMed]
55. Stuetz, W.; Weber, D.; Dolle, M.E.; Jansen, E.; Grubeck-Loebenstein, B.; Fiegl, S.; Toussaint, O.; Bernhardt, J.; Gonos, E.S.; Franceschi, C.; et al. Plasma Carotenoids, Tocopherols, and Retinol in the Age-Stratified (35–74 Years) General Population: A Cross-Sectional Study in Six European Countries. *Nutrients* **2016**, *8*, 614. [CrossRef] [PubMed]
56. Jungert, A.; Neuhauser-Berthold, M. Interrelation between Plasma Concentrations of Vitamins C and E along the Trajectory of Ageing in Consideration of Lifestyle and Body Composition: A Longitudinal Study over Two Decades. *Nutrients* **2020**, *12*, 2944. [CrossRef] [PubMed]
57. Levine, M.; Padayatty, S.J.; Espey, M.G. Vitamin C: A concentration-function approach yields pharmacology and therapeutic discoveries. *Adv. Nutr.* **2011**, *2*, 78–88. [CrossRef] [PubMed]
58. Lykkesfeldt, J.; Tveden-Nyborg, P. The Pharmacokinetics of Vitamin C. *Nutrients* **2019**, *11*, 2412. [CrossRef] [PubMed]
59. Rice, M.E.; Russo-Menna, I. Differential compartmentalization of brain ascorbate and glutathione between neurons and glia. *Neuroscience* **1998**, *82*, 1213–1223. [CrossRef]
60. Vachali, P.P.; Besch, B.M.; Gonzalez-Fernandez, F.; Bernstein, P.S. Carotenoids as possible interphotoreceptor retinoid-binding protein (IRBP) ligands: A surface plasmon resonance (SPR) based study. *Arch. Biochem. Biophys.* **2013**, *539*, 181–186. [CrossRef]
61. Szuts, E.Z.; Harosi, F.I. Solubility of retinoids in water. *Arch. Biochem. Biophys.* **1991**, *287*, 297–304. [CrossRef]
62. Rozanowski, B.; Cuenco, J.; Davies, S.; Shamsi, F.A.; Zadlo, A.; Dayhaw-Barker, P.; Rozanowska, M.; Sarna, T.; Boulton, M.E. The phototoxicity of aged human retinal melanosomes. *Photochem. Photobiol.* **2008**, *84*, 650–657. [CrossRef]
63. Romano, F.; Lamanna, F.; Boon, C.J.; Siligato, A.; Kalra, G.; Agarwal, A.; Medori, C.; Bertelli, M.; Pellegrini, M.; Invernizzi, A.; et al. Clinical, Genotypic and Imaging Characterization of the spectrum of ABCA4 Retinopathies. *Ophthalmol. Retina* **2023**, *in press*. [CrossRef]
64. Maeda, T.; Golczak, M.; Maeda, A. Retinal photodamage mediated by all-trans-retinal. *Photochem. Photobiol.* **2012**, *88*, 1309–1319. [CrossRef] [PubMed]
65. Rozanowska, M.; Sarna, T. Light-induced damage to the retina: Role of rhodopsin chromophore revisited. *Photochem. Photobiol.* **2005**, *81*, 1305–1330. [CrossRef] [PubMed]
66. Marshall, J.; Grindle, J.; Ansell, P.L.; Borwein, B. Convolution in human rods: An ageing process. *Br. J. Ophthalmol.* **1979**, *63*, 181–187. [CrossRef] [PubMed]
67. Bazan, H.E.; Bazan, N.G.; Feeney-Burns, L.; Berman, E.R. Lipids in human lipofuscin-enriched subcellular fractions of two age populations. Comparison with rod outer segments and neural retina. *Investig. Ophthalmol. Vis. Sci.* **1990**, *31*, 1433–1443.
68. Edge, R.; Truscott, T.G. Singlet Oxygen and Free Radical Reactions of Retinoids and Carotenoids-A Review. *Antioxidants* **2018**, *7*, 5. [CrossRef]
69. Różanowska, M.; Różanowski, B. Uptake and Photoprotection in Cultured RPE Cells. In *Carotenoids: Physical, Chemical, and Biological Functions and Properties*; Landrum, J.T., Ed.; CRC Press: Boca Raton, FL, USA, 2010; pp. 309–364.
70. El-Agamey, A.; Lowe, G.M.; McGarvey, D.J.; Mortensen, A.; Phillip, D.M.; Truscott, T.G.; Young, A.J. Carotenoid radical chemistry and antioxidant/pro-oxidant properties. *Arch. Biochem. Biophys.* **2004**, *430*, 37–48. [CrossRef] [PubMed]
71. Edge, R.; Land, E.J.; Rozanowska, M.; Sarna, T.; Truscott, T.G. Carotenoid radical-melanin interactions. *J. Phys. Chem. B* **2000**, *104*, 7193–7196. [CrossRef]
72. Rozanowska, M.; Bakker, L.; Boulton, M.E.; Rozanowski, B. Concentration dependence of vitamin C in combinations with vitamin E and zeaxanthin on light-induced toxicity to retinal pigment epithelial cells. *Photochem. Photobiol.* **2012**, *88*, 1408–1417. [CrossRef]
73. Wrona, M.; Rozanowska, M.; Sarna, T. Zeaxanthin in combination with ascorbic acid or alpha-tocopherol protects ARPE-19 cells against photosensitized peroxidation of lipids. *Free Radic. Biol. Med.* **2004**, *36*, 1094–1101. [CrossRef]
74. Wrona, M.; Korytowski, W.; Rozanowska, M.; Sarna, T.; Truscott, T.G. Cooperation of antioxidants in protection against photosensitized oxidation. *Free Radic. Biol. Med.* **2003**, *35*, 1319–1329. [CrossRef]
75. Gao, Y.L.; Focsan, A.L.; Kispert, L.D. The effect of polarity of environment on the antioxidant activity of carotenoids. *Chem. Phys. Lett.* **2020**, *761*, 138098. [CrossRef]
76. Arunkumar, R.; Gorusupudi, A.; Li, B.; Blount, J.D.; Nwagbo, U.; Kim, H.J.; Sparrow, J.R.; Bernstein, P.S. Lutein and zeaxanthin reduce A2E and iso-A2E levels and improve visual performance in Abca4$^{-/-}$/Bco2$^{-/-}$ double knockout mice. *Exp. Eye Res.* **2021**, *209*, 108680. [CrossRef] [PubMed]
77. Arunkumar, R.; Li, B.; Addo, E.K.; Hartnett, M.E.; Bernstein, P.S. Prenatal Carotenoid Supplementation With Lutein or Zeaxanthin Ameliorates Oxygen-Induced Retinopathy (OIR) in Bco2$^{-/-}$ Macular Pigment Mice. *Investig. Ophthalmol. Vis. Sci.* **2023**, *64*, 9. [CrossRef] [PubMed]
78. Bandara, S.; Moon, J.; Ramkumar, S.; von Lintig, J. ASTER-B regulates mitochondrial carotenoid transport and homeostasis. *J. Lipid Res.* **2023**, *64*, 100369. [CrossRef] [PubMed]

79. Bandara, S.; Ramkumar, S.; Imanishi, S.; Thomas, L.D.; Sawant, O.B.; Imanishi, Y.; von Lintig, J. Aster proteins mediate carotenoid transport in mammalian cells. *Proc. Natl. Acad. Sci. USA* **2022**, *119*, e2200068119. [CrossRef] [PubMed]
80. Li, B.; George, E.W.; Vachali, P.; Chang, F.Y.; Gorusupudi, A.; Arunkumar, R.; Giauque, N.A.; Wan, Z.; Frederick, J.M.; Bernstein, P.S. Mechanism for the selective uptake of macular carotenoids mediated by the HDL cholesterol receptor SR-BI. *Exp. Eye Res.* **2023**, *229*, 109429. [CrossRef] [PubMed]
81. Sun, H.; Nathans, J. ABCR, the ATP-binding cassette transporter responsible for Stargardt macular dystrophy, is an efficient target of all-trans-retinal-mediated photooxidative damage in vitro. Implications for retinal disease. *J. Biol. Chem.* **2001**, *276*, 11766–11774. [CrossRef]
82. Weng, J.; Mata, N.L.; Azarian, S.M.; Tzekov, R.T.; Birch, D.G.; Travis, G.H. Insights into the function of Rim protein in photoreceptors and etiology of Stargardt's disease from the phenotype in abcr knockout mice. *Cell* **1999**, *98*, 13–23. [CrossRef]
83. Maeda, A.; Maeda, T.; Golczak, M.; Palczewski, K. Retinopathy in mice induced by disrupted all-trans-retinal clearance. *J. Biol. Chem.* **2008**, *283*, 26684–26693. [CrossRef]
84. Zhao, J.; Liao, Y.; Chen, J.; Dong, X.; Gao, Z.; Zhang, H.; Wu, X.; Liu, Z.; Wu, Y. Aberrant Buildup of All-Trans-Retinal Dimer, a Nonpyridinium Bisretinoid Lipofuscin Fluorophore, Contributes to the Degeneration of the Retinal Pigment Epithelium. *Investig. Ophthalmol. Vis. Sci.* **2017**, *58*, 1063–1075. [CrossRef]
85. Lenis, T.L.; Hu, J.; Ng, S.Y.; Jiang, Z.; Sarfare, S.; Lloyd, M.B.; Esposito, N.J.; Samuel, W.; Jaworski, C.; Bok, D.; et al. Expression of ABCA4 in the retinal pigment epithelium and its implications for Stargardt macular degeneration. *Proc. Natl. Acad. Sci. USA* **2018**, *115*, E11120–E11127. [CrossRef] [PubMed]
86. Fang, Y.; Tschulakow, A.; Taubitz, T.; Illing, B.; Biesemeier, A.; Julien-Schraermeyer, S.; Radu, R.A.; Jiang, Z.; Schraermeyer, U. Fundus autofluorescence, spectral-domain optical coherence tomography, and histology correlations in a Stargardt disease mouse model. *FASEB J.* **2020**, *34*, 3693–3714. [CrossRef] [PubMed]
87. Pan, C.; Banerjee, K.; Lehmann, G.L.; Almeida, D.; Hajjar, K.A.; Benedicto, I.; Jiang, Z.; Radu, R.A.; Thompson, D.H.; Rodriguez-Boulan, E.; et al. Lipofuscin causes atypical necroptosis through lysosomal membrane permeabilization. *Proc. Natl. Acad. Sci. USA* **2021**, *118*, e2100122118. [CrossRef] [PubMed]
88. Fang, Y.; Taubitz, T.; Tschulakow, A.V.; Heiduschka, P.; Szewczyk, G.; Burnet, M.; Peters, T.; Biesemeier, A.; Sarna, T.; Schraermeyer, U.; et al. Removal of RPE lipofuscin results in rescue from retinal degeneration in a mouse model of advanced Stargardt disease: Role of reactive oxygen species. *Free. Radic. Biol. Med.* **2022**, *182*, 132–149. [CrossRef] [PubMed]
89. Maeda, A.; Maeda, T.; Golczak, M.; Imanishi, Y.; Leahy, P.; Kubota, R.; Palczewski, K. Effects of potent inhibitors of the retinoid cycle on visual function and photoreceptor protection from light damage in mice. *Mol. Pharmacol.* **2006**, *70*, 1220–1229. [CrossRef]
90. Maeda, T.; Maeda, A.; Matosky, M.; Okano, K.; Roos, S.; Tang, J.; Palczewski, K. Evaluation of potential therapies for a mouse model of human age-related macular degeneration caused by delayed all-trans-retinal clearance. *Investig. Ophthalmol. Vis. Sci.* **2009**, *50*, 4917–4925. [CrossRef]
91. Maeda, A.; Maeda, T.; Golczak, M.; Chou, S.; Desai, A.; Hoppel, C.L.; Matsuyama, S.; Palczewski, K. Involvement of all-transretinal in acute light-induced retinopathy of mice. *J. Biol. Chem.* **2009**, *284*, 15173–15183. [CrossRef]
92. Maeda, A.; Golczak, M.; Chen, Y.; Okano, K.; Kohno, H.; Shiose, S.; Ishikawa, K.; Harte, W.; Palczewska, G.; Maeda, T.; et al. Primary amines protect against retinal degeneration in mouse models of retinopathies. *Nat. Chem. Biol.* **2011**, *8*, 170–178. [CrossRef]
93. Shiose, S.; Chen, Y.; Okano, K.; Roy, S.; Kohno, H.; Tang, J.; Pearlman, E.; Maeda, T.; Palczewski, K.; Maeda, A. Toll-like receptor 3 is required for development of retinopathy caused by impaired all-trans-retinal clearance in mice. *J. Biol. Chem.* **2011**, *286*, 15543–15555. [CrossRef]
94. Chen, Y.; Okano, K.; Maeda, T.; Chauhan, V.; Golczak, M.; Maeda, A.; Palczewski, K. Mechanism of all-trans-retinal toxicity with implications for Stargardt disease and age-related macular degeneration. *J. Biol. Chem.* **2012**, *287*, 5059–5069. [CrossRef]
95. Okano, K.; Maeda, A.; Chen, Y.; Chauhan, V.; Tang, J.; Palczewska, G.; Sakai, T.; Tsuneoka, H.; Palczewski, K.; Maeda, T. Retinal cone and rod photoreceptor cells exhibit differential susceptibility to light-induced damage. *J. Neurochem.* **2012**, *121*, 146–156. [CrossRef] [PubMed]
96. Chen, Y.; Sawada, O.; Kohno, H.; Le, Y.Z.; Subauste, C.; Maeda, T.; Maeda, A. Autophagy protects the retina from light-induced degeneration. *J. Biol. Chem.* **2013**, *288*, 7506–7518. [CrossRef] [PubMed]
97. Chen, Y.; Palczewska, G.; Mustafi, D.; Golczak, M.; Dong, Z.; Sawada, O.; Maeda, T.; Maeda, A.; Palczewski, K. Systems pharmacology identifies drug targets for Stargardt disease-associated retinal degeneration. *J. Clin. Investig.* **2013**, *123*, 5119–5134. [CrossRef] [PubMed]
98. Kohno, H.; Chen, Y.; Kevany, B.M.; Pearlman, E.; Miyagi, M.; Maeda, T.; Palczewski, K.; Maeda, A. Photoreceptor proteins initiate microglial activation via Toll-like receptor 4 in retinal degeneration mediated by all-trans-retinal. *J. Biol. Chem.* **2013**, *288*, 15326–15341. [CrossRef] [PubMed]
99. Maeda, A.; Palczewska, G.; Golczak, M.; Kohno, H.; Dong, Z.; Maeda, T.; Palczewski, K. Two-photon microscopy reveals early rod photoreceptor cell damage in light-exposed mutant mice. *Proc. Natl. Acad. Sci. USA* **2014**, *111*, E1428–E1437. [CrossRef] [PubMed]
100. Sawada, O.; Perusek, L.; Kohno, H.; Howell, S.J.; Maeda, A.; Matsuyama, S.; Maeda, T. All-trans-retinal induces Bax activation via DNA damage to mediate retinal cell apoptosis. *Exp. Eye Res.* **2014**, *123*, 27–36. [CrossRef] [PubMed]

101. Wu, X.; Yu, G.; Luo, C.; Maeda, A.; Zhang, N.; Sun, D.; Zhou, Z.; Puntel, A.; Palczewski, K.; Lu, Z.R. Synthesis and evaluation of a nanoglobular dendrimer 5-aminosalicylic Acid conjugate with a hydrolyzable schiff base spacer for treating retinal degeneration. *ACS Nano* **2014**, *8*, 153–161. [CrossRef]
102. Yu, G.; Wu, X.; Ayat, N.; Maeda, A.; Gao, S.Q.; Golczak, M.; Palczewski, K.; Lu, Z.R. Multifunctional PEG retinylamine conjugate provides prolonged protection against retinal degeneration in mice. *Biomacromolecules* **2014**, *15*, 4570–4578. [CrossRef]
103. Puntel, A.; Maeda, A.; Golczak, M.; Gao, S.Q.; Yu, G.; Palczewski, K.; Lu, Z.R. Prolonged prevention of retinal degeneration with retinylamine loaded nanoparticles. *Biomaterials* **2015**, *44*, 103–110. [CrossRef]
104. Parmar, T.; Parmar, V.M.; Arai, E.; Sahu, B.; Perusek, L.; Maeda, A. Acute Stress Responses Are Early Molecular Events of Retinal Degeneration in Abca4$^{-/-}$Rdh8$^{-/-}$ Mice After Light Exposure. *Investig. Ophthalmol. Vis. Sci.* **2016**, *57*, 3257–3267. [CrossRef]
105. Schur, R.M.; Gao, S.; Yu, G.; Chen, Y.; Maeda, A.; Palczewski, K.; Lu, Z.R. New GABA modulators protect photoreceptor cells from light-induced degeneration in mouse models. *FASEB J.* **2018**, *32*, 3289–3300. [CrossRef] [PubMed]
106. Parmar, T.; Parmar, V.M.; Perusek, L.; Georges, A.; Takahashi, M.; Crabb, J.W.; Maeda, A. Lipocalin 2 Plays an Important Role in Regulating Inflammation in Retinal Degeneration. *J. Immunol.* **2018**, *200*, 3128–3141. [CrossRef] [PubMed]
107. Chen, C.; Chen, J.; Wang, Y.; Liu, Z.; Wu, Y. Ferroptosis drives photoreceptor degeneration in mice with defects in all-trans-retinal clearance. *J. Biol. Chem.* **2021**, *296*, 100187. [CrossRef] [PubMed]
108. Yu, G.; Gao, S.Q.; Dong, Z.; Sheng, L.; Sun, D.; Zhang, N.; Zhang, J.; Margeivicus, S.; Fu, P.; Golczak, M.; et al. Peptide Derivatives of Retinylamine Prevent Retinal Degeneration with Minimal Side Effects on Vision in Mice. *Bioconjug. Chem.* **2021**, *32*, 572–583. [CrossRef] [PubMed]
109. Cai, B.; Liao, C.; He, D.; Chen, J.; Han, J.; Lu, J.; Qin, K.; Liang, W.; Wu, X.; Liu, Z.; et al. Gasdermin E mediates photoreceptor damage by all-trans-retinal in the mouse retina. *J. Biol. Chem.* **2022**, *298*, 101553. [CrossRef] [PubMed]
110. He, D.; Tao, L.; Cai, B.; Chen, X.; Wang, Y.; Li, S.; Liao, C.; Chen, Y.; Chen, J.; Liu, Z.; et al. eIF2alpha incites photoreceptor cell and retina damage by all-trans-retinal. *J. Biol. Chem.* **2023**, *299*, 104686. [CrossRef]
111. Chen, C.; Yang, K.; He, D.; Yang, B.; Tao, L.; Chen, J.; Wu, Y. Induction of ferroptosis by HO-1 contributes to retinal degeneration in mice with defective clearance of all-trans-retinal. *Free Radic. Biol. Med.* **2023**, *194*, 245–254. [CrossRef]
112. Maeda, A.; Maeda, T.; Imanishi, Y.; Sun, W.; Jastrzebska, B.; Hatala, D.A.; Winkens, H.J.; Hofmann, K.P.; Janssen, J.J.; Baehr, W.; et al. Retinol dehydrogenase (RDH12) protects photoreceptors from light-induced degeneration in mice. *J. Biol. Chem.* **2006**, *281*, 37697–37704. [CrossRef]
113. Maeda, A.; Golczak, M.; Maeda, T.; Palczewski, K. Limited roles of Rdh8, Rdh12, and Abca4 in all-trans-retinal clearance in mouse retina. *Investig. Ophthalmol. Vis. Sci.* **2009**, *50*, 5435–5443. [CrossRef]
114. Zobor, D.; Bruhwiler, B.; Zrenner, E.; Weisschuh, N.; Kohl, S. Genetic and Clinical Profile of Retinopathies Due to Disease-Causing Variants in Leber Congenital Amaurosis (LCA)-Associated Genes in a Large German Cohort. *Int. J. Mol. Sci.* **2023**, *24*, 8915. [CrossRef]
115. Sen, P.; Srikrupa, N.; Maitra, P.; Srilekha, S.; Porkodi, P.; Gnanasekaran, H.; Bhende, M.; Khetan, V.; Mathavan, S.; Bhende, P.; et al. Next-generation sequencing--based genetic testing and phenotype correlation in retinitis pigmentosa patients from India. *Indian. J. Ophthalmol.* **2023**, *71*, 2512–2520. [CrossRef] [PubMed]
116. Schlottmann, P.G.; Luna, J.D.; Labat, N.; Yadarola, M.B.; Bainttein, S.; Esposito, E.; Ibanez, A.; Barbaro, E.I.; Alvarez Mendiara, A.; Picotti, C.P.; et al. Nationwide genetic analysis of more than 600 families with inherited eye diseases in Argentina. *NPJ Genom. Med.* **2023**, *8*, 8. [CrossRef] [PubMed]
117. Kuo, C.Y.; Chung, M.Y.; Chen, S.J. Pseudocoloboma-like maculopathy with biallelic RDH12 missense mutations. *J. Med. Genet.* **2023**, *60*, 859–865. [CrossRef] [PubMed]
118. Daich Varela, M.; Moya, R.; Azevedo Souza Amaral, R.; Schlottmann, P.G.; Alvarez Mendiara, A.; Francone, A.; Guazi Resende, R.; Capalbo, L.; Gelvez, N.; Lopez, G.; et al. Clinical and Genetic Characterization of RDH12-Retinal Dystrophy in a South American Cohort. *Ophthalmol. Retina* **2023**, *in press*. [CrossRef]
119. Hanneken, A.; Neikirk, T.; Johnson, J.; Kono, M. Biochemical Measurements of Free Opsin in Macular Degeneration Eyes: Examining the 11-CIS Retinal Deficiency Hypothesis of Delayed Dark Adaptation (An American Ophthalmological Society Thesis). *Trans. Am. Ophthalmol. Soc.* **2017**, *115*, T1.
120. Obana, A.; Gohto, Y.; Asaoka, R.; Gellermann, W.; Bernstein, P.S. Lutein and Zeaxanthin Distribution in the Healthy Macula and Its Association with Various Demographic Factors Examined in Pseudophakic Eyes. *Antioxidants* **2021**, *10*, 1857. [CrossRef]
121. Rozanowska, M.; Sarna, T.; Land, E.J.; Truscott, T.G. Free radical scavenging properties of melanin interaction of eu- and pheo-melanin models with reducing and oxidising radicals. *Free Radic. Biol. Med.* **1999**, *26*, 518–525.

MDPI AG
Grosspeteranlage 5
4052 Basel
Switzerland
Tel.: +41 61 683 77 34

International Journal of Molecular Sciences Editorial Office
E-mail: ijms@mdpi.com
www.mdpi.com/journal/ijms

MDPI

www.ingramcontent.com/pod-product-compliance
Lightning Source LLC
Chambersburg PA
CBHW041949260926
51850CB00019B/2256
* 9 7 8 3 7 2 5 8 7 0 3 8 7 *